Drafting and Design with AutoCAD LT

David Dye

Paul Moore

Contributions by:
      Rusty Gesner
      Ronn Lansky

New
Riders

New Riders Publishing, Indianapolis, Indiana

Drafting and Design with AutoCAD LT

By David Dye and Paul Moore

Published by:
New Riders Publishing
201 West 103rd Street
Indianapolis, IN 46290 USA

Copyright ® 1995 by New Riders Publishing

Printed in the United States of America 1 2 3 4 5 6 7 8 9 0

```
Drafting and Design with AutoCAD LT/Rusty Gesner…
      p. cm.
   Includes index.
   ISBN 1-56205-501-1
   1. Mechanical Drawing. 2. Computer-aided design.
3. AutoCAD LT for Windows.  I. Gesner, Rusty.
T353.D723  1995
604.2'0285'5369--dc20                         95-37179
                                                   CIP
```

Warning and Disclaimer

Publisher	Don Fowley
Marketing Manager	Ray Robinson
Acquisitions Manager	Jim LeValley
Managing Editor	Tad Ringo

Product Development Specialist
Kurt Hampe

Acquisitions Editor
Mary Foote

Production Editor
Sarah Kearns

Copy Editor
Margaret Berson

Technical Editor
Alex Lepeska

Assistant Marketing Manager
Tamara Apple

Acquisitions Coordinator
Stacey Beheler

Publisher's Assistant
Karen Opal

Cover Designer
Fred Bowers

Cover Illustrator
Greg Phillips

Book Designer
Sandra Schroeder

Manufacturing Coordinator
Paul Gilchrist

Production Manager
Kelly Dobbs

Production Team Supervisor
Laurie Casey

Graphics Image Specialist
Clint Lahnen

Production Analysts
Angela Bannan
Bobbi Satterfield

Production Team
Angela Calvert
Kim Cofer
Kevin Foltz
Aleata Howard
Erika Millen
Gina Rexrode
Erich Richter
Christine Tyner
Robert Wolf

Indexer
Christopher Cleveland

About the Authors

David Dye graduated from the University of Michigan with a Bachelor of Science and a Master of Architecture degree. He is a registered architect, and formerly practiced the profession in Michigan and Indiana. Mr. Dye is currently an instructor and Autodesk Training Center manager at Grand Rapids Community College in Grand Rapids, Michigan.

Paul Moore currently works for a large publishing company, producing a quarterly CD-ROM for the building industry. He also functions as a consultant and adjunct faculty member at Grand Rapids Community College for Autodesk multimedia products, specifically 3D Studio. Mr. Moore is a graduate of the School of Architecture and Urban Planning at the University of Michigan. He is the co-author of *Learn CAD with AutoSketch*, from New Riders Publishing.

Rusty Gesner, one of the original members of NRP and formerly New Riders' CAD Publishing Director, is a freelance author and AutoCAD expert. Mr. Gesner was born and raised in Indianapolis, and educated at Antioch College and the College of Design, Art, and Architecture at the University of Cincinnati before migrating to Oregon. He is a registered architect, formerly practicing in Oregon and Washington before buying a copy of the original release of AutoCAD in early 1983. Prior to joining New Riders in 1986, he was founder and president of CAD Northwest, Inc., one of the first AutoCAD dealerships. Mr. Gesner is co-author and/or developer of various editions of many New Riders books, including *Maximizing AutoCAD*, *Maximizing AutoLISP*, and *Inside AutoCAD*. He now resides in Boring, Oregon with his wife, Kathy, their two children, and seven pets (cats, dogs, and horses). In his meager spare time, Mr. Gesner skis, bicycles, shoots arrows at unsuspecting bales of excelsior, and rides his tractor in circles.

Ronn Lansky co-authored curriculum for a private technical college in southern California during the 1980s, where he also taught AutoCAD, architecture, and drafting. He is a graduate of the School of Architecture and Urban Planning at the University of Michigan. Currently, Mr. Lansky is an administrator at a technical school, where he also teaches AutoCAD and architecture.

Trademark Acknowledgments

Acknowledgments

David Dye

I would like to thank my wife, Valerie, and daughter, Laura, for their patience and continued encouragement while writing this book. Thanks to my sons, Scott and Chris, for giving Dad priority on the computer while I was writing. At the College, thanks to Tom Boozer and David Piggott for their patience and flexibility, and a thank you to Tom Boersma for presenting me with this challenge. The support and coaching skills of the people at New Riders Publishing was also greatly appreciated, especially the help of Sarah Kearns, Stacey Beheler, and Kurt Hampe.

Paul Moore

I would like to thank my wife, Laurie, and our daughters, Sierra and Summer, for their humor and understanding while I pursued this opportunity. Thanks also to Rusty and Ronn for their help in completing this book. In addition, I also wish to thank the staff at New Riders for their unselfish assistance and guidance, specifically Stacey, Sarah, Kurt, Margaret, Mary, and anyone else who ultimately contributed to the publication of this book.

Contents at a Glance

Table of Contents

Chapter 2   43

An Overview of Drafting Tools and Techniques

Chapter 3   91

CAD Drawing Techniques

Part 3: Viewing and Reproducing Drawings   291

Part 4: Projects to Challenge Your Knowledge   437

Chapter 11   439

Creating a Mechanical Drawing

Chapter 12   505

Architectural Drawing Project

Introduction

Computer-aided drafting (CAD) is a very popular subject in high school, college, and the workplace. Many people want to learn CAD, but more than half of these people have no experience with technical drafting and design. This is a frequent complaint of employers—their employees are trying to learn CAD, but lack the required graphic background. Essentially, most of these employees are unfamiliar with the complexities of technical drawing; they simply do not possess the organization and discipline learned from this type of drawing.

The intent of this book is to give readers a solid foundation in the terminology, tools, and concepts used in technical design and drafting. Hopefully, this knowledge of technical drawing will not only make learning CAD easier, but will also enable readers to successfully converse with others about CAD.

The software used to teach CAD is AutoCAD LT for Windows, an affordable software package that is appropriate for a wide range of people—from the beginning user on a home computer to an experienced technician at an engineering office. AutoCAD LT for Windows is attractive because it has the power and flexibility of CAD and the user friendliness of Windows.

In this book, you learn CAD first by doing a freehand sketch in CAD itself. You then start drawing more precisely by using coordinate information and picking exact points. You'll see a direct comparison between technical drafting tools and techniques and CAD tools. You'll draw simple CAD drawings at first—such as a stick man, a house, and a boat—followed by more complex drawings of a barn, and then a mechanical drawing of a bracket. Ultimately, at the end of the book, you'll be creating architectural and mechanical projects in CAD. The intent of this book is to give you a background in drafting tools and techniques, along with familiarizing you with important drawing skills that you can apply when using CAD.

Who Should Read This Book?

The purpose of *Drafting and Design with AutoCAD LT* is to introduce you to the powerful capability of CAD through the use of AutoCAD LT for Windows. This book is intended for first-time users of CAD, such as the following:

◆ Home computer users who have worked on other computer programs and now want to try CAD

◆ Individuals tired of their existing jobs and wanting a change

◆ Designers, drafters, and technicians who need to learn CAD for their jobs

These different types of first-time CAD users are discussed in the following sections.

Home Computer Users

Home computers are quite prevalent in today's fast-paced world—there are over 90 million home computers in the United States and Canada alone. Home computers offer an opportunity for people to familiarize themselves with various software packages in their spare time at home. Many workplaces are now requiring employees to learn CAD, but not always giving them the time or resources to do so. *Drafting and Design with AutoCAD LT* was created to help readers learn AutoCAD LT at their own pace on their home computers.

Individuals Needing a Change

Many people are tired of their existing jobs and want a change. They have heard about this mysterious and exciting new field of computer-aided drafting, but they aren't familiar with what it entails. This book gives you first-hand experience by teaching you the basics of CAD, and helping you grasp its overall concept. Because it is a beginning to intermediate book on CAD, *Drafting and Design with AutoCAD LT* gives you a realistic taste of this program. It can help you decide if you really want to work in the CAD area.

Designers, Drafters, and Technicians Needing to Learn CAD

Drafting and Design with AutoCAD LT is intended for those designers, drafters, and technicians who have either just started their careers, or are experienced in manual drafting. Both new and experienced employees needing to learn CAD for the first time can use this book as a guide. CAD can essentially be learned by anyone willing to experiment with the software and see the results.

About the AutoCAD LT for Windows Software

This software has proven to be amazingly popular during its first year of introduction—over 125,000 copies were sold. AutoCAD LT has the functionality and power of its parent software, AutoCAD Release 12 for Windows, but it has a simpler interface and is moderately priced (under $500). AutoCAD LT for Windows is thus affordable for the small company, home computer user, or the student. It also can be used in larger companies for training needs, quality assurance review, demonstrations, 2D CAD drafting, and project management.

In addition, this software is completely compatible with AutoCAD Release 12 for DOS and Release 12 for Windows. Drawings can be originally created in AutoCAD Release 12, loaded and changed on AutoCAD LT, and then returned to AutoCAD Release 12 without any loss in data.

How This Book is Organized

Drafting and Design with AutoCAD LT is organized to progressively build your knowledge and understanding of technical drafting and CAD through explanations, examples, and exercises. Depending on your knowledge of technical drawing, CAD, and Windows, you may spend more or less time in each chapter—the chapters do build on each other. You may need to learn the concepts and do the drawings at the beginning of the book to successfully use the tools and create the drawings later in the book.

The book is divided into five major parts, with each part containing several chapters. These are described in the following sections.

Part One: The Basics of CAD and Technical Drawing

Part One introduces you to CAD by using AutoCAD LT for Windows. An overview of tools for technical drawing and how they relate to CAD is included. Several questions that are answered in this part are as follows:

- ◆ How do you start drawing in CAD?

- ◆ What are some drafting tools and techniques that can be used in CAD?

- ◆ What are some CAD drawing techniques?

- ◆ How do you create a basic CAD drawing with multiviews, layers, and a border?

You also learn the basics of CAD and technical drawing by doing different drawing exercises.

Chapter 1, "Welcome to the World of Computer-Aided Drafting," starts you on your journey of CAD by exploring the advantages of CAD over manual drafting. How do you get going in AutoCAD LT? How is the AutoCAD LT graphics window organized? These questions are answered in this chapter when you draw a stick figure and a house using CAD. You're up and running on CAD after finishing Chapter 1.

Chapter 2, "An Overview of Drafting Tools and Techniques," makes a direct comparison between tools in technical manual drawing and the tools you use in CAD. What are some tools you use in CAD? How do you lay out a page? In the process of learning these new tools, you'll draw a boat in CAD.

Chapter 3, "CAD Drawing Techniques," teaches you the techniques that give CAD a tremendous advantage over manual drawing. What are the techniques used to draw precisely in CAD? How do you duplicate lines and circles, and how do you revise drawings? How do you bring in symbols from other drawings or even other computer software within Windows? You'll create a drawing of a barn with trees, hay bales, and clouds, and learn the different techniques in the process of doing so.

Chapter 4, "Multiview Drawings," explains how to take the front view of a barn and create a multiview drawing, complete with front, top, and side views. This is how you can use 2D drawings to describe a 3D object. You'll also learn to draw a mechanical part, complete with a border, title block, multiviews of the part, and text.

Part Two: Notes and Dimensions

Now that you have learned the basics of architectural and mechanical drawing, you next learn how to put notes and dimensions on them. One of the great advantages of using CAD is that you draw full scale; for example, a house is drawn full scale using CAD (11" x 2"). To create a dimension, you simply pick two points on an object you want to measure. CAD will give you the correct dimension measurement and all the appropriate lines and arrows to create the dimension. You'll learn how to use text and how to insert and modify dimensions using a drawing from Part One.

Chapter 5, "Text and Annotation," describes how to insert and modify text for use as notes on your drawings. How do you change text fonts and styles? How do you bring in notes from other Windows programs and insert them into your drawing? In this chapter, you learn how to complete your title block and add important notes to your drawing.

Chapter 6, "How to Start Dimensioning," shows you how to insert dimensions into your drawing. How do you use linear, angular, radial, and circular dimensioning? What is an ordinate dimension? These questions are answered as you explore dimensioning using different drawings.

Chapter 7, "How to Modify Dimensions," demonstrates how to use the Dimension Styles and Settings dialog box to change your dimensions or create a dimension style. How do you move or modify a dimension? Drawing exercises will show you the way.

Part Three: Viewing and Reproducing Drawings

Part Three presents different ways to view a part or building. How do you determine where a cutting plane should be to create a section? Should you use an isometric, dimetric, or trimetric view to show the part? Once the drawing is complete, how do you plot, print, and reproduce the drawing? This part deals with these questions and more.

Chapter 8, "Developing Sectional and Auxiliary Views," presents the different ways to cut an object, and answers questions such as: What are auxiliary views? What are the different ways to section an object and how do you present it in a drawing? This chapter contains examples to explain how it's all done.

Chapter 9, "Exploring Paraline Projection" shows different ways to present an object using isometric drawings or similar methods. What is an assembly drawing? What is the difference between isometric, dimetric, trimetric, and oblique drawings? The different drawing approaches are presented so you can evaluate the best way to make your presentation drawing.

Chapter 10, "Reproducing Drawings," reviews the different ways drawings can be reproduced once they are completed. What are the different methods and mediums for reproducing drawings? How do you set up a printer or plotter? These and other alternative printing methods are discussed.

Part Four: Projects to Challenge Your Knowledge

Once you have completed the examples in *Drafting and Design with AutoCAD LT*, you will need to be challenged to prepare you for day-to-day work with CAD. Two chapters give you this preparation.

Chapter 11, "Creating a Mechanical Drawing," explains the steps to creating a part. How do you determine the layout of the sheet? How do you develop the different views? What is the best way to dimension? These and other questions are answered in this chapter.

Chapter 12, "Architectural Drawing Project," describes the process for creating an architectural drawing. How do you create a prototype drawing? How do you create a floor plan and elevations? What are the settings to plot the drawing? This chapter addresses these questions and more.

Part Five: Appendix

The final part of the book contains Appendix A, "Glossary of CAD and Drafting Terms," which describes tools or references that can be used for CAD and technical drawing.

Getting the Most from This Book

No matter how proficient you are with AutoCAD LT, Windows, or your computer, and no matter how thoroughly you read this book, you will revisit it again and again as a reference manual. AutoCAD LT offers so many ways to accomplish the task of CAD design that you will simply not have the time to explore them all until some need arises. You will then find *Design and Drafting with AutoCAD LT* indispensable as you use it to find explanation and examples of specific commands and techniques. To ensure that you are getting the most from this book, read the following sections, which explain the notational conventions used throughout the book.

Exercises

A sample exercise follows. You do not need to work through it or understand the content, but you should look over the format so that you will know how the book's exercises are presented. Most exercises are accompanied by one or more illustrations, most of which were captured from the screen during an actual exercise. Exercises are arranged in two columns, with direct instructions on the left and explanatory notes on the right. Lengthy instructions or explanations sometimes extend across both columns.

Exercise

Trying a Sample Exercise

Continue in the SAMPLE drawing from the preceding exercise.

`Command:` *Choose the Circle button*	Issues the CIRCLE command
`_Circle 3P/2P/TTR/<Center Point:`**7,4.5** (Enter)	Locates the center of the circle at the coordinates you type
`Diameter/<Radius>:`**3** (Enter)	Gives the circle a radius of 3
`Command:` *Choose* **D**raw, *then* **L**ine	Issues the LINE command
`From point:` *Choose the Center button*	Specifies the CENter object snap

continues

`_center of` *Select circle at* ①	Selecting the edge of the circle will object snap the line to the center of the circle
`To point:` *Choose the Quadrant button*	Specifies the QUADrant object snap
`_quadrant of` *Select circle at* ②	Selecting the edge of the circle will select a nearby circle quadrant
`Command:` *Choose the Polygon button*	Issues the POLYGON command
`Number of sides <4>:`**6**	Specifies six sides for the polygon
`Edge/<center of Polygon>:` *Choose the Midpoint button*	Specifies the MIDpoint object snap
`_midpoint of` *Select line at* ③	Draws polygon at midpoint of line
`Command:` *Choose* **F**ile, *then* **S**ave	Issues the SAVE command

Your drawing should look like figure I.1

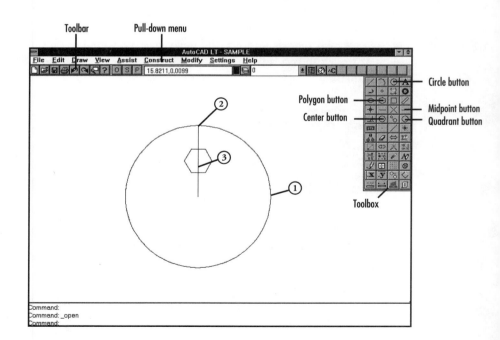

Because AutoCAD LT for Windows is a Windows application, its interface uses many of the elements used by other Windows applications. The toolbar across the top of the graphics window incorporates buttons from a few commonly used commands. The toolbar also contains the current layer name and the coordinate displays. The toolbox at the side of the screen contains buttons for commonly used commands and all of the object snap modes.

You will be using your Windows pointing device (referred to throughout this book as a *mouse*) for much of your input in AutoCAD LT. You should become familiar with the following terms, which are used to describe mouse actions:

◆ **Click.** To press and release the pick button.

◆ **Click on, click in.** To position the cursor on the appropriate user interface object (icon, input box, menu item, and so on) and click the pick button.

◆ **Double-click.** To press and release the pick button twice, in rapid succession. You can double-click on numbers in an input box.

◆ **Pick.** To position the cursor on the appropriate object or point, and then click the pick button.

◆ **Select.** To highlight an object in the AutoCAD LT drawing area by picking it, windowing it, or other object selection methods.

◆ **Choose.** To select a button in a toolbar, toolbox, or dialog box by clicking on it. You can select a command in the pull-down menu by clicking on it.

◆ **Drag.** To move the mouse and cursor, causing lines or objects on the screen to move with the cursor.

◆ **Press and drag.** To press and hold down a mouse button, drag something on screen, and then release the mouse button.

Notes, Tips, and Warnings

Drafting and Design with AutoCAD LT features many special sidebars, which are set apart from the normal text by icons. The book includes three distinct types of sidebars: notes, tips, and warnings. These passages have been given special treatment so that you can instantly recognize their significance and easily find them for future reference.

A note includes "extra" information you should find useful, but which complements the discussion at hand instead of being a direct part of it. A note might describe special situations that can arise when you use AutoCAD LT for Windows under certain circumstances, and might tell you what steps to take when such situations arise. Notes also might tell you how to avoid problems with your software and hardware.

A tip provides you with quick instructions for getting the most from your AutoCAD LT system as you follow the steps outlined in the general discussion. A tip might show you how to conserve memory in some setups, how to speed up a procedure, or how to perform one of many time-saving and system-enhancing techniques.

A warning tells you when a procedure might be dangerous—that is, when you run the risk of losing data, locking your system, or even damaging your hardware. Warnings generally tell you how to avoid such losses, or they describe the steps you can take to remedy these situations.

New Riders Publishing

The staff of New Riders Publishing is committed to bringing you the very best in computer reference material. Each New Riders book is the result of months of work by authors and staff who research and refine the information contained within its covers.

As part of this commitment to you, the NRP reader, New Riders invites your input. Please let us know if you enjoy this book, if you have trouble with the information and examples presented, or if you have a suggestion for the next edition.

Please note, though: New Riders staff cannot serve as a technical resource for AutoCAD LT for Windows or for questions about software- or hardware-related problems. Please refer to the documentation that accompanies AutoCAD LT for Windows or to AutoCAD's Help systems.

If you have a question or comment about any New Riders book, there are several ways to contact New Riders Publishing. We will respond to as many readers as we can. Your name, address, or phone number will never become part of a mailing list or be used for any purpose other than to help us continue to bring you the best books possible. You can write us at the following address:

New Riders Publishing
Attn: Publisher
201 W. 103rd Street
Indianapolis, IN 46290

If you prefer, you can fax New Riders Publishing at (317) 581-4670.

You can send electronic mail to New Riders at the following Internet address:

ddweyer@newriders.mcp.com

NRP is an imprint of Macmillan Computer Publishing. To obtain a catalog or information, or to purchase any Macmillan Computer Publishing book, call (800) 428-5331.

Thank you for selecting *Drafting and Design with AutoCAD LT*!

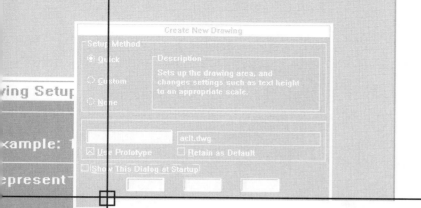

Part One

The Basics of CAD and Technical Drawing

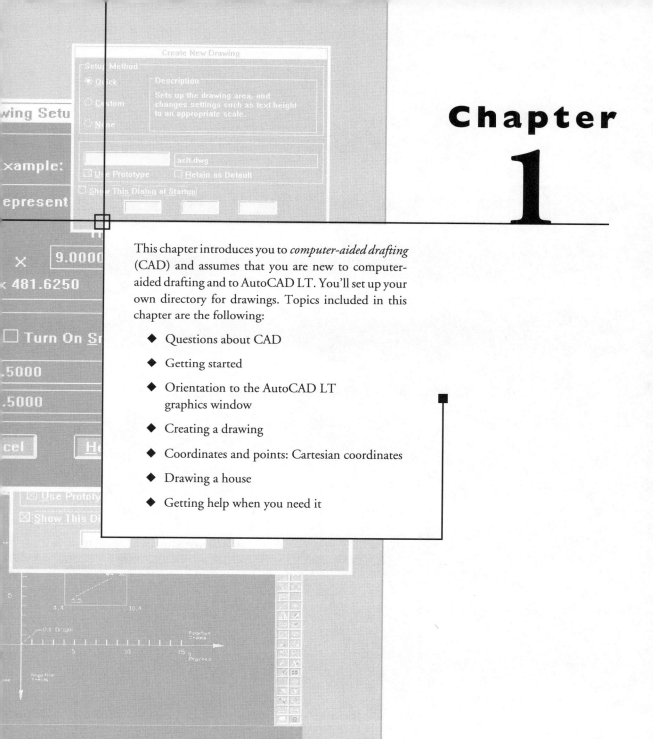

Chapter

1

This chapter introduces you to *computer-aided drafting* (CAD) and assumes that you are new to computer-aided drafting and to AutoCAD LT. You'll set up your own directory for drawings. Topics included in this chapter are the following:

◆ Questions about CAD

◆ Getting started

◆ Orientation to the AutoCAD LT graphics window

◆ Creating a drawing

◆ Coordinates and points: Cartesian coordinates

◆ Drawing a house

◆ Getting help when you need it

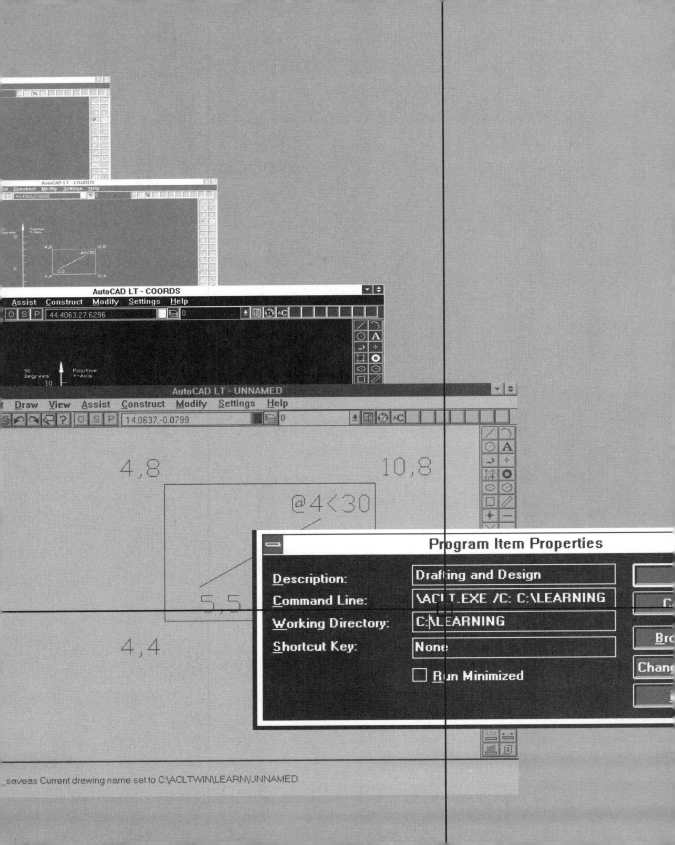

Welcome to the World of Computer-Aided Drafting

In this chapter, you first create some simple drawings. As your skills increase while you explore the book, you will begin to draw more complex drawings. Before you dive into actually drawing, however, let's cover some of the basics of CAD.

Questions about CAD

If you are interested in AutoCAD LT enough to be reading this book, you must have some questions about computer-aided drafting. The following sections explore some questions commonly asked about CAD.

What is CAD?

The term CAD refers to *computer-aided drafting.* Another term that is sometimes used is CADD—*computer-aided design and drafting.* When you draw using CAD, you are operating a computer to draw lines and geometric shapes, insert symbols, and place text and dimensions to create mechanical or architectural drawings.

A typical CAD system of today consists of a personal computer, a monitor, a keyboard, a mouse or tablet for inputting information into the computer, and a plotter and/or printer for outputting information from the computer (see fig. 1.1).

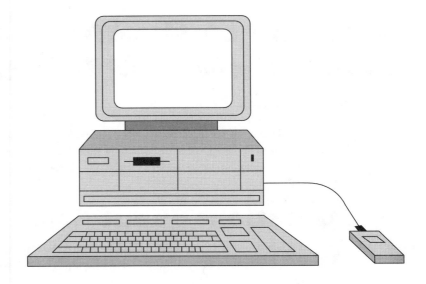

Why Use CAD?

What are some of the advantages of using CAD over manual drafting? Three advantages that are immediately apparent are speed, accuracy, and ease of revision.

Speed

You can draw in CAD faster than in manual drafting, because you can copy lines and geometric shapes, and you can use pre-drawn symbols. With training and practice, you can draw faster than a manual draftsperson. Generally, a draftsperson will match his or her speed on the drafting board within two to six months after learning CAD. Speed improves over time with continued use and learning of a CAD system.

Accuracy

Objects are drawn full-scale in CAD—they are the actual size of the object they represent. The length of objects input into the CAD drawing will have the same accuracy as the actual object. Drawings for a building can be drawn full-scale with the accuracy of .001 of an inch or better. CAD drawings are so precise that the dimension is determined automatically by picking an item to create a dimension.

Ease of Revision

Making changes to a manual drawing can be very time-consuming and messy. Pencil or ink lines can be smeared and the drawing paper can sometimes be damaged. Revisions can be an unwelcome endeavor that reduces the quality of the final print.

In CAD, revisions are not made on the paper; they are made on the computer. Revisions can be made even before a print is made of the computer image. One other advantage of CAD is that when objects or buildings are revised on the computer, the dimensions are automatically updated. Over time, this can translate to dollar savings for a design office because revisions take less time.

Who Uses CAD?

There are many different users of CAD. Software is available for CAD in many different areas: architecture; mechanical, civil, and electrical engineering; interior design; facilities management; mapping; piping and plant layout; manufacturing design and engineering; and virtually any type of drawing needed by industry.

Some professions that use CAD are the following:

◆ Architects or draftspersons who design or draw buildings

◆ Engineers, technicians, or draftspersons in the building trades who design or draw bridges, roads, underground utilities, and other engineered landscapes and structures

◆ Engineers, designers, and technicians who design manufactured items like cars, airplanes, boats, clothing, furniture, electronic equipment, and so on

◆ Interior designers, facility managers, and space planners who design the interiors of buildings

What about AutoCAD LT?

Over 125,000 copies of AutoCAD LT were sold during the first year it was introduced. It can be used on a personal computer and is considered lower-priced software, below $500. AutoCAD LT has much of the functionality of AutoCAD Release 12, but it has a simpler interface. AutoCAD LT also is completely compatible with AutoCAD Release 11 and 12, and its drawings can be opened in AutoCAD Release 13. Essentially, AutoCAD LT is a lot of software for its price.

Drafting and Design with AutoCAD LT was written using the newest release of software, AutoCAD LT Release 2 for Windows. This book can be used with AutoCAD for Windows Release 1 and 2.

Getting Started

The best way for you to learn how to load AutoCAD LT for Windows on your computer is to read Chapter 1 of the *AutoCAD LT User's Guide*. It reviews system requirements, how to install AutoCAD LT, and other fine points of installation.

In this section, you learn how to create your own special learning directory. This directory will be independent of your daily work or anyone else who may be using AutoCAD LT on your computer. It will be your own self-contained workspace in which you should feel comfortable drawing and experimenting.

Setting Up a Workspace for this Book

After you have Windows and AutoCAD LT for Windows installed, you can start to work with the exercises in this book. In Windows, there is a straightforward way to set up a special directory in Program Manager for the exercises you will be doing in AutoCAD LT.

Exercise

Setting Up a Workspace	
In the Program Manager area, double-click on the Main program group icon	Opens the group so program items can be accessed
Double-click on the File Manager program item (see fig. 1.2)	Activates the File Manager

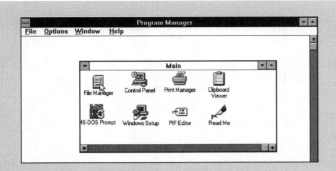

Figure 1.2

*The Main Program
group.*

*Double-click on the C:\ ①
in the corner of the File Manager
until you see files similar to
figure 1.3*	Selects the C:\ directory under
which the new directory will be
created
*Choose **F**ile, then **Cre**ate Directory,
then type **LEARNING** and press Enter*	Creates a new directory called
LEARNING under the root
directory

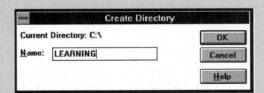

Figure 1.3

*Creating a
LEARNING
directory.*

Double-click on the ACLTWIN
directory ②	Shows the files for this directory
in the right side of the window
Hold the Ctrl key down and choose the
ACLT.INI *and* ACLT.CFG *files
from the right side*	Selects the ACLT.INI and the
ACLT.CFG files
*Choose **F**ile, then **C**opy, type*
C:\LEARNING, *and then press Enter
(see fig. 1.4)*	Copies the selected files from the
ACLTWIN directory to the
LEARNING directory

continues

Figure 1.4

*Copying files to the
LEARNING
directory.*

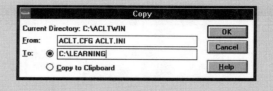

Close the File Manager window by choosing **F**ile, *then* E**x**it	Exits the File Manager window
Double-click on the Program Manager icon	Opens up the Program Manager window
Choose **F**ile, *then* **N**ew, *make sure Program Group is selected, and choose* OK	Opens the Program Group Properties dialog box
In the **D**escription: *box, type* **Learning AutoCAD LT.** ⏎	Gives the program group a name
Choose **F**ile, *then* **N**ew, *make sure Program Item is selected, and choose* OK	Opens the Program Item Properties dialog box

See figure 1.5 to continue this exercise.

Figure 1.5

*Creating a Drafting
and Design
program item.*

In the **D**escription: *box, type*
Drafting and Design, *then press Tab*

In the **C**ommand Line: *box, type*
C:\ACLTWIN\ACLT.EXE /C: C:\LEARNING,
then press Tab

In the **W***orking Directory: box, type*
LEARNING*, then press Tab*

In the **S***hortcut Key: box, press Tab,*   Creates a program item with an
then choose OK   icon with the description Drafting
                                                    and Design

Now when you choose the icon you have created, it will start AutoCAD LT
using LEARNING as the working directory.

Starting AutoCAD LT

Now that you have created a Drafting and Design icon, you can use that icon
to start AutoCAD LT, and it will use the directory you created to contain new
drawings and save existing drawings.

Starting AutoCAD LT

Exercise

Double-click on the Drafting and   Starts AutoCAD LT using the
Design icon (see fig. 1.6)   LEARNING directory

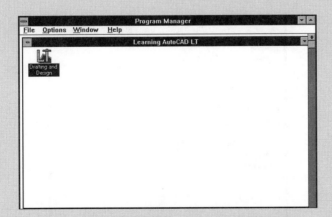

Figure 1.6

*The Drafting and
Design icon.*

In the Create New Drawing dialog   Accepts the Quick Setup method
box, choose OK *(see fig. 1.7)*

continues

Figure 1.7

*The Create New
Drawing dialog box.*

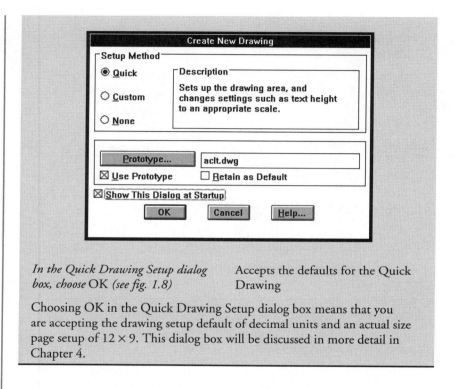

In the Quick Drawing Setup dialog box, choose OK *(see fig. 1.8)*	Accepts the defaults for the Quick Drawing

Choosing OK in the Quick Drawing Setup dialog box means that you
are accepting the drawing setup default of decimal units and an actual size
page setup of 12 × 9. This dialog box will be discussed in more detail in
Chapter 4.

Figure 1.8

*The Quick Drawing
Setup dialog box.*

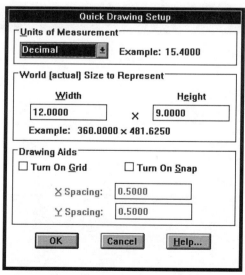

Now you have finally arrived at the AutoCAD LT graphics window. This is a good opportunity for you to become oriented.

Orienting Yourself to the AutoCAD LT Graphics Window

In order to be able to draw effectively in CAD, you need to be familiar with the different parts of the CAD graphics window. The AutoCAD LT window is very similar to that of other Windows applications. The window can be resized—by choosing the arrow in the upper right-hand corner, the window can be made larger or smaller. Other similarities include a title bar across the top, pull-down menus, and a toolbar.

The areas surrounding the graphics area contain tools for communicating with AutoCAD LT, pull-down menus for opening and saving files, and also for altering environmental settings (see fig. 1.9). These areas include the following:

◆ **Title bar.** The title bar displays the name of the current drawing file and also contains the control menu button and the window sizing button.

◆ **Pull-down menu bar.** By using the mouse or keyboard, you can activate the pull-down menus to open dialog boxes or issue commands. You can pick different menus with the left mouse button. You can pick **D**raw, then **L**ine to start the line command. Another way to access menus is with the keyboard. Hold down the Alt key, press D (for **D**raw), and then press L (for **L**ine) to start the LINE command. To cancel the command, press Ctrl+C.

◆ **Window sizing button.** The AutoCAD LT graphics window does not cover the entire terminal screen; press the right window sizing button to make it larger.

◆ **Toolbar.** The toolbar provides direct access to the file and print commands, undo and redo commands, drawing settings, layer controls, and ortho and snap toggles. It also has a coordinate display for the graphics area. There may also be some blank customizable icons on the toolbar, depending on the size of the AutoCAD LT window, the size of the toolbar buttons, and the screen resolution.

◆ **Toolbox.** The toolbox is located in the graphics area and contains icons that you can click on with the mouse to activate commands. The toolbox can be floating, can be in a fixed position on the left or right side of the screen, or can be turned off. The icons in the toolbar and toolbox are also referred to in this book as *buttons*.

◆ **Graphics area.** This is the part of the screen where drawing entities are created, inserted, or selected.

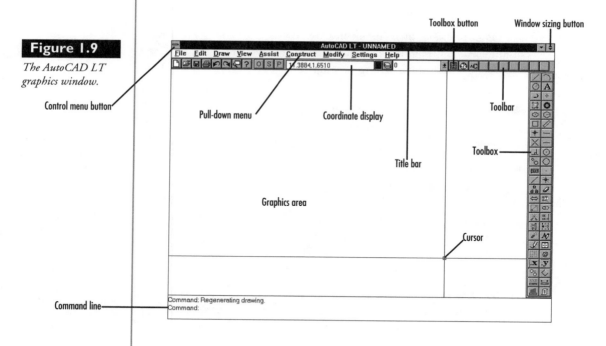

Figure 1.9

The AutoCAD LT graphics window.

Control menu button

Pull-down menu

Coordinate display

Toolbox button

Window sizing button

Toolbar

Toolbox

Title bar

Graphics area

Cursor

Command line

◆ **Command line.** The command line area takes all the input from the keyboard or the mouse, lists the command options, and displays the prompts for required input. It is very important to watch the command line to make sure you are using the command you intended to use.

◆ **Cursor.** The cursor is shown in the graphics area as a cross-shaped cursor. This is the cursor you use to pick points in the graphics area. The appearance of the cursor changes as it moves over the toolbar and toolbox—it becomes an arrow. The cursor can also appear as a small square when it is being used to select objects within the graphics area.

To get the toolbox to appear as shown in figure 1.9, press your right mouse button with your cursor over a button in the toolbox. A Toolbox Customization dialog box will appear. Type **2** (Enter) in the Toolbox Width - Locked input box. Second, choose the Toolbox button on the toolbar until the toolbox appears.

If you hold the cursor for approximately three seconds over one of the icons in the toolbar or toolbox, small labels called ToolTips will suddenly appear. These ToolTips are intended to help you locate tools within the graphics window by calling out the name of the tool as the cursor touches the icon.

It is important to note the effective use of buttons on the mouse. Different buttons have different effects. On a two-button mouse, the left button is the *Pick button,* used to pick or select points on the screen, or select tools on the toolbar or toolbox or on the pull-down menus. The left button is used a lot in AutoCAD LT.

The right button is the *Enter button.* The right button also has other special uses. If you press it while the cursor is over a toolbar or toolbox, you open a dialog box for customization. This box enables you to customize buttons on the toolbar or toolbox. If you ever open this customization dialog box by mistake, just press the Esc key to close the dialog box.

Creating a Drawing

The first figure you will draw will be a very simple stick figure. This will introduce you to the typical methods of creating entities in a CAD environment. You will also be creating other familiar figures you have drawn in the past such as a house, a boat, and so on. The drawings will become more complex and technical as you proceed through the book.

Sketching a Stick Figure

As a way to get you comfortable with drawing in CAD, imagine that you first sketch out the stick figure on a sheet of paper. How would you sketch it? Generally, you would draw either the head or stick body first. You could draw the legs or arms next. Then draw the face next, with a mouth, eyes, and a nose. If you don't like what you're drawing at any time, you can use the eraser on the pencil to remove it. Basically, to draw the stick figure, you are using lines, circles, and arcs to construct it.

Drawing a Stick Figure Using CAD

Now try drawing the stick figure in CAD. You will use commands that get used very often in AutoCAD LT—for example, the **D**raw commands of LINE, CIRCLE, and ARC. You will also pick points for the LINE command, two points for each line. The CIRCLE command will be accomplished by picking points for the center and radius of the circle. With the ARC command, you will draw from a start point, to a second point, and to an end point.

Sometimes when you are drawing, you will have the Ortho mode turned on. When Ortho mode is turned on, it forces the cursor to move just horizontally or vertically, not diagonally. Ortho mode works whenever a second point is required in a command. It can be turned on and off in the middle of commands.

If at any time you make an error, you can always eliminate it. With the ERASE command, you can select objects and press Enter, and the objects will disappear. You can also use the UNDO command by choosing the Undo button, which will undo the most recent command, one command at a time.

Now, see how these commands work by drawing a stick figure. The intent of this exercise is to get you to sketch in AutoCAD LT.

Exercise

Drawing a Stick Figure in AutoCAD LT

Continue from the previous drawing exercise.

Command: *Choose the Ortho button*	Toggles on the Ortho mode setting
Command: *Choose the Line button*	Issues the LINE command
LINE From point: *Pick point at* ① *(see fig. 1.10)*	Places the starting point of the line for the stick body
To point: *Pick* ②	Places the ending point of the line
Command: *Choose the Ortho button*	Toggles off Ortho setting
LINE From point: *Pick* ③	Places the ending point of a second line, the stick leg
To point: (Enter)	Ends the LINE command
Command: (Enter)	Reissues the LINE command
LINE From point: *Pick point at* ④	Places the starting point of the line, the stick arm
To point: *Pick* ⑤	Places the ending point of the line
To point: (Enter)	Ends the LINE command

Pick points in a similar fashion for the right side of the body, pressing Enter to end the LINE command and pressing Enter again to reissue the LINE command.

Command: *Choose the Circle button*	Issues the CIRCLE command
CIRCLE 3P/TTR/<Center point>: *Pick point at* ⑥	Places the center point of the circle directly above the stick body

continues

`Diameter/<Radius>:` *Pick* ①	Specifies the radius of the circle

Now the edge of the circle should be directly at the top of the stick body. If not, choose the Undo button, and try it again.

`CIRCLE 3P/TTR/<Center point>:` *Pick point at* ⑦	Places the center point of the circle for the left eye
`Diameter/<Radius>:` *Pick a second point so the eye will appear as in figure 1.10*	Specifies the radius of the circle and completes the eye

Pick points in a similar fashion for the right eye and the nose.

`Command:` *Choose the Arc button*	Issues the ARC command
`Arc Center/<Start point>:` *Pick point at* ⑧	Starts the left end of the arc
`Center/End/<Start Point>:` *Pick* ⑨	Picks the midpoint of the arc
`End point:` *Pick* ⑩	Picks the end of the arc

Create the upper portion of the mouth in a similar fashion. If at any time you need to delete a portion of a drawing, you can use the ERASE command.

`Command:` *Choose the Erase button*	Issues the ERASE command
`Select objects:` *Pick* ⑪	Selects the arm
`Select objects:` Enter	Erases the arm
`Command:` *Choose the Undo button*	Issues the UNDO command, and brings back the arm
`Command:` **REDRAW** Enter	Issues the REDRAW command and cleans up the screen
`Command:` *Choose the Save button*	Issues the SAVE command
In the File **N**ame *area, type* **STICK** *and choose* OK	Saves the drawing as the file STICK.DWG

Your drawing should now resemble figure 1.10

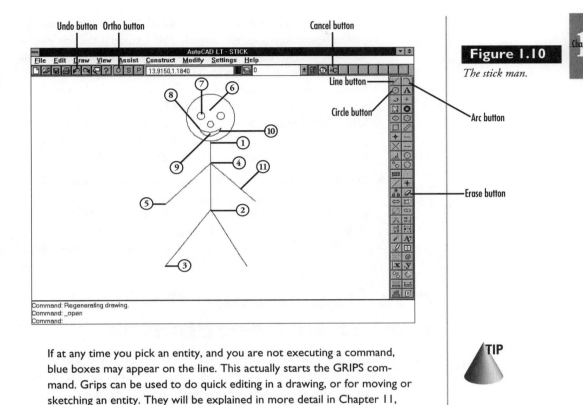

Undo button Ortho button

Cancel button

Line button
Circle button
Arc button
Erase button

Figure 1.10

The stick man.

TIP

If at any time you pick an entity, and you are not executing a command, blue boxes may appear on the line. This actually starts the GRIPS command. Grips can be used to do quick editing in a drawing, or for moving or sketching an entity. They will be explained in more detail in Chapter 11, "Creating a Mechanical Drawing." To get rid of these blue boxes, press the Cancel (^C) button on the toolbar. Also, if you pick an object and just slightly miss it, you may start to create a box. Press the Cancel (^C) button to terminate this box.

There are primarily three ways to issue commands in AutoCAD LT, and you used all three methods in the last exercise, as follows:

◆ **Choosing a button.** Choosing a button from the toolbar or toolbox is the way you issued the LINE command. Although there are many more commands in AutoCAD LT than there are buttons, you can customize buttons for those commands you use most often. Choosing a button is the method that you will use most often in this book.

◆ **Choosing from a pull-down menu.** Choosing commands from a pull-down menu is another method. This method is the way you activated the **F**ile, then **S**ave menu item. There are more commands to choose from in this method, but it's not quite as quick as the buttons method. You can also use your Alt key on the keyboard to activate the pull-down menu commands. For example, if you press Alt+F, and then press **S** on your keyboard, you can save your drawing. The underlined letter is the hot key. Hold down the Alt key on the keyboard at the same time you press the hot key.

◆ **Typing the command.** In this method, you type the command from the keyboard and press Enter. This method is the way you issued the REDRAW command in the last exercise. All commands in AutoCAD LT are available by this method. The disadvantage is remembering the different commands and entering them correctly.

Coordinates and Points: Cartesian Coordinates

When creating a drawing in AutoCAD LT, you are doing it within a 3D Cartesian coordinate system. The CAD program locates a point in space relative to the origin, measured along the perpendicular axes of x, y, and z. The origin is 0, 0, 0, which is where all three axes intersect.

As for the exercises in this book, you will be referring to a 2D area with 0,0 being the origin, X being the horizontal axis, and Y being the vertical axis. As you look at the graphics area, X coordinate values to the right of the origin are positive and to the left are negative. Y values above the origin are positive and below the origin are negative.

Absolute Coordinate Values

The origin 0,0 is generally in the lower left corner of the graphics area of AutoCAD LT. In terms of absolute coordinates, each number on the screen represents the distance from the origin. To give you an idea of how these points are located within the graphics area in relation to the origin, look at the rectangle in figure 1.11. The coordinate 4,8 specifies a point four units in the positive x axis and eight units in the positive y axis, measured from the 0,0 origin. Each point in AutoCAD LT requires coordinate location as part of its definition.

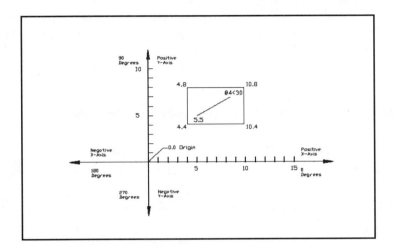

Figure 1.11

The 2D Cartesian coordinate system with x and y axes.

Polar Coordinates: An Alternative

Just as you can enter points for lines based on the 0,0 origin, you can also enter points based on the last point picked. For example, start a line at 5,5 (an absolute point), and from this point, enter a distance of @4<30. A line will be drawn that is drawn from the last point at (@), 4 units long, angled (<), at 30°. In other words, the syntax for polar coordinates is @*distance*<*direction*. So instead of relating to the 0,0 origin, the new point relates to the last point picked (see fig. 1.11).

To clarify how polar coordinates work, see the following. If you start a line at any point in the graphics area, and want to draw it:

◆ Up: in the 90° direction

◆ Left: in the 180° direction

◆ Down: in the 270° direction

◆ Right: in the 0° direction

Typed Coordinates

Coordinate locations for points, lines, circles, and so on can either be entered, or they can be precisely picked on the screen with the snap mode on. If the snap mode is on—the S button—you are restricted to picking points to the nearest .5000 unit. These units can be set in the Quick Drawing Setup dialog box. To see how absolute and polar coordinates work, you can enter some different coordinates in the next exercise.

Exercise

Using Typed Absolute and Relative Coordinates

Use the drawing in figure 1.12.

Choose **F**ile, **N**ew, *then choose* OK *in the Create New Drawing dialog box*	Starts a new drawing and opens the ACAD file
In the Quick Drawing Setup dialog box, choose OK	Accepts the defaults for units of measurement (decimal) and size of the sheet
Command: *Choose the Line button*	Issues the LINE command
LINE from point: **4,4** (Enter)	Starts the line at point 4,4
To point: **4,8** (Enter)	Draws the line up in the y axis
To point: **10,8** (Enter)	Draws the line to the right in the x axis
To point: **10,4** (Enter)	Draws the line down in the y axis
To point: **4,4** (Enter)	Completes the rectangle and ends the LINE command
To point: (Enter)	Ends the LINE command
Command: (Enter)	Reissues the LINE command
LINE from point: **5,5** (Enter)	Starts the line at point 5,5
To Point: **@4<30**	Draws a line 4 units long in a 30° direction
To point: (Enter)	Ends the LINE command

Command : *Choose the Save button,* type **HOUSE**, *and press Enter*	Saves the drawing as HOUSE for the next exercise

Your drawing should now resemble figure 1.12

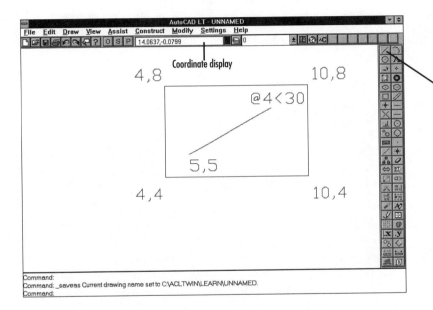

Figure 1.12

Using absolute coordinates to create geometry.

Line

You have learned some important commands and how to type coordinates into the computer to draw a precision drawing. When you first look at AutoCAD LT, it looks as if it has a lot of commands to learn. If you learn about 20 percent of the commands, however, you can do about 80 percent of the work. Some important commands to be familiar with are the following:

◆ Draw commands—LINE, ARC, CIRCLE, RECTANGLE, and PLINE

◆ Construct commands—COPY, OFFSET, and MIRROR

◆ Modify commands—ERASE, MOVE, and TRIM

◆ Settings—drawing aids and layer control

◆ File commands—NEW, OPEN, SAVE, and print/plot

Drawing a House

Now that you've learned how to enter absolute and polar coordinates to create a drawing, you can use the Snap button and the coordinate display to precisely pick points to create the house drawing. To use the coordinate display to pick points, the coordinate display needs to be picked. When you click on the coordinate display, it toggles through three displays, as follows:

◆ **Dynamic Absolute.** X, Y coordinates, continuously updated

◆ **Dynamic Polar.** Polar coordinates during certain commands, like the LINE command

◆ **Static Absolute.** X, Y coordinates, updated only when you pick a point

In the next exercise, you draw a house while watching the coordinate display and picking points. Your Snap button will be highlighted. Remember that this button restricts movement of the cursor to the nearest .5000.

You try out two new commands in this exercise. The POLYLINE command enables you to give thickness to a line, which you cannot do with the LINE command. The POLYLINE command can also be issued by typing **PLINE** (Enter) at the command line. Another command that is quicker than the LINE command is the RECTANG command. By picking two points, you can create a rectangle, instead of having to pick four points with the LINE command. The RECTANG command can also be issued by typing **RECTANG** (Enter) at the command line.

Exercise

Picking Points to Draw a House

Command: *Choose* **F**ile, *then* **O**pen, *then choose the HOUSE file*	Opens the HOUSE file, if it is not already open
Command: *Choose the Erase button*	Issues the ERASE command
Select Objects: *Select the lines of the last exercise, then press Enter*	Erases the lines of the last exercise so you can have a clean graphics area to create a house
Select Objects: (Enter)	Ends the ERASE command

`Command:` *Choose the Snap button*	Activates the snap grid (coordinates should read to the nearest .5000)
`Command:` *Choose the Polyline button*	Issues the PLINE command

`Arc/Close/Halfwidth/Length/`
`Undo/Width/<Endpoint of Line>:`
`W` (Enter)
	Starts drawing a ground line

`Starting Width<0.00>:` `.05` (Enter)	Gives the line a starting width of .05
`Ending Width<0.05>:` `.05` (Enter)	Gives an ending width of .05

`Arc/Close/Halfwidth/Length/`
`Undo/Width/<Endpoint of Line>:`
Pick 1.5000,1.0000 at ① *(see fig. 1.13)*
	Starts drawing a ground line

You can pick the point precisely by watching the coordinate display and picking at the right time.

`Arc/Close/Halfwidth/Length/`
`Undo/Width/<Endpoint of Line>:`
Pick 13.5000,1.0000 at ②
	Finishes drawing the ground line

`Arc/Close/Halfwidth/Length/`
`Undo/Width/<Endpoint of Line>:`
(Enter)
	Ends the PLINE command

`Command:` *Choose the Line button*	Issues the LINE command
`LINE from point:` *Pick* *3.5000,1.0000 at* ③	Places the starting point of the line for the edge of the house
Click on the Coordinate Display area until the dynamic polar coordinates appear	Displays the distance from the last point using the < sign
`To point:` *Pick 4.0000<90 at* ④	Places the ending point of a line
`To point:` *Pick 4.0311<30 at* ⑤	Places the ending point of a line
`To point:` *Pick 4.0311<330 at* ⑥	Places the ending point of a line
`To point:` *Pick 4.0000<270 at* ⑦	Places the ending point of a line
`To point:` (Enter)	Ends the LINE command

continues

Command: `Enter`	Reissues the LINE command
`LINE from point:` *Pick 5.0000, 1.0000 at* ⑧	Places the starting point of the line for the door
`To point:` *Pick 2.5000<90 at* ⑨	Places the ending point of a line
`To point:` *Pick 1.5000<0 at* ⑩	Places the ending point of a line
`To point:` *Pick 2.5000<270 at* ⑪	Places the ending point of a line connecting the door to the bottom of the building
`To point:` `Enter`	Ends the LINE command
Command: *Choose the Rectangle button*	Issues the RECTANG command
`First Corner:` *Pick 7.0000,2.0000 at* ⑫	Starts the rectangle in the lower left corner
`Other Corner:` *Pick 8.0000,3.5000 at* ⑬	Ends the RECTANG command and completes the rectangle and window, using dynamic absolute coordinates as a way to place the other corner
Command: *Choose the Copy button*	Issues the COPY command
`Select Objects:` *Select the rectangle*	Selects the rectangle to copy
`Select Objects:` `Enter`	Ends the selection process
`<Base point or displacement> /Multiple:` *Pick any point*	Selects a base point to copy from anywhere on the screen
`Second point of displacement:` *Pick 1.5000<0 at* ⑭	Copies the rectangle over 1.5 units in the 90° direction
Command: *Choose the Circle button*	Issues the CIRCLE command
`CIRCLE 3P/TTR/<Center Point>:` *Pick 7.0000,5.5000 at* ⑮	Places the center point of the circle
Click on the coordinate display until the dynamic absolute coordinates appear	Displays the distance from the next point to the origin
`Diameter/<Radius>:` *Pick 6.5000, 6.0000*	Picks the radius of the circle

Command: *Choose the Save button*   Saves the drawing

Your drawing should now resemble figure 1.13

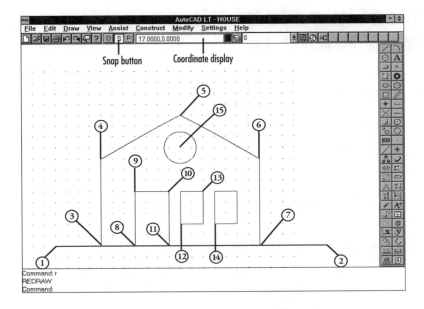

Figure 1.13

Picking points to draw a house.

If you want to change the spacing of the grid or snap in a drawing, adjustments can be made in the Drawing Aids dialog box. To access this dialog box, choose **S**ettings, then **D**rawing Aids from the pull-down menu. Changing the numbers in this dialog box does not affect the existing lines in your drawing. You can also adjust the grid and snap spacing as you enter a new drawing in the Quick Drawing Setup dialog box. In addition, you can use the F7 function key on the keyboard to toggle the grid on and off, and the F9 key to toggle snap on and off.

NOTE

Congratulations—you have now drawn two or more CAD drawings. These drawings will become more challenging as you continue working through the book.

Getting Help When You Need It

Another tool that will make it easier to use AutoCAD LT is the HELP command. The **H**elp pull-down menu gives you a number of choices as far as making use of help. If you choose **H**elp in the pull-down menu, you will see the following options:

- ◆ **Contents.** Defines all the different parts of AutoCAD LT and gives information on basic drawing techniques, menus, toolbar, toolbox, and command line.

- ◆ **Search for Help On.** Information on commands and system variables.

- ◆ **How to Use Help.** Instructions on how to use help.

- ◆ **Orientation.** Information on how to make the transition from manual drafting to using AutoCAD LT Release 2.

- ◆ **What's New.** Information about ease of use, new tools, and productivity in AutoCAD LT Release 2.

- ◆ **Tutorial.** Walkthrough exercises on how to use AutoCAD LT.

- ◆ **About AutoCAD LT.** Important notification for new users.

One way to access **H**elp, **C**ontents is from the pull-down menu. Another way is to type **HELP** at the command line. Still a third way is to press the F1 key at the keyboard. As shown in figure 1.14, when the AutoCAD LT **H**elp **C**ontents window opens, there are a number of options to choose from, which you can activate by clicking on the item:

- ◆ **Quick Help.** Gives graphic information on how to use the buttons across the top of the dialog box—**C**ontents, **S**earch, **B**ack, His**t**ory, **<**<, **>**>, and **G**lossary.

- ◆ **How do I.** Gives basic drawing techniques with instructions.

- ◆ **Toolbar.** Describes commands on the toolbar.

- ◆ **Toolbox.** Describes commands on the toolbox.

- ◆ **Menus.** Describes commands on the pull-down menus.

- ◆ **Command line.** Details commands at the command prompt.

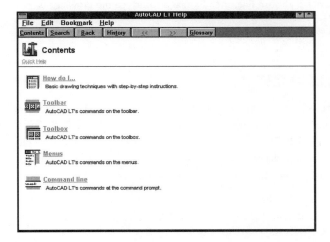

Figure 1.14

The AutoCAD LT Help Window, Contents section.

The buttons across the top of the dialog box are the following:

- ◆ **Contents.** Displays the Contents menu.

- ◆ **Search.** Opens the Search dialog boxes, where you can enter words or key phrases to search for help on.

- ◆ **Back.** Returns to previously displayed window.

- ◆ **History.** Displays the previously displayed help subjects from the current help session; you can choose one to reopen.

- ◆ **<< and >>.** Pages to the previous and next topics in the alphabetical topic list for the current window.

- ◆ **Glossary.** Displays a glossary of terms.

Other options also at the top of the dialog box are **F**ile, **E**dit, Book**m**ark, and **H**elp. You can choose from these pull-down menus to print topics, copy and annotate information, and provide bookmark information so that a topic can be easily found again.

TIP

A quick and easy way to use **H**elp is to choose the command you have a question about; for example, choose the ARC command. After you have issued the command, press F1 on the keyboard, and you will get a complete description of how it works. If you click on the green letters within the description, you will get additional information. This is an easy, less complex way to use **H**elp. When you are done getting the information you need, choose **F**ile, then E**x**it, and return to the drawing.

Summary

In this chapter, you were introduced to the AutoCAD LT graphics window. Some important commands were discussed, such as LINE, CIRCLE, ARC, PLINE, RECTANG, ERASE, and UNDO. You created a sketch in CAD by drawing a stick man, and used precision coordinates to draw a house. You also learned about the different Help options.

Now that you have a taste for how CAD works, you'll look at an overview of manual drafting tools compared to CAD tools in the next chapter.

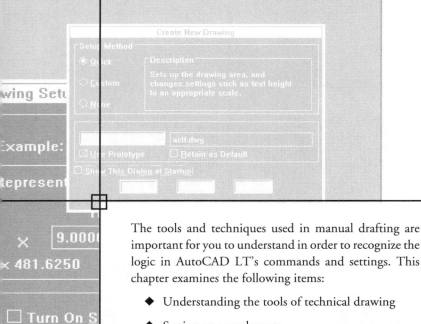

Chapter

2

The tools and techniques used in manual drafting are important for you to understand in order to recognize the logic in AutoCAD LT's commands and settings. This chapter examines the following items:

◆ Understanding the tools of technical drawing

◆ Setting up page layouts

◆ Understanding lettering and text

◆ Working with linetypes within a drawing

The first half of this chapter is designed to provide a comparison between traditional technical drafting tools and those now used within the CAD environment. The second half defines traditional techniques used in manual drafting and their current CAD equivalents.

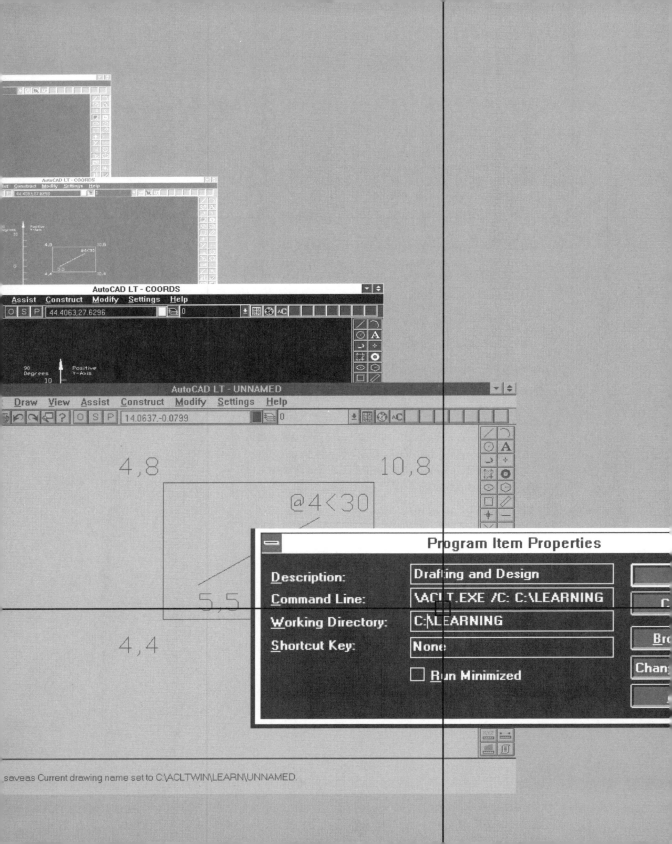

AutoCAD LT - COORDS

Assist Construct Modify Settings Help

O S P 44.4063,27.6296   0

90
Degrees
10

Positive
Y-Axis

AutoCAD LT - UNNAMED

Draw View Assist Construct Modify Settings Help

O S P 14.0637,-0.0799   0

4,8   10,8

@4<30

5,5

4,4

	Program Item Properties
Description:	Drafting and Design
Command Line:	\ACLT.EXE /C: C:\LEARNING
Working Directory:	C:\LEARNING
Shortcut Key:	None
☐ Run Minimized	

.saveas Current drawing name set to C:\ACLTWIN\LEARN\UNNAMED.

An Overview of Drafting Tools and Techniques

Many of the tools and techniques that are about to be discussed have a rich history dating back to 4000 B.C. At that time, draftspeople carved ideas in stone with chisels. Drafting technology has come a long way. Although many draftspeople are still using traditional tools and techniques, the advent of computers has turned the profession into an exciting field dominated by computer technology.

In this chapter, you explore the various tools and drawing aids provided in AutoCAD LT to produce the drawing shown in figure 2.1.

Figure 2.1

The completed sailboat drawing.

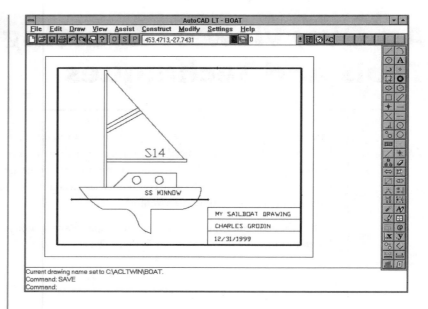

Understanding Drafting Tools

In the past, the tools used to develop drawings were handled with care in order to provide the utmost accuracy in a drawing. Draftspeople had an assortment of tools and equipment to utilize based on the type of drawing that was to be created. Many of the limitations these tools present are no longer an issue in a CAD environment. The purpose of this section is to introduce the tools of manual drafting and their CAD equivalents.

Using a Drafting Board

Drafting boards are smooth tables on which to develop drawings (see fig. 2.2). The boards were traditionally made of soft wood so paper could be tacked down at the corners. After repeated use, however, the boards were filled with holes, so the profession switched to tape as a means to hold the paper down.

Figure 2.2

*A typical
drafting table.*

2
Chapter

These boards are adjustable tables to provide comfort and can grow to considerable size based on the drawings created. Mies VanDerRohe, a famous modern architect, was fond of developing full-scale, building wall sections on the drafting board. Developing a 10- to 12-foot-high drawing can be cumbersome. This problem of drawing size is eliminated with CAD.

Think of the monitor as the drafting board. All lines are drawn to full scale and can be viewed at any distance—with a magnifying glass or from a mile away. The computer is incredibly flexible. In the last chapter, you explored the drawing interface of AutoCAD LT. This interface is the equivalent of the drafting board.

In the following exercise, start a new drawing and draw a small square to explore how the magnification technique works. You will use the ZOOM command to adjust the magnification of the drawing.

Exercise

Changing the Magnification of a Drawing

Start a new drawing.

Command: *Choose* **F**ile, *then* **N**ew	Opens a new drawing
Press **Enter** *to accept all of the defaults within the dialog boxes*	
Command: *Choose the Rectangle button*	Issues the RECTANGLE command
First corner: **1,1** **Enter**	Sets the lower left corner
Other corner: **4,4** **Enter**	Sets the upper right corner
Command: *Choose the Zoom button on the toolbar*	Issues the ZOOM command
All/Center/Extents/Previous/ Window/<Scale(X/XP)>: **.8x** **Enter**	Zooms the view to 80 percent of the original
Command: **Enter**	Reissues the ZOOM command
All/Center/Extents/Previous/ Window/<Scale(X/XP)>: **.05x** **Enter**	Zooms the view to five percent of the previous view

This is a quick exercise to gain insight into the magnification of drawings. The ZOOM command and its options will be discussed throughout the book.

Using a Pencil and Ink Pen

As mentioned earlier, drawings used to be carved in stone with a chisel. Reed pens dipped in ink were the tool of choice to create a straight line until the seventh century, when quills were introduced. Quills were created from bird feathers, particularly goose feathers. Goose-quill pens were used by draftspeople until the eighteenth century, when the graphite pencil was invented.

Pencils and ink pens are the most essential tool for any draftsperson. Pencils provide the drafter with a variety of line weights based on the lead's grade or hardness. Leads range from 9H (hardest) to 7B (softest)—the softer the lead, the larger the lead diameter. The familiar 2H lead, used to fill in the circles of standardized tests, falls in the middle of the grades. Harder leads are used to create light, thin lines in drawings, whereas softer leads, which smudge easily, are used for artwork applications.

Ink pens were used by many draftspeople as a replacement for ink-filled quills. A ruling pen is a refillable pen with an adjustable tip for creating a variety of line thicknesses. Ruling pens were replaced by mechanical pens. Mechanical pens are refillable or come with cartridges. Each pen in a set comes with a nib of varying width. Although pencils are the tool of choice when drafting manually, ink provides crisp, clear, consistent lines.

NOTE

Pen widths will be addressed again in Chapter 10, "Reproducing Drawings," during the discussion of pen plotters.

Simple straight lines created with pencils and pens on a drafting board are translated in AutoCAD LT as the LINE command. Like pencils and pens, this will be the most often-used tool while creating drawings.

The LINE command asks for a start point (where to put the pen on the paper to start the line) and an end point (where to stop the line). By entering coordinates, you can draw lines anywhere, at any distance and angle. The @ symbol represents X,Y coordinates from the last point specified instead of the origin (0,0) of the drawing. The methods of entering coordinates are presented in further detail in Chapter 3, "CAD Drawing Techniques." Also, the ZOOM command is issued at the end of this exercise to display the whole drawing. The ZOOM command will be presented in the following chapters. In this exercise, you create the top and rear of the boat's cabin.

Using the LINE Command

Exercise

Start a new drawing; do not save the previous drawing.

Command: *Choose* **F**ile, *then* **N**ew   Opens a new drawing

continues

Press (Enter) *to accept all of the defaults within the dialog boxes*

Command: *Choose the Line button*	Issues the LINE command
From point: **204,213** (Enter)	Starts a line at the coordinates 204,213
To point: **@120,0** (Enter)	Creates a line 10" long to create the top of the cabin
To point: **@0,-36** (Enter)	Creates a line 3" down to create the back of the cabin
To point: (Enter)	Exits the LINE command
Command: (Enter)	Reissues the LINE command
LINE From point: **84,177** (Enter)	Reissues the LINE command
To point: **@300,0** (Enter)	Creates the deck
To point: (Enter)	Ends the LINE command
Command: *Choose the Zoom button from the toolbar*	Issues the ZOOM command
All/Center/Extents/Previous/ Window/ <Scale(X/XP)>: **A** (Enter)	Zooms view to include all entities in the display
Command: *Choose* **F**ile, **S**ave	Saves the drawing

NOTE

Remember to save the drawing often so you do not lose the work you created in the exercises.

Using a T-Square or Parallel Slide

Another archaic drafting tool is the T-square (see fig. 2.4). A T-square consists of two parts, the blade and the head. The AutoCAD LT opening screen features a T-square, which functions as a T in the logo.

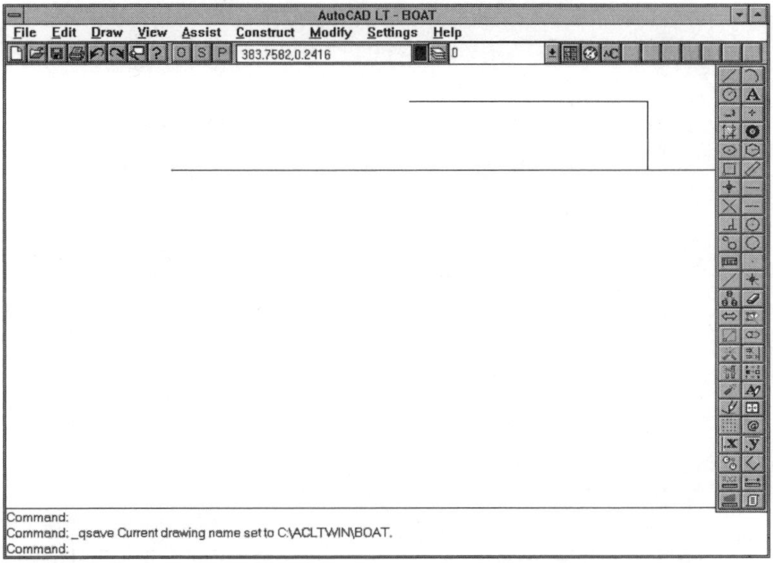

Figure 2.3

*The beginning of
the sailboat.*

The blade is a long strip with clear edges. The head is fastened perpendicular to the blade. By placing the head against the left surface of the drawing board, a drafter is able to produce long straight lines within a drawing. The blade also provides a straight edge to line up the paper before adhering it to the board.

Another tool developed to accomplish this task is the parallel slide (see fig. 2.4). This tool was created to remove the difficulties of keeping a T-square straight. The blade is attached to the drawing board by cables, which allow it to roll up and down. This is the preferred straight-edge in firms still utilizing manual drafting techniques. AutoCAD LT has a variety of aids to keep lines straight and accurate.

Using Ortho Mode

The tool that mimics drawing with a straightedge is called Ortho mode. When Ortho mode is active, the cursor is restricted to vertical and horizontal directions. These movements are described as "orthogonal"—up, down, right, and left. This is helpful not only when creating lines, but also when moving groups of lines or whole drawings on a sheet.

Figure 2.4

A T-square and parallel slide.

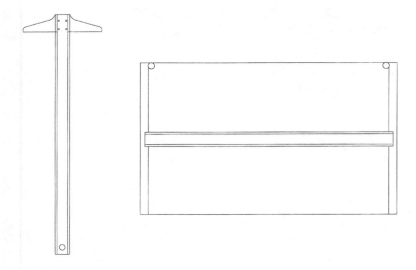

The Ortho button on the toolbar is grayed out and appears raised when Ortho is not activated. When selected, the Ortho button appears highlighted and depressed.

There are a variety of ways to turn on the Ortho mode, as follows:

◆ Click on the **O** in the toolbar

◆ Press F8

◆ Press Ctrl+O

◆ Choose **S**ettings, **D**rawing Aids (or choose the Drawing Aids button), then click in the Ortho box in the Modes section of the dialog box.

All of these methods toggle the Ortho mode on or off. If you issue a command when Ortho mode is on, your cursor will only move horizontally and vertically.

In the next exercise, try the Ortho toggle methods discussed previously.

Setting the Ortho Mode

Within the current drawing, try the following steps.

Command: *Choose the Line button*	Issues the LINE command
From point: **1,1** (Enter)	Sets the lower left corner
To point: *Press F8*	Toggles the Ortho mode on or off
Try the other methods listed to toggle the Ortho mode on and off	
To point: *Press Ctrl+C*	Cancels the LINE command

Using Grid and Snap Modes

When you are generating a drawing, the ability to work within a grid is essential. Graph paper is widely used within the drafting field. The grid is printed as non-photo blue (which means it will not reproduce on a blueprint) or laid under the drawing paper as a guide. You can assign a grid to the drawing, like graph paper, to visually identify points. The grid can be either set up in a traditional square pattern or with unequal X and Y spacing to form a rectangular grid. Like Ortho mode, there are different ways to turn the grid on, as follows:

◆ Press F7

◆ Press Ctrl+G

◆ Choose **S**ettings, **D**rawing Aids, choose the Drawing Aids button, or type **SNAP** at the Command line (see fig. 2.5).

The last option opens the Drawing Aids dialog box. Here, you specify the size of the grid. To set the grid spacing, go through the next exercise.

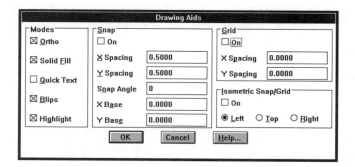

Exercise

Setting the Grid

Proceed within the current drawing.

Command: *Choose* **S**ettings, *then* **D**rawing Aids	Opens the Drawing Aids dialog box
Choose the On box under Grid	Checks the box
Double-click in the X Spacing box	Highlights the text in the box
Type **50**	Sets the X Spacing to 50
Double-click in the Y Spacing box	Sets the Y Spacing to 50
Choose OK	Closes the dialog box

The drawing now displays the grid. Notice that these are just visual points and cannot currently be selected as points. Also notice that the grid does not cover the entire screen. The grid area is defined by the drawing limits. Limits are discussed at length in Chapter 4, "Multiview Drawings."

Setting the snap spacing allows the grid points to be selected. The grid and snap aids work together to form a reliable method of picking points within the drawing.

The snap and grid spacing settings do not have to be equal. Although they are often used together, they operate independently.

To set the snap spacing, follow the steps in the next exercise.

Exercise

Setting the Snap

Proceed within the current drawing.

Command: *Choose* **S**ettings, *then* **D**rawing Aids	Opens the Drawing Aids dialog box
Choose the On box under Snap	Checks the box
Double-click in the X Spacing box	Highlights the text in the box
Type **10**	Sets the X Spacing to 10
Double-click in the Y Spacing box	Sets the Y Spacing to 10
Choose OK	Closes the dialog box

Other options in the Drawing Aids dialog box are explored later in this book.

Within the drawing, the cursor movements are restricted to the grid and snap points because they have equal settings. Again, the snap spacing controls the cursor, but the grid does not.

When turning the Grid and Snap modes on or off, opening the Drawing Aids dialog box is considered inefficient. Use the F key, Ctrl key, or toolbar options.

When you are setting up a preliminary page layout, grid and snap settings will eliminate the guesswork. These aids will be used in exercises throughout this book.

Creating Angled Lines

When supported on a straight-edge, triangles are used to form vertical and angular lines (see fig. 2.6). Historically, they were made of wood and later of lightweight metal. Currently, they are made of clear plastic to ensure drawing visibility. The two most popular triangles are 30°-60° and 45°.

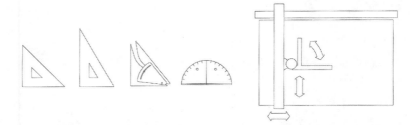

Angles are created every 15° by combining these two triangles. Drawing triangles are available in many colors and sizes, depending on the scale of the drawings to be created.

To form angles other than those that could be generated with the triangles, a protractor is used. Protractors are also made of clear plastic and are circular or semicircular in shape. Degree marks are engraved around the circumference of the protractor, which gives the draftsperson the ability to construct a line at any angle. It is difficult to maintain accuracy with this tool.

For greater accuracy, an adjustable triangle was created (see fig. 2.6). This triangle offers a draftsperson the ability to construct and project lines at any angle and is an invaluable tool. In lieu of all these tools, firms may also employ a drafting machine (see fig. 2.6). This machine is attached to the drafting board and functions as a straightedge, triangle, and protractor.

As with any manual tool, the accuracy of the drawing is in the hands of the drafter. AutoCAD LT eliminates a lot of the guesswork inherent in manual drafting tools, by providing accuracy to eight places.

Continuing with the drawing project, you can use the capability to produce angled lines to create a sail and mast. In order to specify angular coordinates, an angle is added when prompted for To points in the exercise. The methods of entering coordinates are presented in further detail in Chapter 3. Also, the drawing will be cleaned up later in the exercise, so don't be alarmed when lines overlap.

Creating Angled Lines

Proceed within the current drawing.

Command: *Choose the Line button*	Issues the LINE command
From point: **204,213** (Enter)	Starts a line at the top of the cabin
To point: **@60<225** (Enter)	Creates a line 5" long to create the front of the cabin
To point: (Enter)	Ends the LINE command
Command: (Enter)	Reissues the LINE command
LINE From point: **145,174** (Enter)	Starts a line at the top of the deck
To point: **@0,300** (Enter)	Begins the mast
To point: **@6,0** (Enter)	Creates top of mast
To point: **@0,-300** (Enter)	Completes the mast
To point: (Enter)	Ends the LINE command
Command: *Choose the Zoom button on the toolbar*	Issues the ZOOM command
All/Center/Extents/Previous/ Window/ <Scale(X/XP)>: **A** (Enter)	Zooms view to include all entities in the display
Command: *Choose the Line button*	Issues the LINE command
LINE From point: **148,248** (Enter)	Starts the line
To point: **@200,0** (Enter)	Begins the boom
To point: **@0,-6** (Enter)	Creates end of boom

continues

To point: **@-200,0** (Enter)	Completes the boom
To point: (Enter)	Ends the LINE command
Command: (Enter)	Reissues the LINE command
From point: **148,471** (Enter)	Starts the line
To point: **@300<311** (Enter)	Creates the sail
To point: (Enter)	Ends the LINE command
Command: (Enter)	Reissues the LINE command
LINE From point: **148,333** (Enter)	Starts the line
To point: **@110<30** (Enter)	Creates a sail graphic
To point: (Enter)	Ends the LINE command
Command: (Enter)	Reissues the LINE command
From point: **148,323** (Enter)	Starts the line
To point: **@110<30** (Enter)	Creates a sail graphic
To point: (Enter)	Ends the LINE command
Command: (Enter)	Reissues the LINE command
LINE From point: **148,313** (Enter)	Starts the line
To point: **@110<30** (Enter)	Creates a sail graphic
To point: (Enter)	Ends the LINE command
Command: *Choose the Zoom button in the toolbar*	Issues the ZOOM command
All/Center/Extents/ Previous/Window/ <Scale (X/XP)>: **A** (Enter)	Zooms view to include all entities in the display
Command: **SAVE** (Enter) *(see fig. 2.7)*	Saves the drawing

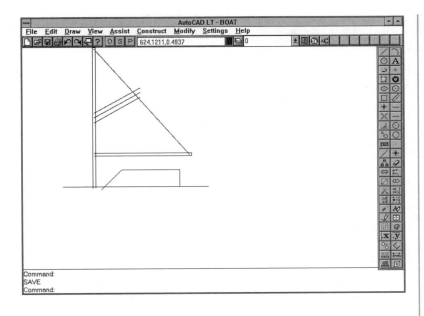

Figure 2.7

The sail, mast, and boom of the boat.

Using a Compass

Along with the pen, the compass was one of the first tools developed for technical drawing. Early compasses were made of bronze and were used to etch or scratch arcs and circles. After graphite lead was introduced, compasses were fitted for pencils and were similar to those still used today.

There are two basic types of compasses: the common bow compass and the beam compass (see fig. 2.8).

The bow compass consists of a needle point and a pen point. The needle point holds the center point, which allows the pen point to rotate and form a circle or arc. Think of the needle point when AutoCAD LT prompts for a Center point. A beam compass operates on the same principle, but the beam allows very large circles to be created.

Figure 2.8

Typical compass types.

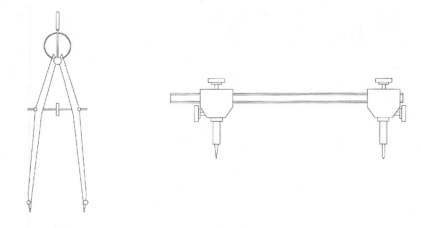

Circles and arcs are important entities to be included in a drawing. All arcs and circles have a center point (the needle point) and a radius (the pen movement). Arcs have different start and end points, while circles share these points. Great care is necessary to produce accurate circles; the pen point needs constant maintenance to keep a consistent line. If large drawings are produced, the physical size of the workspace creates a limitation. AutoCAD LT solves the problem again by taking away limitations of space. The ability to draw lines from a variety of points along or tangent to the circle also eliminates sloppy drafting techniques.

Circles can be created in the following ways:

◆ **3P.** Specify three points, and a circle will be generated through the points.

◆ **TTR.** Pick tangent points on two different curves, and then specify a radius.

◆ **Center point.** Specify a center point (needle point) and set the radius. This is the default option.

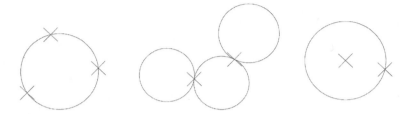

Figure 2.9

Options for specifying circles.

In the next exercise, create a few circles to explore the CIRCLE command. First, you will use a technique that places a mark in your drawing. After performing several commands, you can undo back to the mark and continue with the drawing.

Performing an Undo Back will undo all commands to the last mark. If you have not saved your drawing, all drawing operations will be removed. Remember to save or place a mark.

CAUTION

Start this exercise by creating a mark in the drawing.

Exercise

Drawing Circles

Proceed within the current drawing.

Command: **UNDO** (Enter)	Issues the UNDO command
Auto/Back/Control/End/Group/ Mark/ <number>: **M** (Enter)	Sets a mark in the drawing
Command: *Choose the Circle button*	Issues the CIRCLE command
3P/TTR/<Center point>: *Pick anywhere in the drawing area*	Sets the center point of the circle
Diameter/<Radius>: **2** (Enter)	Sets the radius to 2
Command: (Enter)	Reissues the CIRCLE command
3P/TTR/<Center point>: **3P** (Enter)	Selects the 3 point option

continues

`First point:` *Pick anywhere in the drawing area*	Sets the first point
`Second point:` *Pick anywhere in the drawing area*	Sets the second point
`Third point:` *Pick anywhere in the drawing area*	Sets the third point
Create several circles	
`Command:` **UNDO** (Enter)	Issues the UNDO command
`Auto/Back/Control/End/` `Group/Mark/` `<number>:` `B` (Enter)	Performs the UNDO command until reaching the mark

Arcs are generated by specifying three different points. AutoCAD LT will prompt you for the information.

TIP

Remember to read the prompts when executing the Arc command. Notice that the first letter of each option is capitalized. This is common in all commands with options. By entering the capitalized letter(s) only, you can issue the command without typing its full name. This method will save you time and avoid confusion.

Here are the options for the Arc command:

- ◆ **Start point.** The Start point is the default option. This is the point at which the arc will begin.

- ◆ **Center.** This will specify the center point of the arc.

- ◆ **End.** The endpoint of the arc.

- ◆ **Second point.** The point between the start and end points of the arc. Using the Start, Second, and End option will automatically set the radius.

- ◆ **Angle.** This angle is measured counterclockwise from the base angle established by the center point and start point.

3 pt   S,C,E or   S,E,Angle   C,S,E   C,S,Angle
        S,C,Angle

Figure 2.10

Options for developing arcs.

Chapter

2

Although there are a lot of options, once you use these commands with frequency, the mystery will fade.

Continuing with the drawing project, you use circles and arcs to create portholes, the bow, and the rest of the sailboat.

Exercise

Working with Arcs and Circles

Proceed within the current drawing.

Command: *Choose the Circle button*	Issues the CIRCLE command
3P/TTR/<Center point>: **225, 195** (Enter)	Specifies the center point of the circle
Diameter/<Radius>: **10** (Enter)	Sets radius to 10
Command: (Enter)	Reissues the CIRCLE command
CIRCLE 3P/TTR/<Center point>: **281,195** (Enter)	Specifies the center point of the circle
Diameter/<Radius> <10.0000>: (Enter)	Accepts radius of 10
Command: *Choose the Arc button*	Issues the ARC Command
Center/<Start point>: **84,177** (Enter)	Sets the arc Start point
Center/End/<Second point>: **110,145** (Enter)	Specifies the Second point
End point: **142,121** (Enter)	Specifies the endpoint
Command: *Choose the Line button*	Issues the LINE command

continues

From point: **142,121** (Enter)	
To point: **@60<0** (Enter)	Creates a 5" line
To point: (Enter)	Ends the LINE command
Command: *Choose the Arc button*	Issues the ARC command
Center/<Start point>: **227,92** (Enter)	Sets the arc Start point
Center/End/<Second point>: **E** (Enter)	Specifies the End point option
End point: **202,121** (Enter)	Sets the endpoint
Included angle: **45** (Enter)	Specifies the included angle measured counterclockwise from the start point
Command: (Enter)	Reissues the ARC command
ARC Center/<Start point>: **C** (Enter)	Specifies the Center option
Center: **267,100** (Enter)	Sets the Center point
Start point: **227,92** (Enter)	Specifies the Start point
Angle/<End point>: **261,59** (Enter)	Specifies the endpoint
Command: *Choose the Line button*	Issues the LINE command
From point: **261,59** (Enter)	Specifies the Start point
To point: **@62<90** (Enter)	
To point: **@70<0** (Enter)	
To point: (Enter)	Ends the LINE command
Command: *Choose the Arc button*	Issues the ARC command
Center/<Start point>: (Enter)	Specifies last point entered
End point: **384,177** (Enter)	Specifies the endpoint
Command: *Choose the Zoom button in the toolbar*	Issues the ZOOM command

`All/Center/Extents/`
`Previous/Window/<Scale`
`(X/XP)>:` **A** (Enter)
	Zooms view to include all entities in the display

Command: *Choose the Save button in the toolbar (see fig. 2.11)*	Saves the drawing

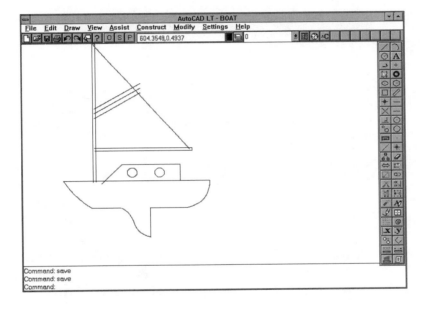

Figure 2.11

The boat's hull.

Using Templates

Repetitive drawing tasks in manual drafting are handled with templates. These templates save a drafter hours of adjusting a compass or drafting individual shapes like triangles, squares, and hexagons (see fig. 2.12).

Figure 2.12

Examples of drawing templates.

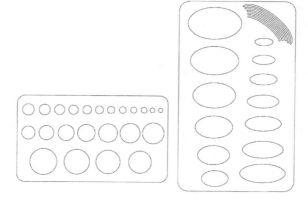

The templates are made of thin plastic in a variety of colors and styles. The shapes in the templates are larger than the noted dimension, to allow for the width of a pencil or pen tip.

All disciplines use templates. Architects use templates for plumbing fixtures such as sinks and toilets. Electrical engineers use symbols for wiring diagrams. Every type of practice can benefit from templates by reducing time and increasing accuracy.

Chapter 3 discusses the use of blocks within a drawing. Blocks are frequently used drawings that can be inserted into your own drawing, quite similar to the template concept. If you are familiar with clip art that is shipped with popular Windows word-processing applications, it can be imported also. This is explained at the end of Chapter 3.

Using an Eraser

Erasing mistakes while drawing is inevitable. Erasers were created specifically for this task and are part of every drafter's tool kit. Erasing is made easier by using an electric eraser (see fig. 2.13). This device contains a length of eraser that can be fed to the tip. The eraser tip rotates quickly, making cleanup and revisions much easier than the time-honored method of using elbow grease. Electric erasers are the tool of choice to make ink drawing revisions, for example.

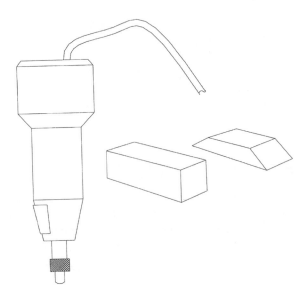

Figure 2.13

Examples of erasers and an electric eraser.

Erasing entities in AutoCAD LT is quite similar to erasing lines on a drawing. To erase any of the lines, do the following:

1. Type **E** at the Command: prompt and press Enter—many of the commands used to this point have keyboard shortcuts. A keyboard shortcut is a letter or two used to issue commands, like L for LINE, C for CIRCLE, and in this case, E for ERASE.

2. At the Select objects: prompt, pick an entity to be erased with a left mouse click.

3. Press Enter; the entity is removed.

4. Type a **U** at the Command: prompt and press Enter. U is the shortcut for UNDO. This will undo the last command or action.

AutoCAD LT is very forgiving. If you make a mistake, utilize the UNDO command. Repeating UNDO will continue to undo any actions that have taken place in the drawing until the last save. UNDO will not undo a save to disk. You can also use the OOPS command, which functions in much the same way.

If a group of entities is to be removed, there are many other options you can use in the ERASE or any other editing command. These options are discussed at length in Chapter 3.

NOTE

Many of the editing commands utilize a selection set prior to adjusting the selected entities. See Chapter 3 for an in-depth discussion.

Using an Erasing Shield

Erasers tend to be quite large in comparison to the detail on a drawing. When small portions of a drawing need to be adjusted, an erasing shield is used (see fig. 2.14). This thin piece of stainless steel will not smudge the drawing and will never wear from repeated eraser use. The openings in the shield form windows around lines to be erased and protect surrounding lines. This is another invaluable tool in every draftsperson's arsenal.

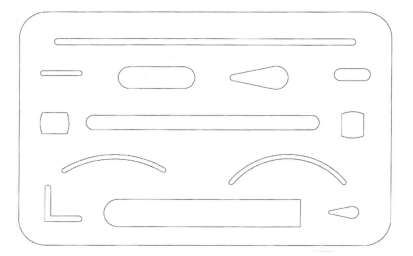

The AutoCAD LT command that emulates an erasing shield is the TRIM command. As the boat drawing nears completion, you will use the TRIM command to clean up the overlapping lines in the mast, boom, and sail. When you are prompted for the "cutting edge(s)," pick the lines that the erasing shield would cover. In order to move in on the drawing, you will utilize the ZOOM command's Window option. By specifying the corners of the window, you can magnify the view to select the appropriate entities.

Cleaning up the Drawing

Proceed within the current drawing.

Command: *Choose the Zoom button in the toolbar*	Issues the ZOOM command
All/Center/Extents/Previous/ Window/ <Scale(X/XP)>: **W** (Enter)	Specifies the Window option
First corner: **134,159** (Enter)	Specifies first corner
Other corner: **359,278** (Enter)	Specifies opposite diagonal corner
Command: *Choose the Trim button*	Issues the TRIM command
Select cutting edge(s):	
Select objects: *Pick the line at* ①, *then press Enter (see fig. 2.15)*	Selects the deck as the cutting edge
<Select object to trim>/ Undo: *Pick the lines at* ② ③ ④	Removes the overlapping lines
Command: (Enter)	Reissues the TRIM command
Select cutting edge(s):	
Select objects: *Pick the line at* ⑤ *and* ⑥, *then press Enter*	Selects the boom and mast as the cutting edges
<Select object to trim>/ Undo: *Pick the lines at* ⑦ ⑧ ⑨	Removes the overlapping lines
Command: *Choose the Zoom button in the toolbar*	Issues the ZOOM command
All/Center/Extents/Previous/ Window/ <Scale(X/XP)>: **A** (Enter)	Zooms view to include all entities in the display

continues

`Command:` *Choose the Zoom button in the toolbar*	Issues the ZOOM command
`All/Center/Extents/Previous/ Window/<Scale(X/XP)>:` **W** (Enter)	Specifies the Window option
`First corner:` **134,307** (Enter)	Specifies first corner
`Other corner:` **264,473** (Enter)	Specifies opposite diagonal corner
`Command:` *Choose the Trim button*	Issues the TRIM command
`Select cutting edge(s):`	
`Select objects:` *Pick the line at* ⑥ *and* ⑩, *then press Enter*	Selects the mast and sail as the cutting edge
`<Select object to trim>/ Undo:` *Pick the lines at* ⑪ ⑫ ⑬ ⑭ ⑮ ⑯ ⑰	Removes the overlapping lines
`Command:` *Choose the Zoom button in the toolbar*	Issues the ZOOM command
`All/Center/Extents/Previous /Window/ <Scale(X/XP)>:` **A** (Enter)	Zooms view to include all entities in the display
`Command:` *Choose the Save button in the toolbar*	Saves the drawing

Using a Drafting Brush

A drafting brush enables a draftsperson to clear away debris that is generated from revising the drawing. If you use your hand, there is a strong danger that the drawing will be smeared. The drafting brush is fine enough to remove the eraser dust and keep the drawing clean.

As you work in AutoCAD LT, there will be occasion to clean up the screen. The command to use is REDRAW. You can issue the REDRAW command quickly by typing **R** and pressing Enter at the command prompt. You will use this command a lot while working on CAD drawings.

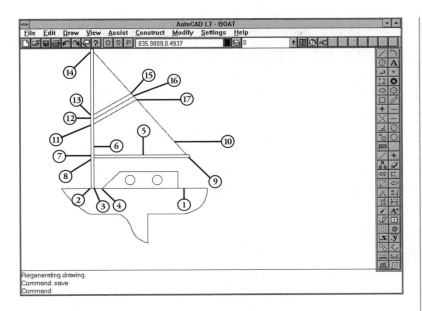

Figure 2.15

Trimming the lines to clean up the sailboat drawing.

Cleaning up the screen can be done in a variety of ways besides REDRAW. You will settle into the best technique as you gain more experience with the program. Try the following techniques:

◆ **REGEN.** This command will review the drawing database and recreate the drawing with the current data and settings. The difference between REDRAW and REGEN is this recalculation. REDRAW merely refreshes the current view.

◆ **F7 key.** Many people who use AutoCAD LT press the F7 key twice. This turns the grid off or on. Redisplaying the grid dots forces a redraw.

◆ **Any ZOOM commands.** All ZOOM commands force a redraw because magnification changes.

As you develop your CAD skills, these techniques will become automatic keystrokes or mouse clicks.

Understanding Techniques of Technical Drawing

Many time-honored techniques are utilized in technical drawing. In the past, the art of graphic language was taught through an apprentice program, as were most trades. In this training, students were taught techniques to improve speed, to remain neat, and to stay consistent and accurate. The most important technique is arguably penmanship. The ability to draw with a pen or pencil and produce a document that explains an idea is the foundation of technical drawing. Penmanship includes both lettering and linework on the drawing. In the following sections, you are introduced to these concepts. Before you begin, however, it is important to understand the media on which the lines and letters will be presented.

Understanding Sheet Sizes

One of the most popular tracing papers is called *vellum paper*. This paper is treated with oil or wax to make it more transparent. The paper used to create drawings is as individual as the tools needed to produce the drawings. As you gain more experience, you will develop a preference as to how you perform a specific task or series of commands like redrawing the screen. The type of paper you like is no exception. The paper used is generally dictated by the firm, and you will quickly adapt to their methods. Also, vellum paper can be printed with a grid like that found on graph paper. This makes measuring and setting up drawings quite easy.

Another paper that is often used is *mylar film*. This polyester film is typically created with at least one matte surface to draw upon. The pencil or pen lines are very sharp for reproduction, and erasing on this type of media leaves very little ghosting or leftover marks. If stored correctly, mylar will resist the aging that most papers endure and is virtually indestructible.

Paper size used throughout the industry is determined in three ways. The first is based on multiples of standard 8-1/2" by 11" paper. The papers are given letters A through E to specify size. If you are working with a laser printer, the paper it holds is probably A-size. Figure 2.16 shows the various standard sheet sizes.

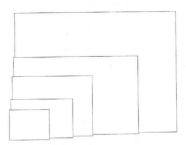 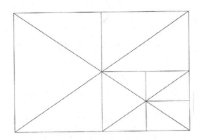

Figure 2.16

Typical paper sizes in standard and metric.

The second method is determined by multiples of 9" by 12" with sizes again specified using letters A to E. The third method is metric. Although metric is not often used in the U.S., many firms are doing more work overseas where metric units are a requirement.

These are the standard sheet sizes used in the industry. Paper can also be cut from rolls and, as mentioned earlier, even printed with reproducible grids.

Designing a Page Layout

Page layout is another skill that is developed by draftspersons from experience. It takes a creative eye and a lot of planning to successfully lay out a drawing. First, you must determine how many drawings are to be included in the sheet, and their typical location. If convention does not dictate paper size (for example, architectural drawings on D-size paper), the drawing and its scale will help you design the page layout. One of the strengths of CAD is the capability to scale drawings effortlessly and to move the drawings as a group or block.

The SCALE command enables you to pick an entity or group of entities and enlarge them or reduce them. In the SCALE command, you have two options: to scale by percentage or by referencing another length.

An entity has an initial scale of 1.0 or 100 percent. Therefore, scaling by .7 will make the entity 70 percent of its original size. If you entered 1.4, the entities would be 140 percent of the original size. You can also specify a reference.

If an entity is 4 feet long and you would like it to be 6 feet, use the reference option. You are prompted for the reference length, in this case 4 (units do not matter). You are then prompted for a new length, in this case 6. All of the entities in the selection set will be scaled according to your inputs. When precise scaling is not necessary, you can drag the entities with the mouse until you are satisfied with the results.

In the next exercise, you draw the outline of an A-size paper to fit around the sailboat. The sheet will be in a landscape orientation, that is, on its side. Remember, you are drawing the boat at full scale and the paper outline will show sheet proportion only. When the drawing is printed, it will be scaled down to fit on the physical sheet. You will also utilize some of the other ZOOM options. These options will be covered more thoroughly in the coming chapters. Begin by drawing the outline of the A-size sheet.

Exercise

Designing the Page Layout

Proceed within the current drawing.

Command: *Choose the Line button*	Issues the LINE command
From point: **0,0** (Enter)	Specifies the lower left corner of the sheet
To point: **@11,0** (Enter)	Specifies the lower right corner of the sheet
To point: **@0,8.5** (Enter)	Specifies the upper right corner of the sheet
To point: **@-11,0** (Enter)	Specifies the upper left corner of the sheet
To point: **C** (Enter)	Closes the lines by creating a line to the first coordinates specified
Command: *Choose the Zoom button in the toolbar*	Issues the ZOOM command

`All/Center/Extents/Previous/`
`Window/ <Scale(X/XP)>:`
`.9x` (Enter)
	Zooms view to 90 percent

`Command:` *Choose the Scale button*	Issues the SCALE command
`Select objects:` **W** (Enter)	Specifies a window selection option
`First corner:` **0,0** (Enter)	Specifies the corner of the window
`Other corner:` **20,20** (Enter)	Specifies opposite diagonal corner
`4 objects found`	Verifies the objects that make up the paper outline
`Select objects:` (Enter)	Accepts the outline as the selection set
`Base point:` **0,0** (Enter)	Specifies the point from which the scaling occurs
`<Scale factor>/Reference:` **60** (Enter)	Scales the paper outline by 60
`Command:` *Choose the Zoom button in the toolbar*	Issues the ZOOM command
`All/Center/Extents/Previous/` `Window/ <Scale(X/XP)>:` **A** (Enter)	Zooms view to include all entities in the display
`Command:` **Z** (Enter)	Issues the ZOOM command

`All/Center/Extents/Previous/`
`Window/ <Scale(X/XP)>:`
`.9x` (Enter)
	Zooms view to 90 percent

`Command:` *Choose the Save button in the toolbar (see fig. 2.17)*	Saves the drawing

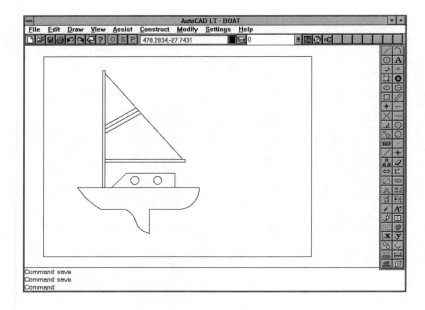

Laying out a page of technical drawings is a simple process once the basic logic of a particular drawing is understood. By performing the exercises in the following chapters, you will be more comfortable with drafting conventions.

Creating a Title Block and Border

No matter what the discipline or the sheet size, all drawings will have a title block. A *title block* is an area on the sheet that gives specific information about the drawing. It is generally located along the bottom of the sheet or in the lower right corner. Architectural firms typically place the title block in a vertical orientation along the left edge of the sheet.

TIP

When loading AutoCAD LT, you are given generic borders for each of the sheet sizes used in the industry, including metric.

The title block usually contains the following information:

- ◆ **The firm's name.** This indicates who is responsible for the drawing.

- ◆ **Drawing title.** This specifies the drawing type or part name.

◆ **Scale.** Drawing sheets can contain different scales; if not, it is indicated in the title block.

◆ **Drawing number.** This helps to organize the assembly of parts.

◆ **Draftsperson's name.** This indicates the creator of the drawing.

◆ **Date completed.** This helps when a drawing is revised.

◆ **Checker's name.** All drawings should be reviewed by another person or a supervisor.

For the purposes of the sailboat drawing, you will create a simple title block.

Another element that is common to drawing sheets is a border. The border defines the drawing area on the sheet. The border is typically 1/2" from the edge of the paper. In some cases, marks are located between the border and the edge of the paper to help a drafter divide the sheet equally.

To develop the border and title block, you will utilize the OFFSET command. This is merely a form of copying entities.

When offsetting arcs and circles, the OFFSET command works wonders. Depending on the side specified for entities to be offset, the entities will maintain a common center point and become smaller or larger.

The OFFSET command prompts for an offset distance or Through point. You enter a distance and select the entities to be offset. Next, you are prompted for a side to offset. This copies the selected entity to the side specified. Begin the following exercise by creating the border of the drawing.

Creating the Border and Title Block

Exercise

Proceed within the current drawing.

Command: *Choose* **C**onstruct, *then* **O**ffset	Issues the OFFSET command
Offset distance or Through <Through>: **30** (Enter)	Specifies the distance

continues

`Select object to offset:` *Pick the bottom edge of the sheet at* ① *(see fig. 2.18)*	Selects bottom edge to be offset
`Side to offset?` *Pick a point near* ②	Copies the edge toward the center of the drawing
`Select object to offset:` *Pick the right edge of the sheet at* ③	Selects right edge to be offset
`Side to offset?` *Pick a point near* ②	Copies the edge toward the center of the drawing
`Select object to offset:` *Pick the top edge of the sheet at* ④	Selects top edge to be offset
`Side to offset?` *Pick a point near* ②	Copies the edge toward the center of the drawing
`Select object to offset:` *Pick the left edge of the sheet at* ⑤	Selects left edge to be offset
`Side to offset?` *Pick a point near* ②	Copies the edge toward the center of the drawing
`Select object to offset:` `Enter`	Ends the OFFSET command
`Command:` *Choose the Line button*	Issues the LINE command
`LINE From point:` **400,30** `Enter`	
`To point:` **400,126** `Enter`	Creates right side of title block
`To point:` **630,126** `Enter`	Creates top of title block
`To point:` `Enter`	Ends the LINE command
Close the toolbox	
`Command:` *Choose the Save button in the toolbar*	Saves the drawing

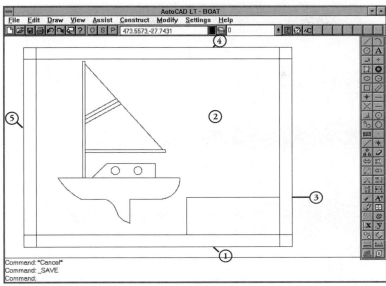

Figure 2.18

The drawing and title block borders.

In the next exercise, you clean up the corners of the border using the FILLET command. Although this command is intended to give straight corners a radius, it is one method used to quickly square off corners. By specifying a radius of 0, the FILLET command is used to clean up the corners of the border. The FILLET command will be discussed at length in the following chapters.

Cleaning Up the Border and Title Block

Exercise

Proceed within the current drawing.

Command: *Choose* **C**onstruct, *then* **O**ffset
Issues the OFFSET command

Offset distance or Through <30.000>: (Enter)
Accepts the current distance

Select object to offset: *Pick* ① *(see fig. 2.19)*
Selects top of title block

continues

`Side to offset?` *Pick near* ②	Offsets the line down
`Select object to offset:` *Pick* ③	Selects the new line
`Side to offset?` *Pick near* ②	Offsets the line down
`Select objects to offset:` (Enter)	Ends the OFFSET command
`Command:` *Choose* **C**onstruct, *then* **F**illet	Issues the FILLET command
`Polyline/Radius/<Select first object>:R` (Enter)	Selects the Radius option
`Enter fillet radius <0.5000>:` `0` (Enter)	Sets the radius to 0
`Command:` (Enter)	Reissues the FILLET command
`FILLET Polyline/Radius/ <Select first object>:` *Pick* ④	Selects the bottom edge of the drawing border
`Select second object:` *Pick* ⑤	Selects right edge of drawing border
`Command:` (Enter)	Reissues the FILLET command
`FILLET Polyline/Radius/ <Select first object>:` *Pick* ⑤	Selects the bottom edge of the drawing border
`Select second object:` *Pick* ⑥	Selects right edge of drawing border
`Command:` (Enter)	Reissues the FILLET command
`FILLET Polyline/Radius/ <Select first object>:` *Pick* ⑥	Selects the bottom edge of the drawing border
`Select second object:` *Pick* ⑦	Selects right edge of drawing border
`Command:` (Enter)	Reissues the FILLET command
`FILLET Polyline/Radius/ <Select first object>:` *Pick* ⑦	Selects the bottom edge of the drawing border
`Select second object:` *Pick* ④	Selects right edge of drawing border
`Command:` *Choose the Save button in the toolbar*	Saves the drawing

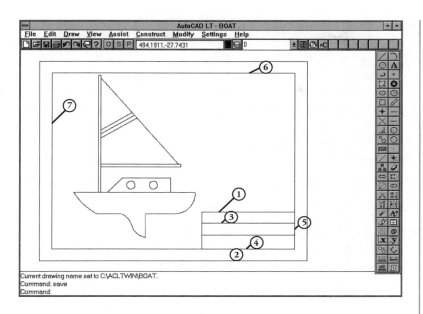

Figure 2.19

The title block and border.

Appreciating the Art of Lettering

Most of the books about manual drafting devote a whole chapter to the subject of lettering. This is a skill that many draftspersons in the past spent hours perfecting. Before Gutenberg invented the printing press, all letters were drawn by hand, each letter carrying the personality of its creator. If you doubt this, find a drawing by the architect Frank Lloyd Wright and study his lettering style. With the introduction of computers, the art of lettering and calligraphy is becoming a lost art.

All draftspeople spend a great deal of time making sure that their letters are straight, have consistent heights, are spaced evenly, and are legible—in addition to staying on a tight time schedule. There are even pens designed specifically to create lettering on technical drawings. Most lettering is done using guidelines that are drawn lightly on the paper to guide the drafter's hand. There are also lettering triangles and instruments. A great deal of pride is reflected in the lettering of most drawing professionals. Again, style is generally dictated by the firm in which you are working, so the ability to adapt to the system is important.

One method to keep the lettering consistent is the use of lettering devices. Similar to the idea of templates, discussed earlier in this chapter, these devices consist of a guide pin that follows the letters on the template, and an ink pen that duplicates the letters on the paper. By adjusting the arm, you can increase or decrease the size of the lettering depending on the application. The most popular lettering device is the Leroy lettering instrument. Although many firms still create drawings by hand, these tools are being outdated by the simplicity and flexibility of the CAD environment and modern reproduction tools.

The art of lettering is replaced in AutoCAD LT by two commands—TEXT and DTEXT. DTEXT displays the text while it is being added to the drawing, whereas TEXT does not. Each command has similar options after you select the point at which the lettering starts. For the benefit of this drawing, you will use the default settings. All of the TEXT options are presented in the exercises in Chapter 5, "Text and Annotation."

Adding Text to a Drawing

You will complete the sailboat drawing by adding text to the title block, adding general notes, and giving the boat a seaworthy name.

In this exercise, you use the DTEXT command, which allows the height of the lettering to be input each time it is used.

Exercise

Adding Lettering to the Drawing

Proceed within the current drawing.

Command: *Choose the Text button*	Issues the DTEXT command
Justify/Style/<Start point>: **247,255** (Enter)	Specifies the starting point for text
Height <0.2000>: **20** (Enter)	Specifies a height of 20
Rotation angle <0>: (Enter)	Accepts a normal orientation
Text: **S14**	Creates text

Text: (Enter)	Ends the DTEXT command
Command: (Enter)	Reissues the DTEXT command
DTEXT Justify/Style/ <Start point>: **247,158** (Enter)	Specifies the starting point for text
Height <20.000>: **10** (Enter)	Specifies a height of 10
Rotation angle <0>: (Enter)	Accepts a normal orientation
Text: *Type your first name or another seaworthy name for the boat*	Creates text
Text: *Pick near* ① *and type* **MY SAILBOAT DRAWING** *(see fig. 2.20)*	Moves the text box and creates the text
Text: *Pick near* ② *and type your name*	Moves the text box and creates the text
Text: *Pick near* ③ *and type the date*	Moves the text box and creates the text
Command: *Choose the Save button in the toolbar*	Saves the drawing

Exploring Linework

Along with lettering, a drafter must acquire the skill of using a pencil or pen to create a variety of linetypes in a drawing. There are two distinguishing factors of a line—its weight or thickness and its linetype. Regardless of the type of line, the techniques associated with it center on reproduction. Should the lines be heavy and crisp to produce clear lines on the blueprint or light, so as to not show? In the past, draftpersons were expected to spend hours on their linework to perfect it.

Figure 2.20

The drawing with text added.

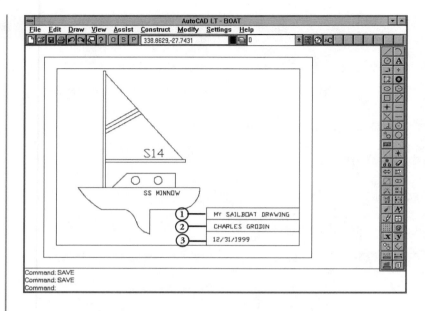

Determining Line Weights

In technical drawing, the thickness of a line can explain a lot about the object that is depicted. These line conventions are used throughout the industry and have evolved to become the integral language of technical drawing. Line weights range from very thin, as in construction lines, to border lines that are drawn very thick. The weight of a line in manual drafting is accomplished by its density and thickness. Within AutoCAD LT, this is achieved in one of two ways.

First, the line weight can be specified at the printing or plotting process. Chapter 10 covers the whole aspect of this subject, but it is important to recognize the advantages at this juncture. A *plotter* is a printing device that is fitted with a series of pens. The ink pens vary in thickness depending on the conventions of the firm. There are usually at least three weights: thin, medium, and thick. These thicknesses are determined by the pen tip and vary from .0001 to .1 or greater. See Chapter 10 for more information on pen assignments.

When the plotting device is set up, colors that will be used in the drawing are assigned pens. Therefore, if the color green is assigned to the .1mm pen, any of the entities drawn in green will be plotted with a line width of .1mm.

If you only have access to a desktop laser printer, you may want to show lines with thickness; they can be changed to polylines with width. Polylines are continuous line segments that are recognized as one entity. These polylines can be assigned a width, which will appear on the screen. Polylines are edited with the PEDIT (Polyline EDIT) command. PLINE and PEDIT are complex commands and will demand more attention in Chapter 3. Changing lines to polylines is not the most efficient way to give a line thickness. While plotting, layers or colors should be assigned a pen width, as described in Chapter 10. Only use the techniques mentioned previously when issues of scale and plotting are understood.

In the last exercise on the sailboat drawing, you will change the border line to a polyline with thickness.

Exercise

Adding Thickness to the Border Line

Proceed within the current drawing.

Command: *Choose* **M**odify, *then* Edit Pol**y**line	Issues the PEDIT command
Select polyline: *Pick* ①	Selects the bottom line of the border
Entity selected is not a polyline	
Do you want to turn it in to one? <Y> (Enter)	Turns the line into a polyline
Close/Join/Width/Edit vertex/ Fit/Spline/Decurve/LType gen/ Undo/eXit <X>: W (Enter)	Selects the Width option
Enter new width for all segments: 3 (Enter)	Changes width to 3

continues

`Close/Join/Width/Edit vertex/`
`Fit/Spline/Decurve/LType gen/`
`Undo/eXit <X>:` (Enter)
	Ends the PEDIT command

`Command:` (Enter)	Reissues the PEDIT command
`Select polyline:` *Pick* ②	Selects the right edge of the border
Entity selected is not a polyline	
`Do you want to turn it in to` `one? <Y>` (Enter)	Turns the line into a polyline

`Close/Join/Width/Edit vertex/`
`Fit/Spline/Decurve/LType gen/`
`Undo/eXit <X>:` **W** (Enter)
	Selects the Width option

`Enter new width for all` `segments:` **3** (Enter)	Changes width to 3

`Close/Join/Width/Edit vertex/`
`Fit/Spline/Decurve/LType gen/`
`Undo/eXit <X>:` (Enter)
	Ends the PEDIT command

`Command:` (Enter)	Reissues the PEDIT command
`Select polyline:` *Pick* ③	Selects the top edge of the border
Entity selected is not a polyline	
`Do you want to turn it in` `to one? <Y>` (Enter)	Turns the line into a polyline

`Close/Join/Width/Edit vertex/`
`Fit/Spline/Decurve/LType gen/`
`Undo/eXit <X>:` **W** (Enter)
	Selects the Width option

`Enter new width for all` `segments:` **3** (Enter)	Changes width to 3

`Close/Join/Width/Edit vertex/`
`Fit/Spline/Decurve/LType gen/`
`Undo/eXit <X>:` (Enter)
	Ends the PEDIT command

`Command:` (Enter)	Reissues the PEDIT command
`Select polyline:` *Pick* ④	Selects the left edge of the border
Entity selected is not a polyline	
`Do you want to turn it in to one? <Y>` (Enter)	Turns the line into a polyline
`Close/Join/Width/Edit vertex/ Fit/Spline/Decurve/LType gen/ Undo/eXit <X>:` `W` (Enter)	Selects the Width option
`Enter new width for all segments:` `3` (Enter)	Changes width to 3
`Close/Join/Width/Edit vertex/ Fit/Spline/Decurve/LType gen/ Undo/eXit <X>:` (Enter)	Ends the PEDIT command
`Command:` *Choose the Polyline button*	Issues the PLINE command
`From point:` `65,146` (Enter)	Starts the polyline
`Arc/Close/Halfwidth/Length/ Undo/Width/ <Endpoint of line>:` `W` (Enter)	Specifies the Width option
`Starting width <0.0000>:` `3` (Enter)	Specifies the line width
`Ending width <3.0000>:` (Enter)	Accepts the current width
`Arc/Close/Halfwidth/Length/ Undo/Width/<Endpoint of line>:` `@336<0` (Enter)	Creates the water line
`Arc/Close/Halfwidth/Length/ Undo/Width/ <Endpoint of line>:` (Enter)	Ends the PLINE command
`Command:` *Choose the Save button in the toolbar*	Saves the drawing

The completed drawing is shown in figure 2.21

Figure 2.21

*The completed
sailboat drawing.*

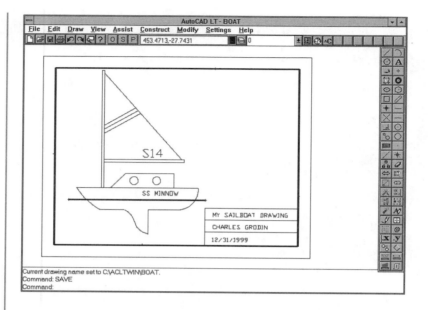

Using Linetypes

Along with the weight of a line, the industry utilizes lines that are drawn in various ways that have definite meaning. In the past, drafters worked excessively to improve their drawing technique. Drawing tools need constant care to draw precisely—the pencil has to be sharp, the pen has to be free of clogs. Many manual drafting textbooks refer to the variety of lines as the *alphabet of lines.* This alphabet of lines will be utilized in all of the future exercises.

AutoCAD LT makes short work of developing the skill to produce multiple linetypes by supplying an array of choices. It is fundamental, however, to understand why a specific linetype is being used.

TIP

Every drawing will load with a continuous or solid linetype as a minimum.

The linetypes are defined in a file called ACLT.LIN. This is a text file that indicates how the lines will be drawn using simple code. Refer to the *AutoCAD LT User's Guide* for information on adjusting this file to create your own linetypes. The following exercise takes you through the steps of loading additional linetypes.

Chapter 2

Loading Linetypes

Proceed within the current drawing.

`Command:` *Choose* **S**ettings, *then* Li**n**etype Style, *then* **L**oad	Issues the LINETYPE command
`Linetype(s) to load: *` (Enter)	Loads all linetypes
Press Enter when the dialog box appears	Loads the file ACLT.LIN
`?/Create/Load/Set:` (Enter)	Ends the LINETYPE command
`Command:` *Choose* **S**ettings, *then* **L**ayer Control	Opens the Layer Control dialog box
In the dialog box, choose Select **A**ll, *then* Set **L**Type	Opens the Select Linetype dialog box
Use the Next and Previous buttons to view the linetypes that are now available	

While in the Select Linetype dialog box, click on any of the linetypes, and it will change all of the lines in the sailboat drawing to that particular linetype. The application of these lines will be examined to set up drawings in the following chapters. As you view the linetypes, review the following section to find out how these lines are used in drawings.

◆ **Continuous.** These lines are used for construction, border, object, and dimension lines. The line weights are usually determined by color.

◆ **Hidden or Dashed.** These lines belong to objects that are hidden in the current view. In manual drafting, hidden lines should start and end while in contact with a visible line, and corners should be touching. Hidden or dashed linetypes are drawn as a series of short dashes.

◆ **Center.** Center lines are used to indicate the center of objects. When circles or arcs are dimensioned, you will surely see center lines used to dimension the center point. Center lines are shown as a dash between adjacent lines.

◆ **Phantom.** These lines are used to show cutting planes and various other alternate lines in a drawing. Phantom lines are drawn as a double dash between two adjacent lines.

Finally, depending on the scale of the drawing, you may need to adjust the scale of the linetype in order to view it properly. To adjust the linetype scale, type **LTSCALE** at the Command: prompt and adjust the number.

NOTE

The LTSCALE will affect all linetypes on the drawing. It may be necessary to make concessions to get the best results.

Summary

This chapter introduced you to the comparisons between manual drafting tools and their AutoCAD LT equivalents. It is important to note that this is just an introduction to the application of the tools. The balance of the book will continue to apply these tools and introduce new commands.

The concept of laying out a page and how to add lettering was also covered in this chapter—these subjects will be discussed at length in future chapters. In addition, you were instructed to load the variety of linetypes that are available while working in AutoCAD LT.

The next chapter, "CAD Drawing Techniques," explores the unique abilities that the CAD environment affords.

In the CAD environment, many commands cannot be duplicated in manual drafting. Revising drawings is thus an inevitable task in manual drawing. These adjustments require tracing or erasing and redrafting for hours to accomplish a change dictated by the client or other unforeseen conditions. AutoCAD LT has a variety of commands that make these revisions much easier to complete. In addition, precision is another advantage while working on the computer. The ability to pinpoint exact coordinates while drawing entities has given the drafter a lot of control and accuracy, and the drafting world has taken considerable notice. This chapter examines the following items:

◆ Methods of coordinate entry

◆ Object snap options for precision

◆ Duplicating a drawing or its individual parts

◆ Revising drawings

◆ Creating free-form shapes

◆ Inserting existing drawings

This chapter is devoted to increasing your knowledge of AutoCAD LT's capabilities and its advantages over manual drafting. By following the examples in this chapter, you should become comfortable duplicating, revising, and inserting drawings.

AutoCAD LT - COORDS
Construct Modify Settings Help
44.4063,27.6296

Positive
Y-Axis

4,8
@4<33
4,4
5,5

AutoCAD LT - COORDS
Assist   Construct   Modify   Settings   Help
44.4063,27.6296   0

90
Degrees
10   Positive
          Y-Axis

AutoCAD LT - UNNAMED
Draw   View   Assist   Construct   Modify   Settings   Help
14.0637,-0.0799   0

4,8   10,8

@4<30

5,5

4,4

	Program Item Properties
Description:	Drafting and Design
Command Line:	\ACLT.EXE /C: C:\LEARNING
Working Directory:	C:\LEARNING
Shortcut Key:	None
☐ Run Minimized	

C
Ca
Brow
Change
He

veas Current drawing name set to C:\ACLTWIN\LEARN\UNNAMED.

CAD Drawing Techniques

Many methods are used to create drawings in AutoCAD LT. As you gain experience, the tips, shortcuts, and tricks of the trade will become second nature. Everyone goes through a period of anxiety as they are learning the logic of CAD and building a new set of skills. It usually lasts several weeks because the complexity continues to build as you learn more about this fascinating aspect of drafting. In this chapter, you explore some of the commands necessary to create the drawing displayed in figure 3.1.

Figure 3.1

The completed barn drawing.

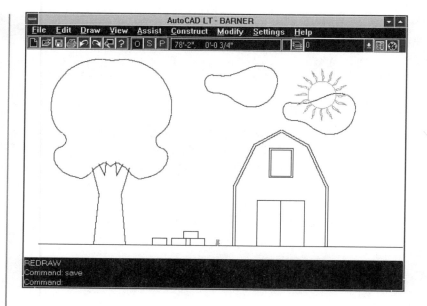

Achieving Precision in AutoCAD LT

To help you understand more about the CAD environment, the next few sections review the methods used to create an accurate drawing. Entering coordinates and setting drawing units are important tasks that are worthy of a review.

Specifying Coordinates

In Chapter 1, "Welcome to the World of Computer-Aided Drafting," you were introduced to the concept of the coordinate system. As you work to create the drawings, there will be circumstances that require an ability to input points more than one way. In Chapter 2, "An Overview of Drafting Tools and Techniques," you utilized these methods briefly, but it is important to understand how they will contribute to your flexibility when drafting in AutoCAD LT.

Setting Up the Units

Until this point, you have been creating drawings using inches as your measurement. If you are creating an architectural floor plan, for example, you must be able to work in feet and inches. Likewise, if you are a civil engineer, you must be able to express lengths and direction as surveying units. This is all accomplished by setting the units in the drawing.

The UNITS command enables you to set the type of measuring unit and its accuracy, in addition to setting the method with which angles are measured. Entering **UNITS** at the Command: prompt will bring up a series of questions related to the units. AutoCAD LT offers a dialog box as an alternative that develops the same settings with greater ease, particularly for a new user. Using the dialog box is recommended for beginning users. As you develop a confidence with the program, you will find secondary methods that are quicker for many commands.

Changing the units will also change the way coordinates are displayed in the coordinate display window in the toolbar.

NOTE

Depending on the type of drawing you are creating, the units will need to be specified. Table 3.1 demonstrates the variety of units and the manner in which the coordinates and measurements are displayed.

Table 3.1
Unit Types

Unit	Display
Scientific	1.55E+01
Decimal	15.50
Engineering	1'-3.50"
Architectural	1'-3 1/2"
Fractional	15 1/2

As you can see, each method varies depending on the discipline or type of drawing. The other option that needs to be specified under units is the precision. The precision display in the dialog box will change based on the unit type specified. AutoCAD LT has the capacity to calculate to sixteen places and display accuracy to eight places. Although it is difficult to create parts, buildings, or dictate property lines within that tolerance, it is important to realize that this level of accuracy exists. If architectural or fractional units are specified, the precision will change to fractional precision. Even though fractions are displayed, AutoCAD LT keeps track of fractions in a decimal form and converts the measurements for display.

TIP

If the units are set to Fractional, you can still input 1-1/2 as 1.5—AutoCAD LT will convert the input value. Also, one unit will always be one inch, foot, meter, or mile, depending on the unit setting.

To access the Units Control dialog box, shown in figure 3.2, do any of the following:

◆ Choose **S**ettings, then **U**nits Style.

◆ Type **DDUNITS**; DD stands for dynamic dialog. All of the dialog boxes can be accessed from the command prompt in this manner.

◆ Type **UNITS** and answer the questions.

◆ Choose **F**ile, then **N**ew; choose **Q**uick, then OK, and set the unit type in the Quick Drawing Setup dialog box.

CAUTION

The last method mentioned will create a new drawing while the other methods are performed within the current drawing. If you are in the middle of an existing drawing, save the changes.

The other unit option is the angle display. Again, the drawing type will dictate the method in which the angles will be displayed. Table 3.2 displays the angle options available.

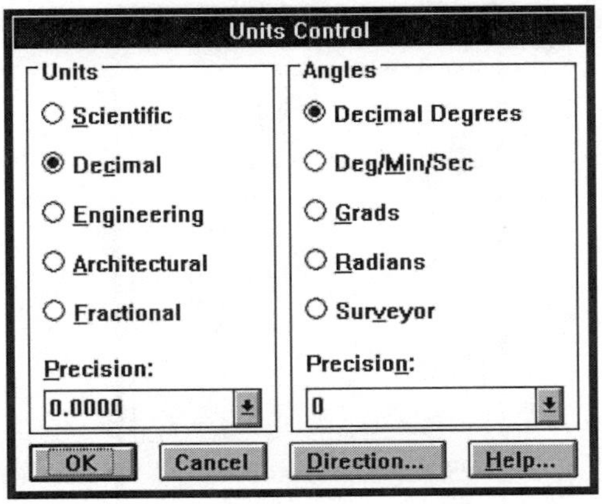

Figure 3.2

The Units Control dialog box.

Table 3.2
Angle Types

Angle Method	Display
Decimal Degrees	30.00
Deg/Min/Sec	60d0'00"
Grads	45.0g
Radians	45.0r

These angle options can also be given a precision. The precision is displayed in the current angle method. Like units precision, it will be accurate to sixteen places and will display up to eight.

The last option in the Units Control dialog box is the direction from which the angles are measured. To set the direction, do the following:

◆ Choose **S**ettings, then **U**nits Style.

◆ Choose the **D**irection button in the Units Control dialog box, which brings up the Direction Control dialog box (see fig. 3.3).

Figure 3.3

*The Direction
Control dialog box.*

Direction Control

Angle 0 Direction

◉ **E**ast   0.0

○ **N**orth   90.0

○ **W**est   180.0

○ **S**outh   270.0

○ **O**ther   Pick/Type

Angle: [0   ]

[**P**ick <]

◉ **C**ounter-Clockwise

○ C**l**ockwise

[OK]   [Cancel]

The top half of the Direction Control dialog box is used to specify the 0 direction. This means you can set the direction from which angles will be measured. It is recommended that this setting should only be adjusted after you have a firm handle on the coordinate system. Most users leave this setting in its default direction; zero is in the 3 o'clock position. Notice that the directions are listed in the current **A**ngle method.

NOTE

Notice the **O**ther option in the Direction Control dialog box. By choosing this option, you can specify any angle for the zero direction datum. By choosing the **P**ick button, you are transferred to the drawing area and prompted to determine an angle. The two points specified form an angle in the XY plane and the measurement is placed in the **A**ngle text box.

The lower half of the box enables you to specify how the angles are measured. **C**ounter-Clockwise is the default and should be left as such. Angles are measured positively from the 3 o'clock position. An angle of 90° is measured from the base line in a counterclockwise motion to end up pointing toward the top of the screen.

Be careful when changing default settings in a drawing. If someone else is planning to work with the drawing, basic settings that have been changed can be a frustrating experience.

Architectural units are unique due to the fact that these units are based on feet and inches, as opposed to decimal numbers.

The following example illustrates how you would input 153 feet 6 and 3/4 inches in architectural units:

153'6-3/4"

TIP

It is important to remember that whenever you input points, pressing the spacebar to add a space will perform the same function as pressing Enter and will end the command prematurely. Spaces are not available in coordinate input.

Architectural measurements are developed with an apostrophe representing the foot mark and quotes representing the inch marks. If a fraction of an inch is to be added, type a dash followed by the fraction. Although the previous example displays inch marks, it is not necessary to add them after the foot mark is used when specifying points. In the following exercise, set the units of the current drawing to architectural units.

Setting the Drawing Units

Exercise

Depending on how the preferences are set, the next few steps may change. Simply open a new drawing, but cancel the Quick Drawing setup.

Command: *Choose* **F**ile, *then* **N**ew, *then* OK	Begins a new drawing and opens the Create New Drawing dialog box
Click on the Cancel *button*	Closes Quick Drawing Setup dialog box

continues

Command: *Choose* **S**ettings, *then* **U**nits Style	Opens the Units Control dialog box
Choose **A**rchitectural *under* Units	Sets the unit type to Architectural
Click in the **P**recision *list box and scroll until* **0'-0 1/4"** *appears*	Displays the fractional precision options
Click on **0'-0 1/4"**	Sets the precision
Choose OK	

Understanding Absolute Coordinates

To help you understand all of the coordinate input options thoroughly, the following is a brief review of absolute coordinates.

The Cartesian coordinate system or rectangular system is based on a grid. The grid is derived from X and Y coordinates whose distances are offset from the location of the origin or 0,0 point. In order to simplify these coordinates, the lower left corner of the drawing should be located at the origin to keep all of the entities in the positive quadrant of the grid (see fig. 3.4).

It is essential that you develop an understanding of the Cartesian coordinate system because all CAD programs operate on this principle.

In *absolute coordinates*, the horizontal distance is listed first, followed by the vertical distance from the origin, as follows:

45,67

You have used this method extensively to this point in the book, but it is important to reiterate its value. In the following exercise, you will utilize absolute coordinates to create the ground line and sides of the barn.

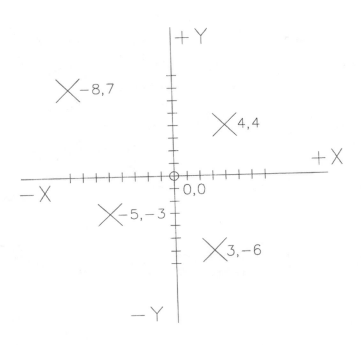

Figure 3.4

The Cartesian grid.

Chapter

3

Exercise

Using Absolute Coordinates

Now that the Units are set, continue with the current drawing.

Command: *Choose the Line button*	Issues the LINE command
LINE From point: 3',3' (Enter)	Specifies the first point with architectural coordinates
To point: 80',3' (Enter)	Specifies a point 80' to the right
To point: (Enter)	Ends the LINE command
Command: *Choose the Zoom button in the toolbar*	Issues the ZOOM command
All/Center/Extents/Previous/ Window/<Scale(X/XP)>: A (Enter)	Zooms the view to include the origin and all of the entities

continues

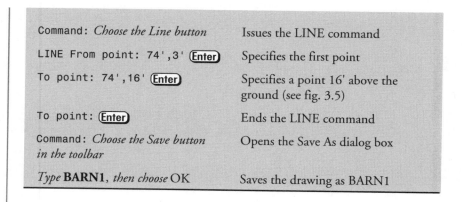

Command: *Choose the Line button*	Issues the LINE command
LINE From point: 74',3' (Enter)	Specifies the first point
To point: 74',16' (Enter)	Specifies a point 16' above the ground (see fig. 3.5)
To point: (Enter)	Ends the LINE command
Command: *Choose the Save button in the toolbar*	Opens the Save As dialog box
Type **BARN1**, *then choose* OK	Saves the drawing as BARN1

Figure 3.5

Creating lines with absolute coordinates.

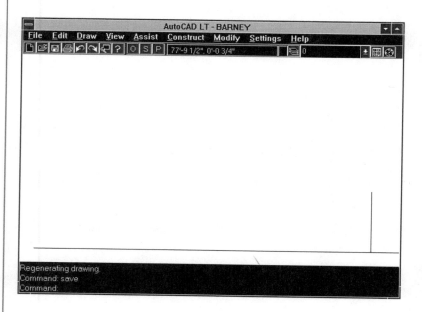

Understanding Relative Coordinates

Relative coordinates rely on the last point picked or typed. By temporarily storing this information in a buffer, AutoCAD LT enables you to specify points in relation to the last point. In effect, the last point becomes a temporary origin. After a new point is selected with relative coordinates, that new point becomes the last point, and so on. This is the most popular method of keyboard input. It is much easier to determine relative coordinates in an object than absolute coordinates. Figure 3.6 illustrates how relative coordinates are determined.

Figure 3.6

Relative coordinates.

Chapter

3

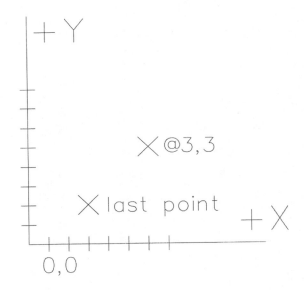

To enter relative coordinates, type an **@** (Shift+2) symbol before the X, Y coordinates. The following example literally means create the next point 28 feet, 6 inches in the x axis, and 10 feet, 8 1/2 inches in the y axis from the last point entered:

@28'6,10'8-1/2

You have already used this method in the Chapter 2 exercises, so now you are aware of the logic. In this exercise, you use relative coordinates to create the barn door and hayloft opening.

Using Relative Coordinates

Continue with the BARN1.DWG.

Command: *Choose the Line button*	Issues the LINE command
LINE From point: **64',3'** (Enter)	Specifies the first point with architectural coordinates
To point: **@0,10'** (Enter)	Specifies a point 10' up from the ground

Note that the foot mark is not necessary when entering a coordinate of zero.

To point: **@5',0** (Enter)	Specifies a point 5' to the right of the last point
To point: **@0,-10'** (Enter)	Specifies a point 5' to the right of the last point
To point: (Enter)	Ends the LINE command
Command: (Enter)	Reissues the LINE command
LINE From point: **66'6,18'** (Enter)	Specifies the first point

Note that the inch mark is not needed for inches and that there are no spaces.

To point: **@-5',0** (Enter)	Specifies the second point
To point: **@0,6'3-1/2** (Enter)	Specifies the next point
To point: **@5',0** (Enter)	Specifies the next point
To point: **C** (Enter)	Uses the first point entered while using the LINE command to close the box (see fig. 3.7)
Command: *Choose the Save button on the toolbar*	Saves the drawing

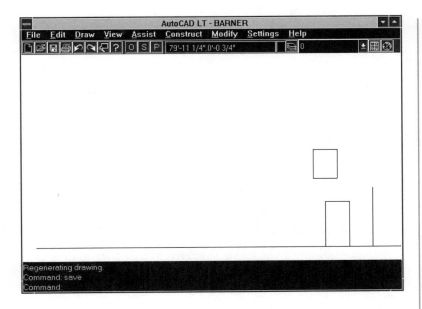

Figure 3.7

Creating lines with relative coordinates.

Understanding Polar Coordinates

Another type of coordinate input, called *polar coordinates*, involves specifying the distance and angle that the line will follow (see fig. 3.8). This is similar to relative coordinates using the @ symbol. The following example displays the format to enter a polar coordinate:

@24'<45

This creates a line 24 feet long at an angle of 45° from the last point input. Angles are measured using the method determined by the unit settings, as mentioned previously. The angle must be input using the method set by the units. The current setting is decimal.

The angle is measured counterclockwise from the 3 o'clock position. This is the default and should be left as such until you are more familiar with the consequences of changing this setting.

Figure 3.8

Polar coordinates.

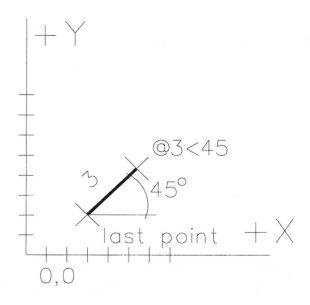

In the next exercise, you use polar coordinates to create half of the gambrel roof on the barn.

Exercise

Using Polar Coordinates	
Continue with the BARN1.DWG.	
Command: *Choose the Line Button*	Issues the LINE command
LINE From point: **74',16'** (Enter)	Specifies the first point
To point: **@10'<115** (Enter)	Specifies a point 10' long at an angle of 115°
To point: **@10'<150** (Enter)	Specifies a point 10' long at an angle of 150°
To point: (Enter)	Ends the LINE command (see fig. 3.9)
Command: *Choose the Save button on the toolbar*	Saves the drawing

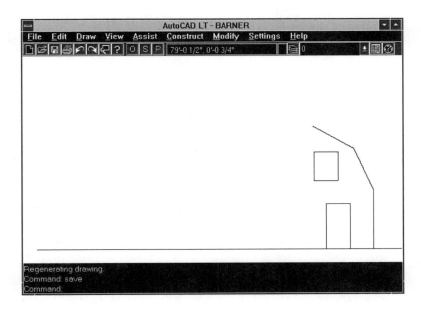

Figure 3.9

Creating lines with polar coordinates.

These three methods of coordinate input should satisfy all of your needs while learning the art of technical drawing. There are other input methods and even a third distance called the Z coordinate. The Z coordinate is used while creating 3D drawings.

Comprehending Object Snap Options

As mentioned in Chapter 2, technical drawing has always been considered an art form. Draftspeople spent years studying descriptive geometry in order to develop the skills that are necessary to create complex drawings. Creating geometric shapes, bisecting angles, and finding midpoints were all necessary to stay accurate within drawings.

There is another set of tools available to you in AutoCAD LT called *object snaps*. These tools enable the drafter to find points on entities with accuracy, precision, and consistency. All of the entities in the CAD environment have at least one snap point. Most of the entities you have been creating have several. Up to this point, the only means of accuracy has been created by the snap and grid settings; object snaps override the current snap, grid, and ortho settings.

Using Object Snap Choices

Object snaps are points on an object that can be selected. It is often necessary to start a line from the end of another line, for example. You can do this by setting your object snap to ENDPoint, which will find the object's endpoint. When activated, object snap settings change the cursor from a plain crosshair to a crosshair with a small target box. This target box is called the *aperture,* and its size can be set based on the type of drawing in which you are working. Any of the entities that fall within the aperture are subject to selection. A simple drawing can benefit from a large aperture, whereas a complex drawing might need a smaller target box.

There are times when you will need an object snap only once. Object snaps can be entered from the keyboard before you pick an entity or picked from the toolbox. This is a benefit of the toolbox, which has a button for each object snap.

Many beginners make the mistake of cautiously surrounding the endpoint of a line with the target box. This target box is the aperture, discussed later in the chapter. It is not necessary to pick an object at the object snap point. If your object snap is set to MIDpoint, you can pick anywhere on a line or arc and the MIDpoint will be selected. These object snaps are designed to save time.

If the toolbox is hidden, choose the Toolbox button on the right side of the toolbar until the toolbox is floating in the drawing area as a window.

In the second row of the default toolbox, each of the object snaps has a button. Hold the cursor over each button until the object snap name is displayed. Over time, you will become quite familiar with these tools and their location in the toolbox:

◆ **Object_Snap.** This opens the Running Object Snap dialog box. The term *running* means that the object snap options are available when selecting entities and will remain active until they are turned off. Several object snap options can be active at one time. This dialog box also enables you to set the size of the aperture box.

◆ **ENDPoint.** This option selects an object's endpoint. When you enter the From and To points for a line, you are actually setting the endpoints, and AutoCAD LT is generating the line between them.

Lines and arcs have endpoints, so this object snap will be quite useful.

◆ **INTersection.** When any two entities overlap, they form an intersection. This spot can be picked by using this option.

◆ **MIDpoint.** Lines and arcs have a point halfway between the endpoints called the *midpoint*.

◆ **PERpendicular.** The PERpendicular option creates an intersection on the entity perpendicular to the last point or next point specified.

◆ **CENter.** All arcs and circles have a center point from which the radius is calculated.

◆ **TANgent.** Locates a point along an arc or circle that is tangent to the last point or next point specified.

◆ **QUAdrant.** Circles and arcs have points at 0°, 90°, 180°, and 270° that can be selected. These are called *quadrants*.

◆ **INSert.** Text strings or lines of text have an insertion point that can be selected. When using the TEXT or DTEXT commands, this is the first point you enter. Later in the chapter, you will explore blocks, which are a group of entities. They also have insertion points.

◆ **NODe.** You will be introduced to points later in the book. They are unique because they can be selected by nodes.

◆ **NEArest.** This selects a point that is nearest to the crosshairs intersection. This option should be used with caution.

◆ **NONe.** This option temporarily removes running object snaps if they are active.

The object snaps can also be entered quickly from the keyboard by using the uppercase options shown in the preceding list.

Create a circle within a square during the next exercise to experiment with these object snap options.

Exercise

Using Object Snaps

Within the BARN1.DWG, try the following.

Command: *Choose* **D**raw, *then* Rectangle	Issues the RECTANGLE command
First Corner: **10',10'** (Enter)	Specifies the lower left corner
Other corner: **@10',10'** (Enter)	Sets opposite diagonal corner
Command: *Choose the Circle button*	Issues the CIRCLE command
CIRCLE 3P/TTR/<Center point>: **15',15'** (Enter)	Sets the center of circle
Diameter/<Radius>: **2'6** (Enter)	Specifies 30" radius
Command: *Choose the Line button*	Issues the LINE command
LINE From point: *Choose the Endpoint button*	Selects ENDPoint object snap
of *Pick* ① *(see fig. 3.10)*	
To point: *Choose the Center button*	Selects CENter object snap
of *Pick* ②	
To point: *Choose the Midpoint button*	Selects MIDpoint object snap
of *Pick* ③	
To point: *Choose the Nearest button*	Selects NEArest object snap
of *Pick* ④	
To point: *Choose the Tangent button*	Selects TANgent object snap
of *Pick* ⑤	
To point: *Choose the Perpendicular button*	Selects PERpendicular object snap
of *Pick* ①	
To point: *Choose the Intersection button*	Selects INTersection object snap
of *Pick* ②	

To point: *Choose* **A**ssist, *then* **O**bject Snap	Opens the Running Object Snap dialog box
Click on the **Q**uadrant *option, then choose* OK	Sets the object snap to QUAdrant
To point: *Pick*⑥	Selects the circle QUAdrant
To point: *Pick*⑦	Selects the circle QUAdrant
To point: Enter	Ends the LINE command

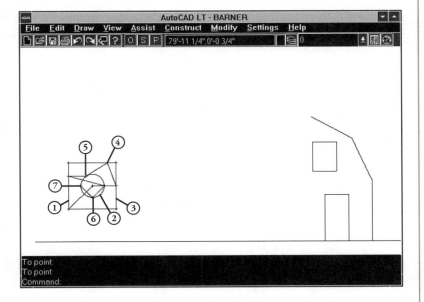

Figure 3.10

The object snap exercise.

Notice that when you open the Running Object Snap dialog box, the bottom of the box is devoted to aperture size. The aperture defines the size of the target box, on the crosshairs, when utilizing object snaps. While working with highly detailed drawings, a smaller box may be necessary; on simpler drawings, a large aperture may speed your object snap selection.

Object snaps are used extensively in the exercises throughout the rest of the book. You will quickly find that they save time and produce a level of accuracy that is difficult to match in manual drafting. When dimensioning a drawing, you will be thankful that you took the time to utilize object snaps. It will save hours of editing the dimension text.

Cleaning Up and Adjusting the Drawing

There are several ways to improve your efficiency while working in the CAD environment. In the next two sections, you will be introduced to commands that will help in this effort. When working in manual drafting, you typically draw in several phases. First, the construction lines are laid out. Second, the object lines are created by tracing over the construction lines. Third, the drawing is edited and cleaned up. Although the same process occurs in the CAD environment, it is not as linear. The process skips around and construction lines can become object lines using a few simple editing commands. These commands will display the power of computer-aided drawing over its manual counterpart by creating as you are editing—a new concept. By utilizing existing entities, you can generate and manipulate drawings with ease.

First, you will explore new ways of selecting entities by creating selection sets.

Selecting Entities

Before you begin to duplicate and revise entities, it is important to introduce the concept of selection sets. These are methods that enable you to duplicate or edit groups of entities as opposed to one at a time.

The default method in AutoCAD LT is to issue a command first, and then select the entities on which to perform the command. This is called the *Verb-Noun method*; for example, Move-Cup is to move a cup, as opposed to selecting the cup first and then moving it. This is actually contrary to the way you live, work, and play. You pick something first and then decide what to do with it. If you accept this new way of thinking, CAD will be easier.

When using duplicating or editing commands, AutoCAD LT prompts you to select the entities with the `Select objects:` prompt. The cursor will change to a small square called the *pick box*. Like the aperture, the pick box size can be set based on the intricacy of the drawing.

To adjust the size of the pick box, choose **S**ettings, then Sele**c**tion Style, which opens the Entity Selection Settings dialog box (see fig. 3.11).

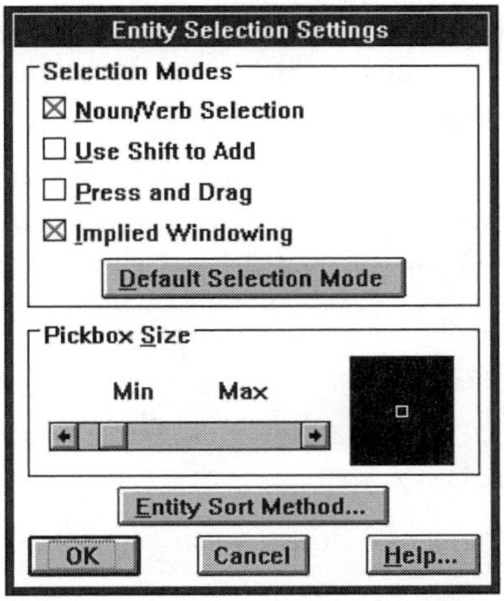

Figure 3.11

The Entity Selection Settings dialog box.

Most of these settings will be helpful as you become more experienced. At this point, use this dialog box to adjust the pick box only and refer to the *AutoCAD LT User's Guide* for further information regarding the other settings. Understanding these settings is not essential to create the drawings in this book.

When objects are selected, they are highlighted or changed temporarily to a dashed line. As you continue to pick objects, you are creating a selection set. There are many options to use when creating a selection set, but unlike other commands, they are not listed each time. After all of the entities are selected, press Enter to end the selecting process and continue with the command.

Make sure to read the command prompt. At times, it will ask you to "Select objects:," which means define a selection set, and other times it will ask for one object with "Select object:." This depends on the type of command you are using.

Here are a few of the options available at the `Select objects:` prompt:

◆ **Window.** This option creates a selection set of all entities that fall *completely* within the window. By picking two opposite diagonal corners, you can construct a solid window box.

◆ **WPolygon.** This option creates a selection set of all entities that fall *completely* within the polygon. By picking the endpoints of a many-sided polygon, you can construct a solid polygon.

◆ **Crossing.** This option creates a selection set of all entities that are crossed or touched by the window. By picking two opposite diagonal corners, a dashed crossing box will be constructed.

◆ **CPolygon.** This option creates a selection set of all entities that are crossed or touched by the polygon. By picking the endpoints of a many-sided polygon, a dashed polygon will be constructed.

◆ **Last.** Picks the last entity created to be included in the current selection set.

◆ **Previous.** Picks all of the entities in the previous selection set to be included in the current set.

◆ **ALL.** Picks all of the entities that can be selected as the selection set.

◆ **Remove.** Removes entities from the current selection set. This can also be accomplished by holding down the Shift key and selecting entities to be removed.

◆ **Add.** In order to continue adding to the selection set after issuing a Remove, you must use this option. The Add option is not necessary if the Shift key is used to remove entities.

There are other options also, but the ones just mentioned will be used in the exercises throughout the rest of the book. If you are interested in researching the additional options, see the *AutoCAD LT User's Guide*.

The keyboard shortcuts are capitalized in the option names above. For example, type **P** for Previous or **WP** for WPolygon at the Select objects: prompt. You should also be aware that Window and Crossing are available by default (see fig. 3.12). If you do not pick on an entity, a window or crossing window will be formed depending on the direction you drag the mouse. If you specify a window by selecting left-to-right, a window will be formed. Likewise, if you specify a window from right-to-left, a crossing window will be formed.

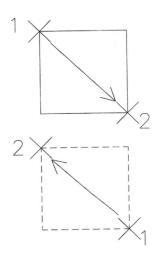
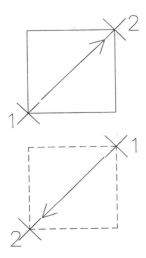

Figure 3.12

Default selection windows.

Duplicating and Moving Existing Entities

You can copy entities in a variety of ways to continue developing the barn drawing. The COPY and MOVE commands are used frequently during a drawing session. It is often easier to copy or move entities and edit them than to redraw them. This is a technique that may or may not be part of your repertoire as you gain experience with AutoCAD LT.

Using the COPY Command

The COPY command enables you to duplicate the selection set. The COPY command needs a point of reference in order to duplicate the selected entities. This point is called the *base point*. After the base point is established, it becomes the *last point*, which was mentioned earlier in this chapter, and new coordinates are given to the base point, which creates the copy.

You can issue the command using the following methods:

◆ Type **COPY** or **CP** at the Command: prompt.

◆ Choose **C**onstruct, then **C**opy.

◆ Choose the Copy button in the default toolbox.

There is also another option, which can be specified before the copy is made—Multiple. This option enables you to duplicate the entity as many times as necessary. After an M (multiple) is typed, you are prompted for a base point. In the next exercise, you will use the COPY command to create a second barn door and explore the multiple copy option.

Exercise

Using the COPY Command

Continue the BARN1.DWG.

Command: *Choose* **C**onstruct, *then* **C**opy	Issues the COPY command
Select objects: W (Enter)	Specifies the Window selection option
First corner: *Pick near* ① *(see fig. 3.13)*	Specifies first window corner
Other corner: *Pick near* ②	Picks opposite diagonal corner
3 found	Returns number of objects selected
Select objects: (Enter)	Continues COPY command
<Base point or displacement>/ Multiple: *Choose the Endpoint button*	Sets object snap to ENDPoint
of *Pick* ③	Sets the base point
Second point of displacement: *Choose the Endpoint button*	Sets object snap to ENDPoint
of *Pick* ④	Creates copy of the door
Command: *Choose the Save button on the toolbar*	Saves the drawing
Command: *Choose the Copy button*	Reissues the COPY command
COPY Select objects: W (Enter)	Specifies the Window selection option

First corner: *Pick near* ⑤	Specifies first window corner
Other corner: *Pick near* ⑥	Picks opposite diagonal corner
4 found	Returns number of objects selected
Select objects: (Enter)	Continues COPY command
<Base point or displacement>/ Multiple: M (Enter)	Selects the Multiple option
Base point: *Pick near* ⑤	Sets the base point
Turn Ortho mode off if it is on by pressing F8	
Second point of displacement: *Pick anywhere on the screen a number of times*	Creates a number of copies
Second point of displacement: (Enter)	Ends the COPY command
Command: U (Enter)	Removes the multiple copies with the UNDO command
Command: R (Enter)	Redraws the screen
Command: *Choose the Save button on the toolbar*	Saves the drawing

Using the MOVE Command

The MOVE command is very similar to the COPY command. The difference is that the entities are placed in a new position as opposed to just duplicating them. MOVE does not have any additional options and simply relocates the selection set. The base point is also used as the last point in order to move the entities correctly. The MOVE command can be accessed in the following ways:

◆ Type **MOVE** or **M** at the Command: prompt.

◆ Choose <u>M</u>odify, then <u>M</u>ove.

◆ Choose the Move button in the default toolbox.

Figure 3.13

Using the COPY command.

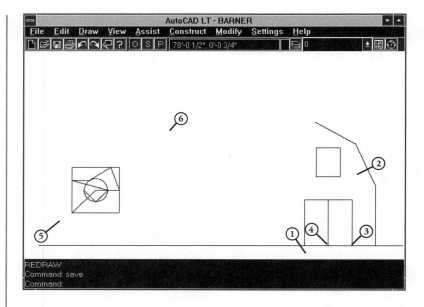

In the next exercise, imagine you have been told that the barn is located 10 feet over the property line, and the neighbor has complained. Use the MOVE command to remedy the situation.

Using the MOVE Command

Continue the BARN1.DWG.

Command: *Choose* **A**ssist, *then* **O**bject Snap	Opens the Running Object Snap dialog box
Deselect any of the checked boxes, then choose OK	Disables any running object snaps
Command: *Choose* **M**odify, *then* **M**ove	Issues the MOVE command
Select objects: W (Enter)	Specifies the Window selection option

`First corner:` *Pick near* ① *(see fig. 3.14)*	Specifies first window corner
`Other corner:` *Pick near* ②	Picks opposite diagonal corner
`13 found`	Returns number of objects selected
`Select objects:` (Enter)	Continues the COPY command
`<Base point or displacement>/ Multiple:` **END** (Enter)	Sets object snap to ENDPoint
`of` *Pick* ③	Sets the base point
`Second point of displacement:` **@10'<180** (Enter)	Moves the entities 10 feet to the left
`Command:` *Choose the Save button on the toolbar*	Saves the drawing

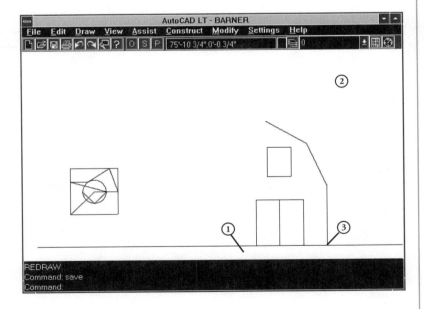

Figure 3.14

Using the MOVE command.

Using the OFFSET Command

Although the OFFSET command will copy existing entities, it operates quite differently than MOVE or COPY. The OFFSET command prompts for the offset distance first; this is different than the other commands. Next, you are asked to select an object. OFFSET will not work with a selection set, which is another difference. When the object is selected, the point at which you picked the object becomes the last point. Finally, you are prompted for the side to offset. AutoCAD LT then applies the specified distance to the copied object based on the last point. To issue the OFFSET command, do one of the following:

◆ Type **OFFSET** or **OF** at the Command: prompt.

◆ Choose **C**onstruct, then **O**ffset.

◆ Choose the Offset button in the default toolbox.

In Chapter 2, you used the OFFSET command to create a border located 1/2" from the paper outline. The result of the OFFSET command depends on the type of entity that is to be copied. If arcs or circles are to be offset, the result will be smaller or larger depending on the side of the offset. Lines that are offset will create parallel lines that are the same length as the original. Polylines can give unexpected results based on their geometry.

In the next exercise, you add a wood trim to the barn around the edge of the building and the hayloft opening.

Exercise

Using the OFFSET Command

Continue the BARN1.DWG.

Command: *Choose* **C**onstruct, *then* **O**ffset	Issues the OFFSET command
Offset distance or Through point <Through>: **6"** (Enter)	Sets offset distance to 6"
Select objects to offset: *Pick* ① *(see fig. 3.15)*	Selects side of barn

`Side to offset:` *Pick near* ②	Copies the line to the inside
`Select objects to offset:` *Pick* ③	Selects roof of barn
`Side to offset:` *Pick near* ②	Copies the line to the inside
`Select objects to offset:` *Pick* ④	Selects roof of barn
`Side to offset:` *Pick near* ②	Copies the line to the inside
`Select objects to offset:` **⟨Enter⟩**	Ends the OFFSET command
`Command:` **⟨Enter⟩**	Reissues OFFSET command
`Offset distance or Through` `<0'-6">:` **4** **⟨Enter⟩**	Sets offset distance to 4"
`Select objects to offset:` *Pick* ⑤	Selects side of hayloft
`Side to offset:` *Pick near* ②	Copies the line to the inside
`Select objects to offset:` *Pick* ⑥	Selects top edge of hayloft
`Side to offset:` *Pick near* ②	Copies the line to the inside
`Select objects to offset:` *Pick* ⑦	Selects side of hayloft
`Side to offset:` *Pick near* ②	Copies the line to the inside
`Select objects to offset:` *Pick* ⑧	Selects bottom edge of hayloft
`Side to offset:` *Pick near* ②	Copies the line to the inside
`Select objects to offset:` **⟨Enter⟩**	Selects bottom edge of hayloft
`Command:` *Choose the Save button on the toolbar*	Saves the drawing

3
Chapter

Figure 3.15

Using the OFFSET command.

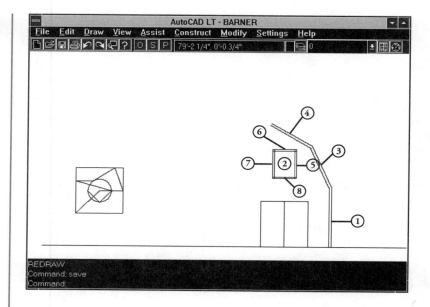

Using the MIRROR Command

Another type of command that involves duplicating entities is MIRROR. Many of the drawings that you will create have groups of objects that are symmetrical or mirrored images of each other. This command enables you to define a line that functions as the axis across which the selection set will be mirrored. The axis line can be at any angle. You can even mirror text without the risk of the text becoming reversed. The only other option within the command is whether or not to delete the selection set.

It is good practice to turn on Ortho mode when using the MIRROR command. This will ensure that the entities are mirrored correctly.

In order to use the MIRROR command, do one of the following:

◆ Type **MIRROR** or **MI** at the Command: prompt.

◆ Choose **C**onstruct, then **M**irror.

In the next exercise, you use the MIRROR command to create the left side of the barn.

Using the MIRROR Command

Exercise

Continue the BARN1.DWG.

Press F8 if Ortho is not currently active

`Command:` *Choose* **C**onstruct, *then* **M**irror   Issues MIRROR command

`Select objects:` **C** `Enter`   Selects crossing option for selection set

`First corner:` *Pick near* ① *(see fig. 3.16)*   Forms lower right corner of crossing window

`Other corner:` *Pick* ②   Highlights all entities touched by the window

`Select objects:` *Hold the Shift key down and select the two lines of the door that are highlighted*   Deselects the door of the barn

`Select objects:` `Enter`   Accepts the selection set

`First point of the mirror line:` **END** `Enter`   Sets object snap to ENDPoint

`of` *Pick* ③

`Second point:` *Pick near* ② *or* **@10'<90** `Enter`   Sets other point of the mirror line

`Delete old objects? <N>:` `Enter`   Accepts the default of not deleting old objects

`Command:` **R** `Enter`   Redraws the view

`Command:` *Choose the Save button on the toolbar*   Saves the drawing

Figure 3.16

Using the MIRROR command.

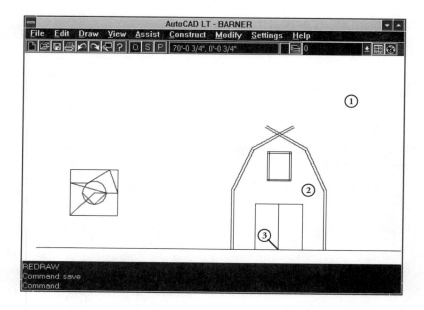

Revising Existing Entities

One of the advantages of working in AutoCAD LT is the ability to lay out drawings quickly. It is essential to understand that the drawing is in a constant state of flux while it is being created. Lots of construction lines, overlapping lines, and even notes on the drawing can be quite useful, but will make the drawing look cluttered. It is necessary that drawings go through this phase, and you will discover that the drawing will look refined rather quickly by applying the following commands. This section describes the methods used to clean up the drawing.

Using the ERASE Command

The ERASE command is another self-explanatory tool that will be used extensively during a drawing session. It is used to remove entities singularly or by selection set. You have used this command during previous exercises in this book. The UNDO and OOPS commands are used as a fail-safe to reverse the results of the ERASE command. You can access the command in one of the following ways:

◆ Type **ERASE** or **E** at the Command: prompt.

◆ Choose **M**odify, then **E**rase.

◆ Choose the Erase button in the default toolbox.

Exercise

Chapter

3

Erasing Objects

Continue the BARN1.DWG.

Command: *Choose the Erase button*	Issues the ERASE command
Select objects: **C** (Enter)	Selects the Crossing option
First corner: *Pick near* ① (see fig. 3.17)	Sets the lower left corner
Other corner: *Pick near* ②	Sets the opposite diagonal corner
Select objects: (Enter)	Erases the entities
Command: **R** (Enter)	Redraws the view
Command: *Choose the Save button on the toolbar*	Saves the drawing

---

OOPS will reverse the effects of the last ERASE command even if other commands have been used in the meantime. Later on in this chapter, the BLOCK and WBLOCK commands will be discussed. These commands erase the entities that are grouped, and the OOPS command will restore them. You can issue the OOPS command in the following two ways:

◆ Type **OOPS** or **OO** at the Command: prompt.

◆ Choose **M**odify, then **O**ops.

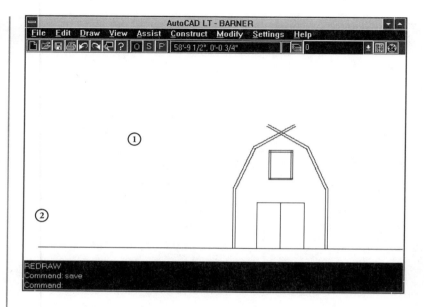

UNDO or U will reverse the effects of any command that has produced unsatisfactory results. UNDO has a variety of options that can be used to your benefit. These options are outlined and used in Chapter 11, "Creating a Mechanical Drawing." In order to quickly reverse the results of the last command, type a U. Typing **U** at the command prompt is the most preferred method. You can continue to type a U at the command prompt, which will reverse the actions of each command you have entered.

CAUTION

Using some of the UNDO options will remove all of your work to the last save. Type **U** at the command prompt or avoid the options to eliminate this risk.

You will utilize the ERASE command many times in the exercises throughout the rest of the book.

Using the FILLET Command

You were introduced to the FILLET command in Chapter 2. When creating mechanical drawings in particular, you can use the FILLET command to add a radius to a corner. If you pick two entities, the FILLET command will attempt to apply the current radius setting to the corner they form. If the radius is larger than the line segments selected, an error message will be displayed, and the radius must be adjusted. The FILLET command will only work on two entities at one time. It will not fillet a polyline to a line. The command has several options, as follows:

◆ **Polyline.** Enables you to set the radius of all corners in a 2D polyline. If you want to fillet only one corner in a polyline, just pick the two lines and ignore the fact that the whole polyline is highlighted.

◆ **Radius.** This will set the radius length.

The FILLET command can be invoked using the following methods:

◆ Type **FILLET** or **F** at the Command: prompt.

◆ Choose **C**onstruct, then **F**illet.

As mentioned earlier, this command is designed to add a radius to an angled corner. In the next exercise, you use the FILLET command to clean up the trim that was added to the barn and hayloft opening. The ZOOM command is invoked to magnify the view in order to help with the editing process.

Using the FILLET Command

Exercise

Continue the BARN1.DWG exercise.

Command: *Choose the Zoom button on the toolbar*	Issues the ZOOM command
All/Center/Extents/Previous/ Window/<Scale(X/XP)>: **W** (Enter)	Selects the Window option
First corner: *Pick near* ① (*see fig. 3.18*)	Sets lower left corner of window
Other corner: *Pick near* ②	Sets upper right corner of window
Command: *Choose* **C**onstruct, *then* **F**illet	Issues FILLET command

continues

`Command: _fillet Polyline/`
`Radius/<Select first object>:`
`R` (Enter)
	Selects the Radius option

`Enter fillet radius <0'-0` `1/2">:` **0** (Enter)	Sets the radius to zero
`Command:` (Enter)	Reissues the FILLET command
`Fillet Polyline/Radius/<Select` `first object>:` *Pick* ③ *(see fig. 3.19)*	Highlights the first object
`Select second object:` *Pick* ④	Cleans up the peak of the roof
`Command:` (Enter)	Reissues the FILLET command
`Fillet Polyline/Radius/` `<Select first object>:` *Pick* ⑤	Highlights the first object
`Select second object:` *Pick* ⑥	Cleans up the trim at the peak of the roof
`Command:` (Enter)	Reissues the FILLET command
`Fillet Polyline/Radius/` `<Select first object>:` *Pick* ⑥	Highlights the first object
`Select second object:` *Pick* ⑦	Cleans up the trim
`Command:` (Enter)	Reissues the FILLET command
`Fillet Polyline/Radius/` `<Select first object>:` *Pick* ⑦	Highlights the first object
`Select second object:` *Pick* ⑧	Cleans up the trim
`Command:` (Enter)	Reissues the FILLET command
`Fillet Polyline/Radius/` `<Select first object>:` *Pick* ⑨	Highlights the first object
`Select second object:` *Pick* ⑩	Cleans up the trim

Command: (Enter)	Reissues the FILLET command
Fillet Polyline/Radius/ <Select first object>: *Pick* ⑪	Highlights the first object
Select second object: *Pick* ⑫	Cleans up the trim
Command: (Enter)	Reissues the FILLET command
Fillet Polyline/Radius/ <Select first object>: *Pick* ⑫	Highlights the first object
Select second object: *Pick* ⑬	Cleans up the trim
Command: (Enter)	Reissues the FILLET command
Fillet Polyline/Radius/ <Select first object>: *Pick* ⑬	Highlights the first object
Select second object: *Pick* ⑭	Cleans up the trim
Command: (Enter)	Reissues the FILLET command
Fillet Polyline/Radius/ <Select first object>: *Pick* ⑭	Highlights the first object
Select second object: *Pick* ⑪	Cleans up the trim
Command: **R** (Enter)	Redraws the view
Command: *Choose the Save button on the toolbar*	Saves the drawing

Although there are many other commands that could fall into this section, they should be applied when needed. The other exercises in this book continue to introduce commands that will help while editing a drawing. The rest of this chapter is devoted to adding additional elements to the scene with the previously described tools and a few new ones.

Figure 3.18

The zoom window points.

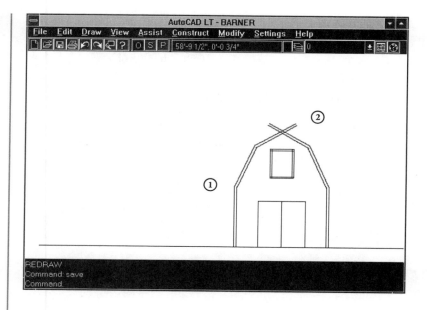

Figure 3.19

Using the FILLET command.

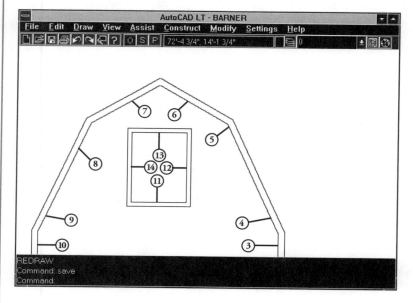

Understanding the **CHAMFER** command

Like FILLET, the CHAMFER command performs a clean-up function at the intersection or the implied intersection of two lines. *Implied intersection* means that if the lines were extended or trimmed to intersect, an intersection point would be formed. A new line will be formed between the two existing lines. There are three options in the CHAMFER command, as follows:

◆ **Polylines.** This option creates a chamfer at the intersection of each segment in the polyline.

◆ **Distances.** This option prompts for the distance from the intersection on each of the two lines. Although it is typical for these distances to be the same, they do not need to be.

◆ **Select first line.** This prompt asks you to choose the first line of the chamfer. After the selection, you are prompted for the next line

If you would like to experiment with the CHAMFER command, replace the FILLET command in the last exercise with the CHAMFER command. Remember not to save these changes if you elect to explore this command.

Adding to the Drawing

When creating a drawing, you will undoubtedly sway from creating to editing and back again as portions of the drawing come to fruition. The next section of this chapter introduces you to other commands that will enable you to create interesting shapes and insert existing drawings.

In order to complete the scene, you will utilize some of the commands reviewed in this chapter to create some bales of straw and the trunk of the tree.

Creating the Hay Bales

Exercise

Continue with the BARN1.DWG.

Command: *Choose the Zoom button on the toolbar*	Issues the ZOOM command

continues

`All/Center/Extents/Previous/`	Zooms the view to include the origin and all of the entities
`Window/<Scale(X/XP)>:` **A** (Enter)	
`Command:` *Choose* **D**raw, *then* **R**ectangle	Issues the Rectangle command
`First corner:` **35',3'** (Enter)	Sets lower left corner of bale
`Other corner:` **@3',1'6** (Enter)	Sets upper right corner of bale
`Command:` *Choose the Copy button*	Issues COPY command
`Select objects:` **L** (Enter)	Selects the Last option
`1 found`	
`Select objects:` (Enter)	Accepts the selection set
`<Base point or displacement>/` `Multiple:` **M** (Enter)	Selects the Multiple option
`Base point:` *Choose the Endpoint button*	Selects the ENDPoint object snap
`of` *Pick* ① *(see fig. 3.20)*	Picks the lower right corner of the bale
`Second point of` `displacement:` *Choose* **A**ssist, *then* **O**bject Snap, *choose the Nearest box, then choose* OK	Sets the running object snap to Nearest
Press F8 if Ortho mode is active	Toggles Ortho mode on/off
`Second point of` `displacement:` *Choose several points similar to figure 3.20*	Places bales in a stack
`Second point of displacement:` (Enter)	Ends the COPY command
`Command:` *Choose the Save button on the toolbar*	Saves the drawing

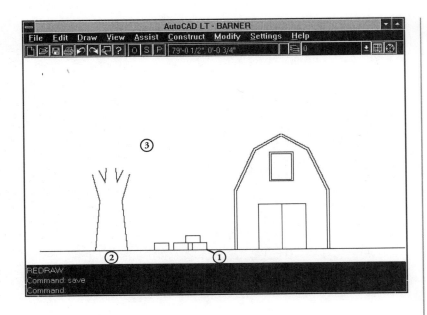

Figure 3.20

Creating the hay bales and tree trunk.

Next, you create a tree trunk using the MIRROR command.

Exercise

Creating the Tree Trunk

Continue the BARN1.DWG.

Command: *Choose the Line button*	Issues the LINE command
From point: **21'4,3'** (Enter)	Sets first point
To point: **@11'<95** (Enter)	
To point: **@6'<75** (Enter)	
To point: (Enter)	Ends the LINE command
Command: (Enter)	Reissues the LINE command
From point: **19',21'** (Enter)	Sets first point
To point: **@3'<275** (Enter)	
To point: **@3'<65** (Enter)	

continues

To point: (Enter)	Ends the LINE command
Command: *Choose the Save button on the toolbar*	Saves the drawing
Command: *Choose* **C**onstruct, *then* **M**irror	Issues the MIRROR command
Select objects: **W** (Enter)	Selects the Window option
First corner: *Pick near* ② *(see fig. 3.20)*	
Other corner: *Pick near* ③	Highlights the trunk
4 found	
Select objects: (Enter)	Accepts the selection set
First point of mirror line: **18',3'** (Enter)	Sets first point
Press F8 to turn on Ortho mode	
Second point: *Pick a point straight up from the first point*	Sets other point
Delete old objects? <N>: **N** (Enter)	Retains selection set
Command: **R** (Enter)	Redraws the view
Command: *Choose the Save button on the toolbar*	Saves the drawing

Creating Free-Form Shapes

The most common complaint about working with CAD tools is a computer's inability to mimic a freehand drawing. Believe it or not, on occasion, precision can be detrimental. AutoCAD LT offers a couple of commands that will create free-form shapes that are difficult to develop with regular geometry. At times, as you use these commands, the objects that are created will spark the imagination and may lead to better ideas or at least interesting shapes. One of these commands is PLINE.

Using Polylines to Create Shapes

The PLINE command creates a series of line segments or arcs or a combination of the two. This series of segments is joined at the endpoints and is recognized as a single entity referred to as a polyline. In Chapter 2, a polyline was used to create thickness at the border line because polylines can be given width. Use one of these methods to invoke the command:

◆ Type **PLINE** or **PL** at the `Command:` prompt.

◆ Choose **D**raw, then **P**olyline.

◆ Choose the Polyline button in the default toolbox.

You will be prompted to begin a line with a `From point:` prompt, just as with the LINE command. A series of options will then be issued, as follows:

`Arc/Close/Halfwidth/Length/Undo/Width/<Endpoint of line>:`

◆ **Arc.** The PLINE command has two modes, either a Line mode or an Arc mode. Each segment can be assigned either property and selecting the endpoint will set the size.

◆ **Close.** This option functions the same way as the LINE command. This will select the first point of the polyline as the final point and close the polyline. This is a closed polyline.

◆ **Halfwidth.** Enables you to specify half of the current thickness.

◆ **Length.** Enables you to enter the length of the line. This is the same as specifying the endpoint.

◆ **Undo.** If a mistake is made while entering points, use the Undo option and the last vertex will be removed.

◆ **Width.** Sets the thickness of the polyline, the starting and ending width of a segment are set separately. You have the option of entering different values for the starting and ending widths.

◆ **Endpoint of the line.** This is the default. This option accepts any of the coordinate inputs, as well as a picked point on the screen.

The next exercise uses the PLINE command to create the vertices for the leaves in the tree and a cloud in the sky.

Chapter

3

Exercise

Utilizing Polylines

Command: *Choose the Polyline button*	Issues the PLINE command
From point: **18',20'** (Enter)	Sets first point of the leaves
Arc/Close/Halfwidth/Length/Undo /Width/<Endpoint of line>: **19'9,20'8** (Enter)	
Arc/Close/Halfwidth/Length/Undo /Width/<Endpoint of line>: **21'2,19'9** (Enter)	
Arc/Close/Halfwidth/Length/Undo /Width/<Endpoint of line>: **23',17'10** (Enter)	
Arc/Close/Halfwidth/Length/Undo /Width/<Endpoint of line>: **27'3,18'3** (Enter)	
Arc/Close/Halfwidth/Length/Undo /Width/<Endpoint of line>: **30',22'8** (Enter)	
Arc/Close/Halfwidth/Length/Undo /Width/<Endpoint of line>: **27'6,25'6** (Enter)	
Arc/Close/Halfwidth/Length/Undo /Width/<Endpoint of line>: **30'2,29'7** (Enter)	
Arc/Close/Halfwidth/Length/Undo /Width/<Endpoint of line>: **18',43'** (Enter)	
Arc/Close/Halfwidth/Length /Undo/Width/<Endpoint of line>: (Enter)	Ends the PLINE command
Command: *Choose the Save button on the toolbar*	Saves the drawing

Command: *Choose* **C**onstruct,   Issues the MIRROR command
then **M**irror

MIRROR Select objects: *Pick* ①   Selects the leaves
(see fig. 3.21)

Select objects: (Enter)

First point of mirror line:
END (Enter)

of *Pick* ③

Press F8 if Ortho mode is not on

Second point: *Drag the cursor straight
down and pick a point*

Delete old objects? <N>: (Enter)

Command: *Choose the Save button*   Saves the drawing
on the toolbar

Command: *Choose the Polyline*   Issues the PLINE command
button

From point: **45',33'** (Enter)   Sets first point

Arc/Close/Halfwidth/Length/Undo
/Width/<Endpoint of line>:
@6'8<2 (Enter)

Arc/Close/Halfwidth/Length/Undo
/Width/<Endpoint of line>:
@8'<72 (Enter)

Arc/Close/Halfwidth/Length/Undo
/Width/<Endpoint of line>:
@6'<156 (Enter)

Arc/Close/Halfwidth/Length/Undo
/Width/<Endpoint of line>:
@7'4<216 (Enter)

Arc/Close/Halfwidth/Length/Undo
/Width/<Endpoint of line>:

continues

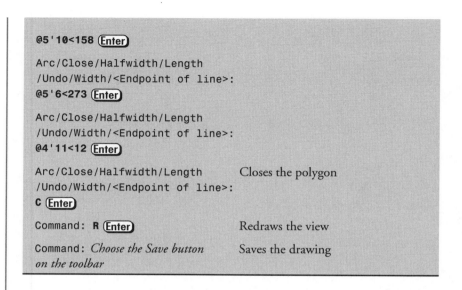

```
@5'10<158 (Enter)

Arc/Close/Halfwidth/Length
/Undo/Width/<Endpoint of line>:
@5'6<273 (Enter)

Arc/Close/Halfwidth/Length
/Undo/Width/<Endpoint of line>:
@4'11<12 (Enter)

Arc/Close/Halfwidth/Length          Closes the polygon
/Undo/Width/<Endpoint of line>:
C (Enter)

Command: R (Enter)                  Redraws the view

Command: Choose the Save button     Saves the drawing
on the toolbar
```

Figure 3.21

The polyline vertices.

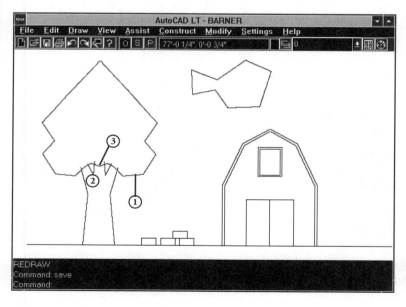

Using PEDIT to Edit Polylines

Each of the points or *vertices* in a polyline can be edited. This is done using the PEDIT command. The PEDIT command enables you to move, straighten, curve, and insert vertices. This option makes polylines very flexible and can create shapes that are quite interesting. Issue the PEDIT command using the following methods:

◆ Type **PEDIT** or **PE** at the Command: prompt.

◆ Choose **M**odify, then Edit Pol**y**line.

Editing Polylines with PEDIT

Exercise

Continue with the BARN1.DWG.

Command: *Choose* **M**odify, *then* **E**dit Polyline	Issues the PEDIT command
PEDIT Select polyline: *Pick the cloud at* ① *(see fig. 3.22)*	Highlights the cloud
Open/Join/Width/Edit vertex /Fit/Spline/Decurve/LType gen /Undo/eXit <X>: **F** (Enter)	Reshapes the polyline with the Fit option
Open/Join/Width/Edit vertex /Fit/Spline/Decurve/LType gen /Undo/eXit <X>: **S** (Enter)	Reshapes the polyline with the Spline option
Open/Join/Width/Edit vertex /Fit/Spline/Decurve/LType gen /Undo/eXit <X>: **D** (Enter)	Reshapes the polyline with the Decurve option
Open/Join/Width/Edit vertex /Fit/Spline/Decurve/LType gen /Undo/eXit <X>: **S** (Enter)	Reshapes the polyline with the Spline option
Open/Join/Width/Edit vertex /Fit/Spline/Decurve/LType gen /Undo/eXit <X>: (Enter)	Exits the PEDIT command
Command: (Enter)	Reissues the PEDIT command

continues

`PEDIT Select polyline:` *Pick the leaves at* ②	Highlights the leaves
`Open/Join/Width/Edit vertex /Fit/Spline/Decurve/LType gen /Undo/eXit <X>:` **F** (Enter)	Reshapes the polyline with the Fit option
`Open/Join/Width/Edit vertex /Fit/Spline/Decurve/LType gen /Undo/eXit <X>:` **X** (Enter)	Ends the PLINE command
`Command:` (Enter)	Reissues the PEDIT command
`PEDIT Select polyline:` *Pick the leaves at* ③	Highlights the leaves
`Open/Join/Width/Edit vertex /Fit/Spline/Decurve/LType gen /Undo/eXit <X>:` **F** (Enter)	Reshapes the polyline with the Fit option
`Open/Join/Width/Edit vertex/Fit/ Spline/Decurve/LType gen/Undo/ eXit <X>:` **X** (Enter)	Ends the PLINE command
`Command:` *Choose the Save button on the toolbar*	Saves the drawing

Figure 3.22

Reshaping the cloud and leaves with PEDIT.

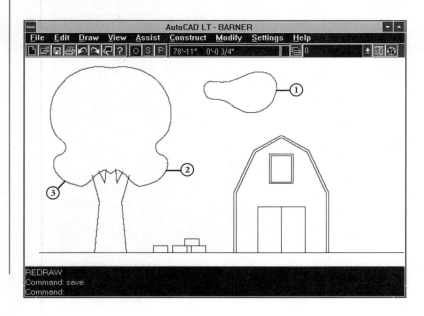

The PEDIT command has several options in addition to those used in the exercise. The following list gives the descriptions of those options:

- **Open or Close.** Depending on whether the polygon to be edited is opened or closed, this option adds or deletes the last segment of the polyline.

- **Join.** Used to create a single polyline from multiple polylines. Endpoints must be shared to use this option. If you join the leaves from the preceding exercise, the two polylines will become one, and any adjustments to the polyline affect the whole "leaves" polyline.

- **Width.** Changes the thickness of the polyline.

- **Edit vertex.** Enables you to edit individual vertices of the polyline.

- **Fit.** Creates a continuous curve through all of the points in a polyline, even if it is open.

- **Spline.** Uses points of the polyline as "pulling forces" to create a curve among the points. The major difference from Fit is the fact that the curve does not pass through the points—it is only influenced by them.

- **Decurve.** Removes the effect of Fit or Spline.

- **Ltype gen.** Enables you to apply a linetype to the whole polyline or each segment independently.

- **Undo.** Reverses the effects of the last option.

- **eXit.** Ends PEDIT options. This is the default.

Use these options to increase your drawing power and flexibility. Rarely will you use these options when creating working drawings, but they may come in handy when working on civil drawings or for sketching ideas.

Inserting Existing Drawings

The ability to use stored drawings or drawings in other Windows formats is an asset in using AutoCAD LT. A drawing can be stored in AutoCAD LT format as a .DWG file on the hard drive of the computer or it can be imported from the Windows Clipboard. The Clipboard is a temporary storage buffer for text or graphics. Good Windows programs will take advantage of this ability, and AutoCAD LT is no exception. This technique will save you a lot of time as your drawing library grows. There are many elements in a drawing that should be repeated within a drawing or from drawing to drawing, and it is to your benefit to always think along these lines. The following sections discuss the types of files that can be inserted into your drawings to save time and increase efficiency.

Inserting Blocks

Blocks are groups of entities that can be saved as part of a drawing's database of information or directly to the hard drive of the computer as a separate file.

A *block* is an object or group of objects within a drawing. The block is made of a selection set and saved within the database of the drawing to be used as needed. This is called *defining a block*. Each block has an insertion point that is created as it is defined. The complexity of blocks is discussed later in the book. At this time, it is important to be aware of the ability to create a block or group of entities.

The WBLOCK command creates a new drawing file of a selection set and saves it to the hard drive. This file can then be inserted into any drawing when necessary.

TIP

The BLOCK command saves a block to the current drawing's database, whereas the WBLOCK command saves a block to the hard drive, so that it can be used in any drawing.

To create a block, AutoCAD LT asks you for the following information:

◆ **Block name.** This is the name that is stored in the database. If you type a question mark (?), you will see a list of available blocks. If the list is blank, no blocks have been defined.

- ◆ **Insertion point.** This can be a valuable and strategic decision. Pick a point on or near the selection set.

- ◆ **Select objects.** This asks about which entities are to be included in the block. Use the selection set options to pick the objects.

After the objects are selected and saved, they are erased. This may not be the result you want, so the OOPS command is appropriate to bring back the entities. Remember, the block will still be defined even though the OOPS command was used. When creating a block, do not use UNDO to bring the entities back. You will reverse the effects of the BLOCK command and will be forced to redefine the block.

The WBLOCK command works in the same way with one exception. WBLOCK asks for a file name to be saved on disk. After a name is specified, you will be prompted for the `Block name:` that is to be saved. The block created by WBLOCK does not have to have the same name as the block in the drawing, nor does it have to be defined yet. If the block is already defined, type in the name. If the block is yet to be defined, press Enter and you will be prompted with the same prompts as the BLOCK command. After the block is defined, the entities selected will be erased and can be returned with the OOPS command.

In the next exercise, you create a block of the cloud and insert it back into the drawing.

Creating a Block of the Cloud

Exercise

Continue with the BARN1.DWG.

Command: *Choose* **C**onstruct, *then* Make **B**lock	Issues the BLOCK command
Block name (or ?): **CLOUD** (Enter)	Gives the block a name
Insertion base point: *Pick in the middle of the cloud*	Sets the insertion point
Select objects: *Pick the cloud*	Highlights cloud
Select objects: (Enter)	Accepts selection set

continues

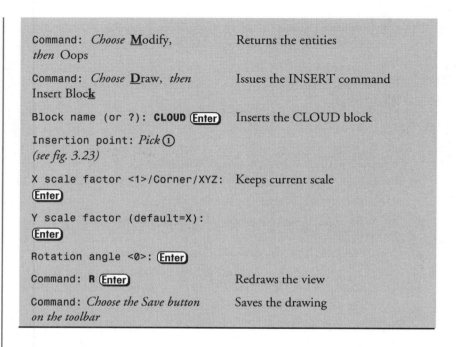

Command: *Choose* **M**odify, *then* Oops	Returns the entities
Command: *Choose* **D**raw, *then* Insert Bloc**k**	Issues the INSERT command
Block name (or ?): **CLOUD** (Enter)	Inserts the CLOUD block
Insertion point: *Pick* ① *(see fig. 3.23)*	
X scale factor <1>/Corner/XYZ: (Enter)	Keeps current scale
Y scale factor (default=X): (Enter)	
Rotation angle <0>: (Enter)	
Command: **R** (Enter)	Redraws the view
Command: *Choose the Save button on the toolbar*	Saves the drawing

Figure 3.23

Adding clouds as blocks.

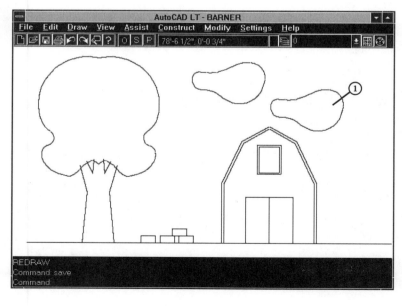

In the previous exercise, in which you created copies of the hay bales, you could have easily created a block and inserted it multiple times. Inserted blocks take up less space within the drawing database. This reduces the byte size of the drawing. Blocks are complex. They can contain hidden text information called *attributes*, which could be extracted from the drawing to create parts lists and pricing. At this point, that would be a confusing discussion, but will be easier to understand later in the book.

Importing DXF and WMF Files

In addition to DWG files, two other types of files can be inserted into a drawing in AutoCAD LT. These files are DXF and WMF files, referred to as vector-based graphics files. This means that the entities are defined by points and the entities in between are given attributes to specify their linetype, color, layer, radius, font, and so on. For example, a line is defined as two endpoints. The actual line that is in between the points is generated with the information, given in the file, about the line.

This is different from raster-based files, which operate strictly on static pixel color. The computer screen is made up of a large grid or *resolution*. A resolution of 640 × 480 means that the current display is showing a grid 640 pixels wide by 480 pixels tall. In raster-based graphics, the only information provided is the pixel's location in the grid and its color.

Data eXchange Format (DXF) files are ASCII text files that contain information directing AutoCAD LT about how the drawing should be created. The DXF format is the file format of choice for most CAD packages when they need to share information with one another. The type of information that is contained in the DXF can be problematic because every CAD package uses a different method to construct entities.

Never assume that a DXF file is bulletproof. Before sending anyone a final copy of a drawing, send a quick test to make sure the CAD packages being used are compatible.

The *Windows MetaFile* (WMF) format is also vector-based. This is the means by which Windows programs can utilize each others files. AutoCAD LT loads several of these files also known as *clip art*. Most Windows programs will support the WMF format. When inserting a WMF file in AutoCAD LT, the entities will appear as gray. Don't be alarmed—they will print correctly.

CAUTION

Inserting WMF files will make your DWG file quite large. Experiment to decide if these file sizes are acceptable.

In the next exercise, you will insert a couple of WMF files, adding detail to the barn drawing.

Exercise

Inserting WMF Files

Command: *Choose* **F**ile, **I**mport/Export, **W**MF In	Opens the Import WMF dialog box
Double-click on the CLIPART *folder under the* ACLTWIN *folder in the* **D**irectories *box*	Lists WMF files in CLIPART folder
Scroll to the bottom of the list and click on SUN.WMF	Displays the file in the Preview box
Double-click on SUN.WMF	
Insertion point: *Click between the clouds in the sky (see fig. 3.24)*	
X scale factor <1>/Corner/XYZ: **.5** (Enter)	Inserts at 50 percent of original size
Y scale factor (default=X): (Enter)	
Rotation angle <0>: (Enter)	
Command: **SAVE** (Enter)	Saves the drawing
Command: *Choose* **F**ile, *then* **I**mport/ Export, **W**MF In	Opens the Import WMF dialog box
Double-click on the CLIPART *folder under the* ACLTWIN *folder in the* **D**irectories *box*	Lists WMF files in CLIPART folder

Scroll in the list and   Displays the file in the Preview box
click on KITTY.WMF

Double-click on KITTY.WMF

Insertion point: *Click near the barn*
to insert the file and press Enter to
accept all the defaults

Figure 3.24

Inserting WMF files.   Chapter

Using the SCALE Command

The ability to resize entities with the SCALE command in AutoCAD LT is
another advantage over manual drafting.

After a selection set is determined, you are prompted for a base point. The base
point can be located anywhere on the drawing, and all entities will be resized
from this point (see fig. 3.25).

Figure 3.25

Scaling results from base point placement.

The SCALE command multiplies all of the dimensions in the selection set by the scale factor that is input. If the number is between 0 and 1, for example .3, all of the dimensions are reduced to 30 percent of their original size. If the scale factor is 1 or greater, for example 6, all objects are scaled to 6 times their original size.

The other option available in the SCALE command is Reference. If an entity has a dimension of 4" and the desired length is 7-3/4", the Reference option will eliminate the risk of improperly calculating the scale factor. Here is the command sequence after the selection set is established:

> <Scale factor>/Reference: **R** (Enter)
>
> Reference length <1>: **4** (Enter)
>
> New length: **7-3/4** (Enter)

The entity will be resized accordingly. If more than one object was included in the selection set, they too will be scaled according to the reference.

Although this book stresses the importance of accuracy, you also can drag an object to scale it. The drag option uses the cursor position as a means of input. As you move the mouse, the objects will be resized. This is not a very accurate method of scaling, but it can be used in this and other commands to quickly assess an idea. In order to scale a selection set unequally in the x or y axis, you need to create a block and insert it. During the insertion process, you are prompted for the X and Y scale factors.

All entities can be scaled, even WMF files. These files are inserted as one object; scaling them is easy.

The KITTY.WMF file that was inserted in the last exercise is too large. Its height is currently 20'7". Using the SCALE command, you will resize the kitty to normal scale.

Scaling WMF Files

Continuing the BARN1.DWG.

Command: *Choose* **M***odify, then* **S***cale*	Issues the SCALE command
_SCALE Select objects: *Select the kitty image (see fig. 3.26)*	Highlights the kitty
Select objects: **Enter**	Accepts the selection set
Base point: *Choose the Nearest button*	Selects the NEArest object snap
of *Pick the ground line under the kitty*	Sets the base point
<Scale factor>/Reference: **R**	Selects the Reference option
Reference length <1>: **20'7**	Sets current height
New length: **15**	Establishes new height
Command: **R**	Redraws the view
Command: *Choose the Save button on the toolbar*	Saves the drawing

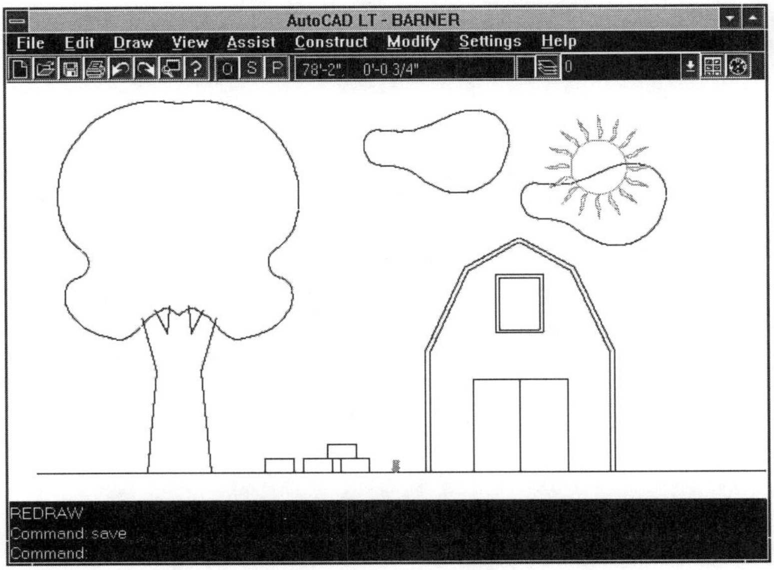

Figure 3.26

The completed drawing.

Summary

In this chapter, you have explored many of the tools available to you in AutoCAD LT. New users often experience a lot of discomfort over the number of commands and options that are available. This is a common feeling—all CAD operators go through a period of information overload. Hang in there while you continue through the rest of the chapters in the book. Practice will make this feeling subside, and you will probably find yourself making a mistake in the physical world only to mumble, "undo..." This chapter has covered the following items:

◆ The three methods of coordinate input: Absolute, Relative, and Polar

◆ The object snap options and selection set definition

◆ Duplicating and moving drawings or their parts

◆ Editing and revising drawings

◆ Creating and editing polylines

◆ The methods of inserting DXF, WMF, and other types of drawings

The following chapters will explore essential drafting concepts while continually introducing new commands. Be prepared to use the object snap options.

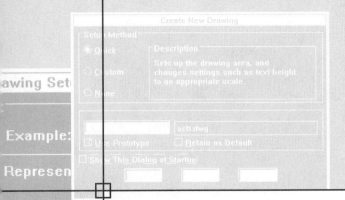

Chapter

4

The intent of multiview drawings is to accurately portray the size and shape of a 3D object on flat 2D surfaces. You will start off with a standard multiview drawing, and develop your skills from there. Topics included in this chapter are the following:

- ◆ Understanding multiview drawings
- ◆ Creating a prototype drawing
- ◆ Creating a real world orthographic drawing

To draw well in CAD, you need to understand the drafting skill and then apply it to working in CAD. One skill you will gain in this chapter is the understanding of how to portray a 3D object with 2D views. You will also learn to create a prototype or template drawing, with the appropriate CAD settings. You can use this prototype drawing again and again. Finally, you will use this prototype drawing to create a mechanical orthographic drawing of a bracket.

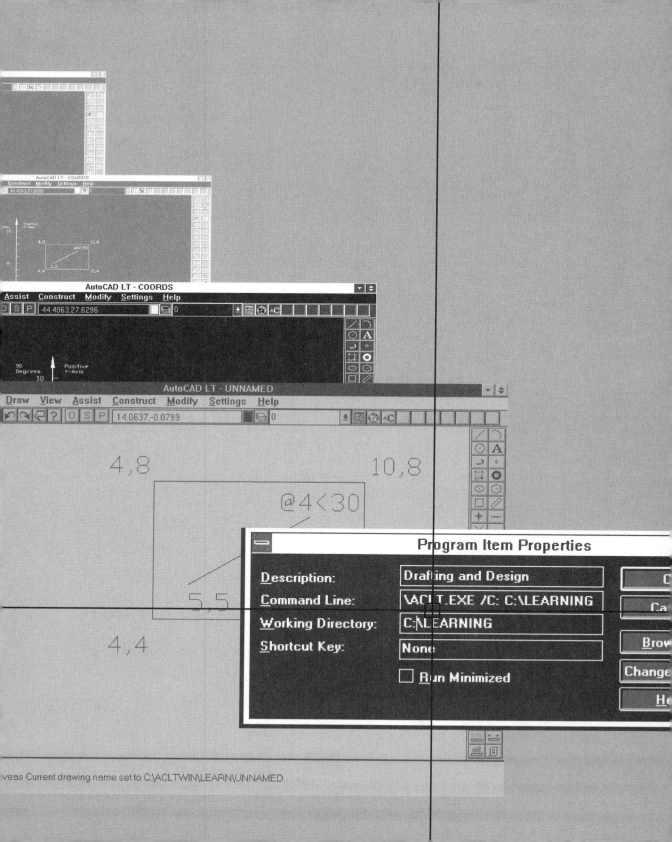

Multiview Drawings

To be effective in drawing, you must be able to draw 2D views of a 3D object. To draw 2D views, you need to be able to visualize these views while looking at the actual object; then you can create the multiview drawing. This drawing can be either architectural plans and elevations, or it can be a mechanical orthographic drawing. Let's look at visualizing a multiview drawing.

Understanding Multiview Drawings

How do you effectively portray a 3D object on a flat 2D sheet of paper? Generally, to represent a 3D object, it normally takes six standard views: top, rear, left side, front, right side, and bottom. What the draftsperson is doing is representing the shape and size of a 3D element on a flat sheet of paper. A *multiview drawing* is one in which a draftsperson uses a flat two-dimensional plane to accurately describe a three-dimensional object. It shows the width, depth, and height of the object.

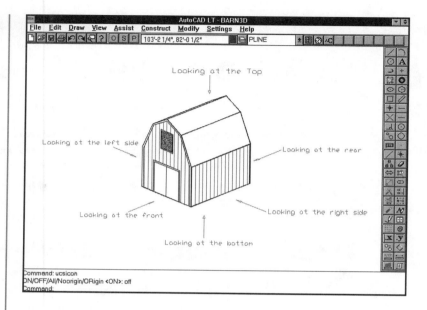

Standard Views

You are looking at a 3D barn (see fig. 4.1). How would you come up with six standard views? If you were a bird overhead and looked down at the barn, that would be your top view. If you were just an ordinary person and started walking around from the rear to the front of the barn, that would give you the four views: rear, left side, front, and right side. Finally, if you were a mole under the ground looking up at the barn, that would give you the bottom view, a concrete slab surrounded by foundation walls. This is a short description of how to create six different views from a single 3D object (see fig. 4.2).

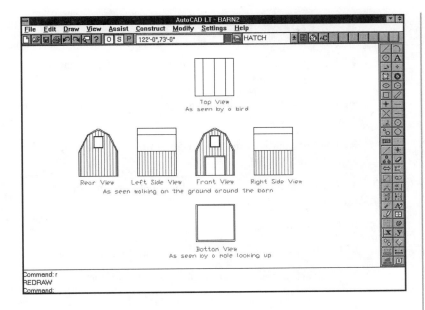

Figure 4.2

Six views of a 3D object.

Mechanical Orthographic Drawings vs. Architectural Plan and Elevations

Multiview drawings can be represented in a number of different ways. Mechanical solid objects are best presented in an orthographic drawing. Buildings are best represented by plans and elevations.

Mechanical Orthographic Drawings

As an example, think of the barn as a 3D solid object to be represented as a mechanical orthographic drawing. Maybe it's a brass model of the barn 4" wide, 4" deep, and 6" high. The window in the front of the barn is a hole in the object that projects all the way through the object (see fig. 4.3). Some guidelines to draw this solid 3D object as a 2D orthographic drawing are as follows:

◆ The front view needs to be the one the draftsperson thinks is the most important view. All other views are projected from it. It is the starting point for the other views. The front view shows width and height of the object, and can also be the view with the most detail.

◆ The top view is directly above the front view, and it shows width and depth.

◆ The right side view is directly to the right of front view, and it shows height and depth.

◆ The left side view is directly to the left of front view and it shows height and depth.

◆ The rear view is projected directly from the left side and it shows width and height.

◆ The bottom view is directly below the front view and it shows width and depth of the object.

◆ Hidden lines and other special lines should be shown.

◆ Required views usually are top, front, and right side.

◆ Units are usually in decimal for inches or millimeters.

◆ The drawing is usually plotted full (1:1), half (1:2), or quarter scale (1:4). (Objects are always drawn full scale.)

◆ Each view shows details specific to that view.

◆ Sectional and auxiliary views may also be needed. See Chapter 8, "Developing Sectional and Auxiliary Views," for more information.

Figure 4.3

Orthographic views of a barn model.

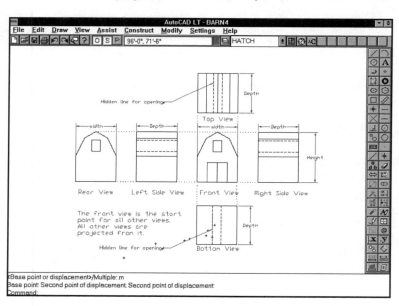

Architectural Plan and Elevations

A more realistic way to look at the barn is as a building, showing floor plans and elevations. When you draw this as a building, it becomes important to know the location of North. A site plan may also be required to locate the building exactly to comply with local zoning and building codes (see fig. 4.4). Some important aspects of an architectural plan for this building are the following:

◆ The plan view or floor plan is the primary view. All other views are created from that view. A North arrow is generally shown on the floor plan.

◆ Elevations are what you see as you walk around the building. They are generally referred to by the direction they face.

◆ Foundation plans may be shown, bottom plans, but you are looking down at them, not up at them.

◆ Roof plans are sometimes shown to show how the roof drains.

◆ Site plans are often used to locate the building on the site.

◆ Elevations are usually shown on a different sheet than the floor plan.

◆ Units used for the drawing are generally architectural, feet and inches, or meters. (Buildings are always drawn full scale.)

◆ Drawings are usually plotted at 1/4"=1(1:48), 1/8"=1"(1:96)

◆ Sectional and auxiliary views may also be needed. See Chapter 8 for more information.

Chapter 4

Figure 4.4

*An architectural
drawing of a barn.*

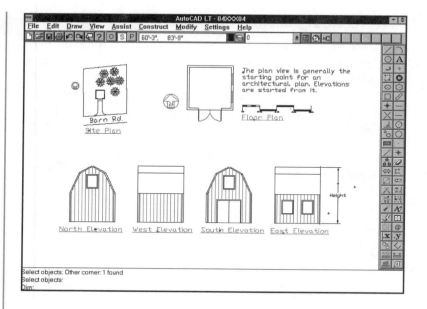

Required Views

How many views are necessary to effectively portray a 3D object on a 2D sheet of paper? One idea is to create as many views as necessary to completely describe the 3D object and clearly present it. How many is that? A spherical baseball may need only one view to effectively portray it. A symmetrical pipe needs only two views, one for the side view and one for the front view (see fig. 4.5). How many more views would you need for a pipe?

For more complex objects, like a bracket, you will need more than two views. Generally, you can have up to six views, but there's no reason to draw two views that are exactly alike. Symmetrical objects like the bracket often look the same from different views: front to back, top to bottom, and right to left. For the bracket, then, the draftsperson could decide on three views: front, top, and right side. The front view will be the one that is considered the most important by the draftsperson. All views are projected from the front view. The front view should also have a minimum number of hidden lines so as not to be confusing (see fig. 4.6).

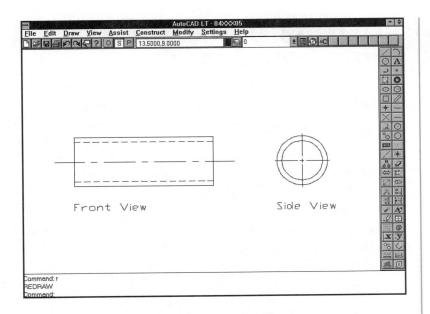

Figure 4.5

Two views of a pipe.

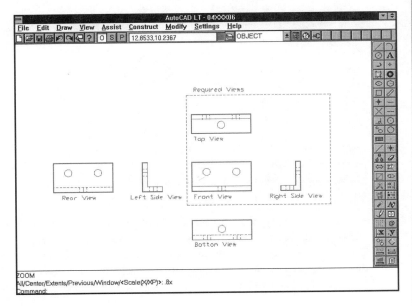

Figure 4.6

Three required views of a bracket.

Creating Your Own Prototype Drawing

Now that you are familiar with the drafting skill of visualizing a 3D object with 2D views, let's look at the CAD skill of creating a prototype drawing. The advantages of such a drawing is that you create the settings for the typical drawing, and use it as a template for future drawings—you can use this template again and again. You will later create a prototype drawing, and then use it as a template for your mechanical orthographic drawing.

Prototype drawings contain standardized settings and drawing elements that can be used as a template for creating new drawings. There is a real advantage to creating your own prototype drawings if you have numerous drawings using the same units, layers, and paper size. Some settings and elements in your prototype drawing that will be discussed are as follows:

- Unit type and level of precision
- Drawing limits
- Snap and grid settings
- Layers
- Linetypes
- Borders
- Title blocks

Other items you can include in your prototype drawing are dimension and text styles. You will not be including these styles in this prototype drawing, but you will learn about dimension styles and text styles in the chapters to follow.

An important detail to keep in mind when you are creating prototype drawings is that they need to be consistent with your office standards. They can be used to support these standards because individuals using these prototypes will have the same units, drawing limits, layering conventions, borders, and title blocks.

Units, Drawing Limits, and Grid

As a way for you to keep working on the same project, you will be creating a bracket in this chapter. It will be similar to the bracket you saw in figure 4.6. You will be showing just the required views.

You always need to start off with your units setting, because it defines the units of measurement that define the limits of your drawing and the border size. Units may also affect some other factors in your drawing—the size of the text and the size of the object. It's important to set the units first. Because you are doing a real-world orthographic drawing of a part to be manufactured, you'll use decimals as your units for this drawing. The linear units' accuracy in this drawing will be to .0001 units accuracy, and angular accuracy will be to 0°.

The drawing limits of the prototype drawing are important because they give you an idea of your drawing area. Knowing your drawing area, you can organize your drawing effectively, without crowding any elements. In this project, you will be printing/plotting the project full scale to an 11×8.5 sheet of paper. An 11×8.5 drawing area will work fine. It becomes more complex, when you are drawing a building full scale, and have to plot to an 11×8.5 sheet of paper. The process for setting up limits for a building is discussed in Chapter 12, "Architectural Drawing Project."

Grid is helpful because you have a sense of scale within the drawing: it's like working on graph paper with your drawing on a sheet of paper. If you set your snap spacing to match the grid and toggle snap on, your cursor will snap to the grid points. Grid and snap can be very helpful for organizing text on a drawing.

Exercise

Creating a Prototype Drawing

Choose the New button	Begins a new drawing
Choose **C***ustom and then* OK *in the Create New Drawing dialog box (see fig. 4.7)*	Opens the Custom Drawing Setup dialog box
In the World (actual) Size to Represent *area, double-click on* 12.0000 *and type* **11***, press Tab, type* **8.5,** *press Tab, then choose* OK	Adjusts sheet size to 11 by 8.5, an American National Standard A-size sheet

continues

In the Drawing Aids area, *choose Drawing Aids button, then choose the box below grid to toggle on the grid*	Accepts snap and grid spacing of 0.5000 and turns on grid
Choose OK	
In the Title Block *area, choose the Title Block button, choose* No title block *under description, then choose* OK	
You will be creating your own simpler title block for this drawing later.	
Choose OK *in the Custom Drawing dialog box*	Finishes this setup task and opens the drawing

Figure 4.7

Figure 4.7

The Custom Drawing Setup dialog box.

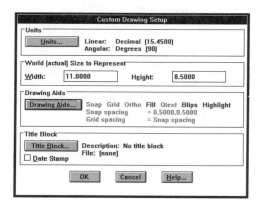

Now you have set up your units and limits for your prototype drawing. Next, take a look at organizing layers within your prototype drawing, complete with names, color, and linetypes.

Layers

Up to this point, you have been working on one layer—layer 0, the default layer. In AutoCAD LT, you can create multiple layers. For the mechanical orthographic drawing, you will be creating seven layers on which to work.

To get an idea of how multiple layers can work, let's look at an architectural drawing. In this type of drawing, you can put the building on one layer, notes on another layer, the border on another, and then dimensions on even another. If you want to show the client the floor plan without all the notes and dimensions, then you simply turn off those layers. The same principle would also work in a mechanical drawing: the dimensions and notes layers could be turned off, and the object and border would be the only layers showing.

In addition to just turning off layers you don't want to show to the customer, there are other advantages to layers. For example, in the OFFICE drawing within the \ACLTWIN directory, the different disciplines in a drawing are shown on different layers. Walls, furniture, phones, doors, hardware, and roomtags are all shown on different layers. The layers can be worked on individually, not affecting the other layers. Also, if you are doing a lighting plan of the ceiling, you might want to turn off or freeze certain layers if they don't directly affect that ceiling lighting drawing. You might want to turn off the phones, doors, hardware, and furniture layers, and show just the walls, ceilings, and roomtags (see fig. 4.8).

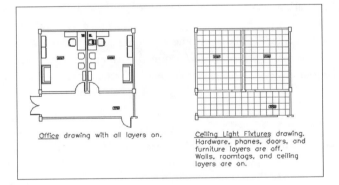

Office drawing with all layers on.

Ceiling Light Fixtures drawing. Hardware, phones, doors, and furniture layers are off. Walls, roomtags, and ceiling layers are on.

Figure 4.8

An office drawing with different layers on.

Creating Layers

To work with layers effectively, you need to use the Layer Control dialog box in AutoCAD LT. For this project to create the bracket, you will use a number of layers, as follows:

◆ **0.** Always has to be in every AutoCAD LT drawing. You cannot delete layer 0.

◆ **OBJECT.** You will draw the object on this layer.

- ◆ **BORDER.** The border goes on this one.

- ◆ **TITLE.** You will put the title block on this layer.

- ◆ **NOTES.** All annotation will go on this layer.

- ◆ **DIM.** Dimensioning will go on this layer.

- ◆ **HIDDEN.** The hidden lines go on this layer.

- ◆ **CENTER.** The center lines go here.

Before you start creating layers, there are some other characteristics that you can add to layers: color and linetypes.

Color on Layers

Color can be added to individual circles or lines, or a layer can be assigned a color. When a color is assigned to a layer, then anything drawn on that layer will take on that color. When you assign a color, you may want to take into consideration what the ultimate width of the lines are going to be on that layer. If you are using a plotter, you could think of it this way: the black lines could be used to denote heavy lines (.50mm pens) for an object, green could be used to denote medium lines (.35mm pens) for text and dimensioning, and yellow could be used to denote fine lines (.25mm pens) for hidden and center lines.

Linetypes on Layers

Another characteristic that can be assigned to a layer is a linetype. There are many linetypes you can use, some of which are BORDER, CENTER, HIDDEN, DOT, and CONTINUOUS. For the bracket project, you will use three linetypes: CONTINUOUS—which are solid lines—for most of the lines, HIDDEN for the details of the objects you are unable to see from a view, and CENTER for the center of the drilled holes.

Two commands are important for you to know when using linetypes. First, you must load the different linetypes using the LINETYPE command, before you open the Layer Control dialog box. The second important command is LTSCALE, (Linetype SCALE), which is also an AutoCAD system variable. The default scale factor for LTSCALE in AutoCAD LT is 1. You need to increase it if you are plotting at a much smaller scale than the real size of the actual object. You will adjust the LTSCALE later when creating the mechanical drawing.

A good rule of thumb for determining the scale factor for LTSCALE is that it should be approximately half your plot scale. If you are plotting full scale, the LTSCALE scale factor should be .5. If you are plotting at a quarter scale (1:4), your LTSCALE scale factor should be approximately 2. If you are plotting at a 1/4"=1' (1:48), your LTSCALE scale factor should be approximately 24.

When you change your LTSCALE scale ratio, your drawing automatically regenerates, and you can see the change in your lines.

In the following exercise, you complete the prototype drawing with layers having colors and linetypes.

Creating Layers for the Prototype Drawing

Continue with the drawing from the previous exercise.

Command: *Choose* **S**ettings, *then* **Li**netype Style, *then* **L**oad	Issues the LINETYPE command with the Load option
Linetype(s) to Load: * (Enter)	Loads all the linetypes into the drawing
Choose ACLT.LIN *from the file name dialog box, and choose* OK	Loads all the linetypes from that file into the drawing
?/Create/Load/Set: (Enter)	Ends the LINETYPE command

If you want to see all the lines that were loaded into your drawing after choosing the ACLT.LIN file, press the F2 function key. You will note that there are three types of hidden lines: HIDDEN, HIDDEN2, HIDDENX2. The HIDDENX2 linetype has the longest dashes and spaces, and the HIDDEN2 has the shortest dashes and spaces. You may want to choose the HIDDENX2 for building plans, and HIDDEN2 for drawings that are plotted full scale.

continues

Figure 4.9

The Layer Control dialog box.

Now start to create layers.

Command: *Choose the Layers button*   Opens the Layer Control dialog box
(see fig. 4.9)

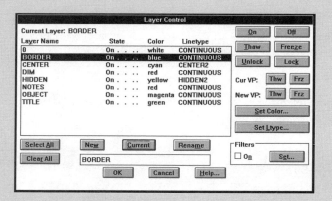

Type **BORDER,CENTER,TITLE,**   Creates seven new layers and adds
OBJECT,DIM,NOTES,HIDDEN *in*   them to the Layer Name list box
the input box, and choose Ne**w**

TIP

> When typing the layer names—BORDER, CENTER, TITLE and so forth—
> make sure there are no spaces after the commas. This is the only number
> of new layers that can be created at once.

Click on BORDER *in the* Layer   Selects a layer
Name *list box*

Choose **S**et Color   Sets a color of blue for the layer

Type Blue *in color input box,*
then choose OK

Click on BORDER *in the* Layer   Unselects one layer and selects
Name *list box, then click on* CENTER   the next layer

Choose **S**et Color, *type* **Cyan** *in the*   Sets a color of cyan for the
color input box, then choose OK   center layer

Choose the Set **L**type *button, click on the line next to* CENTER2, *then choose* OK	Sets the linetype to CENTER2

When you click on the line next to CENTER2, the name CENTER2 appears in the Linetype box (see fig. 4.10).

Figure 4.10

The Select Linetype dialog box.

Click on CENTER *in the* Layer Name *list, then click on* TITLE	Unselects one layer, then selects *box,* the next layer for subsequent action
Choose **S**et Color, type **Green** *in color input box, then choose* OK	Sets a color of green for the title layer
Click on TITLE *in the* Layer Name *list box, then click on* OBJECT	Unselects one layer and then selects the next layer
Choose **S**et Color, type **Magenta** *in color input box, then choose* OK	Sets a color of magenta for OBJECT layer
Click on OBJECT *in the* Layer Name *list box, then click on* NOTES *and* DIM	Unselects one layer and then selects two new layers
Choose **S**et Color, type **Red** *in color input box, choose* OK	Sets a color of red for the notes and dim layers
Click on NOTES *and* DIM *in* Layer Name *list box, then click on* HIDDEN	Unselects two layers and then selects the next layer
Choose **S**et Color, type **Yellow** *in color input box, then choose* OK	Sets a color of yellow for the hidden layer

continues

Choose Set **L**type, *click on the line next to* HIDDEN2, *then choose* OK	Sets the linetype to HIDDEN2
Click on HIDDEN *in the* La**y**er Name *list box, then click on* BORDER, *choose* **C**urrent, *and choose* OK	Unselects one layer and then selects the BORDER layer and makes it current
The current layer of BORDER is the layer on which you will be drawing	

All the layers for your prototype drawing are now complete. In the following exercise, you add a border and a simple title block to your drawing.

Exercise

Adding a Border and Title Block

Now you will draw the border and title block shown in figure 4.11.

Command: *Choose the Rectangle Button*	Creates a rectangular border
First Corner: **1,.5** (Enter)	Places first corner
Second Corner: **10.5,8** (Enter)	Completes the rectangle
Command: *Click on the arrow on Current Layer drop-down list box and choose* TITLE	Makes TITLE the current layer
Command: *Choose the Line Button*	Issues the LINE command
Line From Point: **6,.5** (Enter)	Places starting point of a line
To Point: **@0,1.25** (Enter)	Places ending point of a line
To Point: **@4.5,0** (Enter)	Places ending point of a line
To Point: (Enter)	Ends the LINE command
Command: *Choose* **C**onstruct, *then* **O**ffset	Issues the OFFSET command
Offset Distance or through <0.0000>:**.5** (Enter)	Sets an offset distance
Select Object to Offset: *Pick line at* ① *(see fig. 4.11)*	Selects line to offset

Side to Offset: *Pick* ②	Offsets line
Select Object to Offset: (Enter)	Ends the OFFSET command

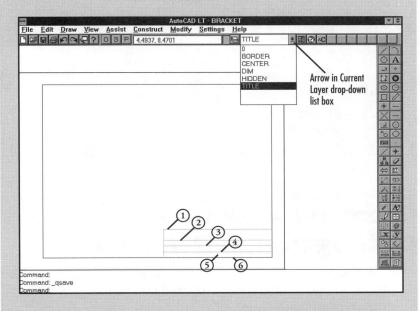

Figure 4.11

The prototype drawing, complete with border and title block.

Command: (Enter)	Reissues the OFFSET command
Offset Distance or through <0.0000>:.25 (Enter)	Sets an offset distance
Select Object to Offset: *Pick line at* ②	Selects line to offset
Side to Offset: *Pick* ③	Offsets line
Select Object to Offset: *Pick line at* ③	Selects line to offset
Side to Offset: *Pick* ④	Offsets line
Select Object to Offset: (Enter)	Ends the OFFSET command
Command: *Choose the Line button*	Issues the LINE command
LINE From point: *Choose the Midpoint button*	Specifies the MIDpoint object snap

continues

_MID of *Pick* ④	Places the starting point of a line
To point: *Choose the Perpendicular button*	Specifies the PERpendicular object snap
_PER to *Pick* ⑤	Places the ending point of a line
To point: (Enter)	Ends the LINE command
Command: *Choose* **F**ile, Save **A**s, *type* **ORTHO-O** *and press Enter*	Saves the drawing to the file ORTHO-O.DWG
You will use the ORTHO-O drawing as the prototype for your BRACKET.DWG	

Creating a Real-World Orthographic Drawing

You have learned about multiview drawings and how to view them. Two types of multiview drawing are orthographic projection drawings and architectural plans and elevations.

There are six standard views in an orthographic drawing and generally three required views for a drawing. The number of required views depends on the object. When you start an orthographic drawing, you start with the most important view, the front view. From the front view, you will project to the other views. See figure 4.12 for a review.

You just finished creating a prototype drawing that you can use for your real world drawing or any drawing. It is set up complete with decimal units, drawing limits, a grid, a border, and a title block.

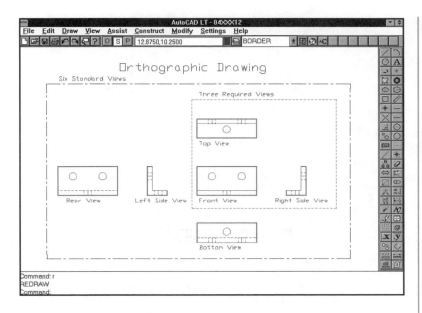

Figure 4.12

The standard and required views of an orthographic drawing.

Constructing the Front View

Now you have the tools to construct a real-world orthographic drawing. You know how to create an orthographic or multiview drawing, and how to create a prototype drawing in AutoCAD LT. As you construct the front view of a bracket, you will also learn some new commands in the process. Let's start an exercise by drawing a rectangle.

Creating a Front View of the Orthographic Drawing

Exercise

Command: *Choose the New Button*	Opens the Create New Drawing dialog box
In the Create New Drawing dialog box, choose **Q**uick *under Setup Method, then choose the* **P**rototype *button*	
Choose the ORTHO-O *drawing (see fig. 4.13)*	Uses ORTHO-O as the prototype drawing

continues

Choose OK *twice*	Exits the Create New Drawing dialog box

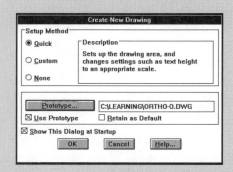

Choose OK *in the Quick Drawing Setup box*	Exits the dialog box and starts the drawing
Click on the arrow on Current Layer name box and choose the OBJECT *layer drop-down list*	Makes the object layer the current layer

Start drawing the object shown in figure 4.14.

Command: *Choose the Rectangle button*	Issues the RECTANG command
From Point: **2.5,2.5** (Enter)	Starts the first corner of the rectangle
To Point: **@3,1.5** (Enter)	Completes the second corner
Command: *Choose the Zoom button*	Issues the ZOOM command
All/Center/Extents/Previous /Window/Scale>(X/XP) **1.4** (Enter)	Brings you closer to the bracket to work on it
Command: *Choose the Explode button*	Issues the EXPLODE command
Select Objects: *Select the rectangle*	Selects the rectangle to explode
Select Objects: (Enter)	Ends the EXPLODE command
Command: *Choose the Copy button*	Issues the COPY command
Select Objects: *Select line at* ① *(see fig. 4.14)*	Selects bottom line to copy

Select Objects: (Enter)	Ends object selection
<Base point or displacement> /Multiple: **0,.25** (Enter)	
Second point of displacement: (Enter)	Copies line up .25
Click on the arrow on Current Layer name box and choose the CENTER *layer*	Makes the center layer the current layer
Command: *Choose the Line button*	Issues the LINE command
LINE From point: *Choose the Midpoint button*	Specifies MIDpoint object snap
_MID of *Select the line at* ②	Places starting point of line
To Point: **@0,.25** (Enter)	Places ending point of line
To Point: (Enter)	Ends the LINE command

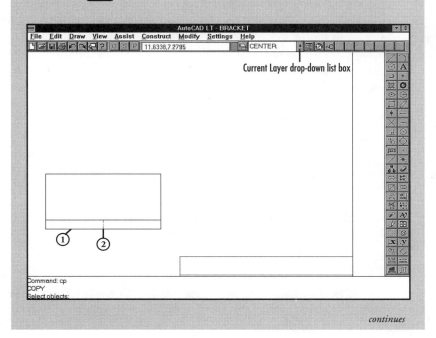

Figure 4.14

Starting to draw the front view.

continues

```
Command: LTSCALE (Enter)              Accesses the LTSCALE system
                                      variable

New Scale Factor<1.000>:              Adjusts the linetype scale to be
.2 (Enter)                            smaller because this is a small object
                                      drawn full scale
```

NOTE

Earlier, it was recommended that you use approximately .5 as the linetype scale factor for objects being plotted full scale. You will be adjusting it to .2 for this drawing.

In the next exercise, you use a new feature called TRACKING mode. This is a method used to locate a point relative to other points on the drawing without having to use construction lines. You will find TRACKING mode in your toolbox.

First, you need to be in a drawing command that prompts you for a point. The idea is that after you select the exact point for tracking to start—for example, by using an ENDpoint object snap—you move the cursor in the direction you want the next point to be placed. For instance, if you want the next point to be four units to the right from the last point, move your cursor to the right, type **4,** and press Enter. The tracking command will continue to move the point around until you press Enter to stop it. It will then place the point and prompt with the next drawing command prompt. The command enables you to place an element relative to another element without having to use construction lines. Another way to use TRACKING mode is to type in polar coordinates. For example, **@3<180** (Enter) will move the TRACKING point three units to the left.

Exercise

Using TRACKING Mode with the CIRCLE Command

Continue the drawing as shown in figure 4.14.

```
Command: Choose the Circle button     Issues the CIRCLE command

Circle 3P/TTR/<Center point>:         Places the center point of the circle
Choose the Tracking button
```

`_TRACKING First Tracking point:` *Choose the Endpoint button*	Uses ENDPoint object snap for exact point to track from
`_ENDP of` *Select* ③ *(see fig. 4.15)*	Selects the upper left corner of the rectangle

Figure 4.15

Constructing the front view.

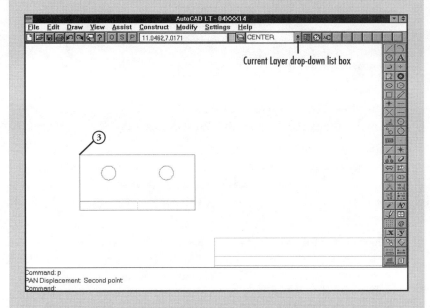

`Next Point(Press RETURN to end tracking):` *type* `@.75<0` `Enter`	Places a point .75 to the right from the corner of rectangle
`Next Point(Press RETURN to end tracking):` *type* `@.5<270` `Enter`	Places a point .5 down from the last point placed
`Next Point(Press RETURN to end tracking):` `Enter`	Ends tracking and places the center of the circle @.75,-.5 from the upper right corner
`Diameter/Radius: D` `Enter`	Specifies the Diameter option
`Diameter: .375` `Enter`	Specifies diameter of .375
`Command:` *Choose the Copy button*	Issues the COPY command

continues

Select Objects: *Select the circle*	Selects the circle to copy
Select Objects: (Enter)	Ends object selection
<Base point or displacement> /Multiple: **1.5,0** (Enter)	Copies circle over 1.5 in horizontal
Second point of displacement: (Enter)	Ends the COPY command

When complete, drawing should look like figure 4.15

In the next exercise, you use the OFFSET, DDCHPROP, and CHAMFER commands, described in the following paragraphs.

DDCHPROP (CHange PROPerties) is another command important after you offset the center lines of the holes. This command is very effective when you have drawn or copied lines and suddenly you determine that they are on the wrong layer. You can change the color, layer, and linetype of a line, circle, or any entity after it is drawn.

The CHAMFER command is similar to the FILLET command except that it creates a beveled edge instead of a rounded edge. The CHAMFER command creates an angled line between two intersecting lines. The command extends or trims lines as necessary and then adds a line to create a beveled edge. You specify the distance from the intersection on each line where the bevel starts and ends. To initiate the command, choose **C**onstruct, C**h**amfer. Continue on through the next exercise to see how this works.

Exercise

Using CHAMFER on the Front View

Continue drawing the bracket in figure 4.16.

Command: *Choose* **C**onstruct, *then* **O**ffset	Issues the OFFSET command
Offset distance or through <0.1875>: **.1875** (Enter)	Sets the offset distance to .1875
Select Object to offset: *Pick line at* ④ *(see fig. 4.16)*	Selects line to offset

`Side to Offset:` *Pick point to right*	Offsets line
`Select Object to offset:` *Pick line at* ④	Selects line to offset
`Side to Offset:` *Pick point to left*	Offsets line
`Select Object to offset:` (Enter)	Ends the OFFSET Command

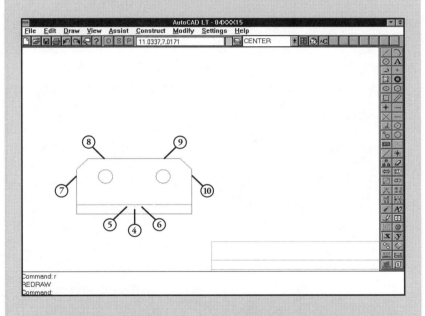

Figure 4.16

The final touches on the front view.

`Command:` *Choose the Change Properties button*	Issues the DDCHPROP command
`Select Objects:` *Select line at* ⑤	Selects the line
`Select Objects:` *Select line at* ⑥	Selects the line
`Select Objects:` (Enter)	Ends object selection
Choose the **L**ayer *button in the Change Properties dialog box*	Opens the Select Layer dialog box
Choose the HIDDEN *layer in the Select Layer dialog box and choose* OK	Selects the HIDDEN layer as a new layer for the selected lines

continues

Command: *Choose* **C**onstruct, *then* **Ch**amfer	Issues the CHAMFER command
_chamfer Polyline/Distances /<Select first line>: **D** (Enter)	Specifies the Distance option
Enter first chamfer distance <0.4722>: **.3** (Enter)	Sets the first distance
Enter second chamfer distance <.2000>: **.3** (Enter)	Sets the second distance
Command: (Enter)	Reissues the CHAMFER command
CHAMFER Polyline/Distances /<Select first line>: *Select line at* (7)	Selects first line to chamfer
Select second line: *Select line at* (8)	Selects second line to chamfer
Command: (Enter)	Reissues the CHAMFER command
CHAMFER Polyline/Distances /<Select first line>: *Select line at* (9)	Selects first line to chamfer
Select second line: *Select line at* (10)	Selects second line to chamfer
Command: *Choose* **F**ile, Save **A**s, *type* **BRACKET**, *and press Enter*	Saves the drawing to the file BRACKET.DWG

Your drawing should now resemble figure 4.16

Constructing the Top View of the Orthographic Drawing

To construct the top view of the bracket, you will first use the COPY command with the Multiple option to define the top and bottom of the top view. The COPY Multiple command can be very effective if you have to copy a number of parallel lines at various distances. To issue the COPY command, choose the Copy button from the toolbox.

To define the sides of the top view, you will use the LINE command and use running object snaps. The advantage of a running object snap over a regular object snap is that it stays on until you turn it off. To turn on running object snaps, choose the Object Snap button from the toolbox and toggle on the appropriate running object snaps. The important point to remember is to turn the running object snap off when you are finished.

You will continue to use the DDCHPROP command and the TRACKING mode to construct the top view.

Exercise

Creating the First Part of the Top View

Continue drawing the bracket in figure 4.17.

Command: *Choose the Copy button*	Issues the COPY command
Select Objects: *Select bottom line at* ① *(see fig. 4.17)*	Selects the bottom line
Select Objects: (Enter)	Ends object selection
<Base point or displacement> /Multiple: M (Enter)	Specifies Multiple copy option
<Base point or displacement>: @ (Enter)	Specifies from this point
Second point of displacement: @0,3.0 (Enter)	Copies a line 3.0 units up
Second point of displacement: @0,3.75 (Enter)	Copies a line 3.75 unit up
Second point of displacement: @0,4.0 (Enter)	Copies a line 4.0 units up
Second point of displacement: (Enter)	Ends the COPY command

continues

Figure 4.17

*Copying and drawing
lines for the top view.*

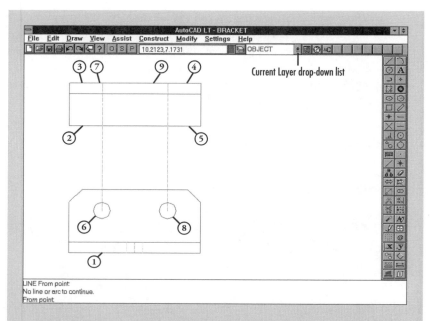

Current Layer drop-down list

LINE From point
No line or arc to continue.
From point

Click on the arrow on Current Layer drop-down list, choose the OBJECT *layer*	Makes the object layer the current layer
`Command:` *Choose the Object Snap button, choose* **E**ndpoint, *then choose* OK	Sets a running object snap of ENDPoint
`Command:` *Choose the Line button*	Issues the LINE command
`LINE From Point:` *Select line at* ② *(see fig. 4.17)*	Places starting point of the line
`To Point:` *Select line at* ③	Places ending point of the line
`To Point:` (Enter)	Ends the LINE command
`Command:` (Enter)	Reissues the LINE command
`LINE From Point:` *Select line at* ④	Places starting point of the line
`To Point:` *Select line at* ⑤	Places ending point of the line
`To Point:` (Enter)	Ends the LINE command

Command: *Choose the Object Snap button, then choose* **E**ndpoint *to toggle off*	Toggles off Endpoint
Choose **C**enter *and* **P**erpendicular *to toggle on, then choose* OK	Toggles on Center, Perpendicular running object snap
Click on the arrow on Current Layer drop-down list, then choose the CENTER *layer*	Makes the CENTER layer the current layer
Command: *Choose the Line button*	Issues the LINE command
Line From Point: *Select circle at* ⑥	Places starting point of line
To Point: *Select line at* ⑦	Places ending point of line
To Point: (Enter)	Ends LINE command
Command: (Enter)	Reissues the LINE command
Line From Point: *Select circle at* ⑧	Places starting point of line
To Point: *Select line at* ⑨	Places ending point of line
To Point: (Enter)	Ends the LINE command
Command: *Choose the Object Snap Button, choose* **C**enter *and* **P**erpendicular *to toggle off, then choose* OK	Toggles off CENter and PERpendicular running object snap
Your drawing should now look like figure 4.17	

In this part of the exercise, you use the OFFSET command to copy the lines that you projected to the top view.

To clean up the view, you use the TRIM command—the Trim button is found in the toolbox. When you issue the TRIM command, you are prompted with Select cutting edges...Select objects:. Respond to the prompt by selecting a single line and then pressing Enter—this is the line that cuts all the other lines. Your next prompt is Select objects to Trim/ UNDO:—you pick the portions of the lines you want removed. You can select objects individually, or you can use F (for Fence) for selecting a number of objects at a time. Whatever the fence crosses is selected and trimmed off. Press Enter, and you complete the TRIM command.

Try these two commands in the next exercise.

Exercise

Using OFFSET and TRIM in the Top View

Continue with the Bracket drawing.

Command: *Choose* **C**onstruct, *then* **O**ffset	Issues the OFFSET command
Offset distance or through <0.1875>: **.1875** (Enter)	Sets offset distance to .1875
Select Object to offset: *Select line at* ⑩ *(see fig. 4.18)*	Selects line to offset
Side to Offset: *Select to the right of line*	Offsets line
Select Object to offset: *Select line at* ⑩	Selects line to offset
Side to Offset: *Select to the left of line*	Offsets line
Select Object to offset: *Select line at* ⑪	Selects line to offset
Side to Offset: *Select to the right of line*	Offsets line
Select Object to offset: *Select line at* ⑪	Selects line to offset
Side to Offset: *Select to the left of line*	Offsets line
Side to Offset: (Enter)	Ends the OFFSET command
Command: (Enter)	Issues the OFFSET command
Offset distance or through <0.1875>: **.3** (Enter)	Sets offset distance to .3
Select Object to offset: *Select line at* ⑫ *(see fig. 4.18)*	Selects line to offset

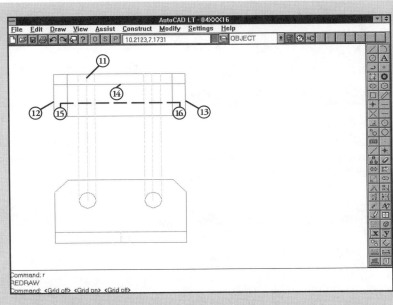

Figure 4.18

The top view under construction.

Side to Offset: *Select to the right of line*	Offsets line
Select Object to offset: *Select line at* ⑬	Selects line to offset
Side to Offset: *Select to the left of line*	Offsets line
Side to Offset: ⏎ Enter	Ends the OFFSET command
Command: *Choose the Trim button*	Issues the TRIM command
Select cutting edges Select objects: *Select line at* ⑭ *and then press Enter*	Selects the line that will cut the other lines
Select objects: ⏎ Enter	Ends selection process
<Select Objects to Trim>/UNDO: F ⏎ Enter	Specifies the Fence option

continues

First fence point: *Pick* ⑮   Selects first point of fence

Undo/<Endpoint of line>: *Pick* ⑯   Selects second point of fence

Undo/<Endpoint of line>: **Enter**   Ends the TRIM command

The bottom portions of the lines should be trimmed

To finish up the top view, you use the DDCHPROP command. To place a circle for a hole in the bracket, you use the TRACKING mode. The top view will then be complete, as demonstrated in the next exercise.

Exercise

Completing the Top View

Continue drawing the bracket in figure 4.19.

Command: *Choose the Change Properties button*	Issues the DDCHPROP command
Select Objects: *Select line at* ① *(see fig. 4.19)*	Selects the line to change properties
Select Objects: *Select line at* ②	Selects the line to change properties
Select Objects: *Select line at* ③	Selects the line to change properties
Select Objects: *Select line at* ④	Selects the line to change properties
Select Objects: **Enter**	Ends the selection process
Choose the **L**ayer *button in the Change Properties dialog box*	Opens the Select Layer dialog box
Choose the HIDDEN *layer in the Select Layer dialog box and choose* OK	Selects the hidden layer as a new layer for the selected lines
Command: *Choose the Circle button*	Issues the CIRCLE command
CIRCLE 3P/TTR/<Center point>: *Choose the Tracking button*	Initiates TRACKING
TRACKING First tracking point *Choose the Endpoint button*	Uses ENDPoint object snap
ENDP of *Pick point at* ⑤	Starts tracking at lower left corner

`Next Point(Press RETURN to end tracking): type` `@1.5<0` `Enter`	Picks a point 1.5 to the right of the corner
`Next Point(Press RETURN to end tracking): type` `@.4<90` `Enter`	Picks a point .4 up from the previous point
`Next Point(Press RETURN to end tracking):` `Enter`	Ends tracking, resumes CIRCLE command
`Diameter/<Radius><0.1875>:` `.1875` `Enter`	Specifies same radius as other circle
`Command:` *Choose the Save button*	Saves drawing to BRACKET.DWG

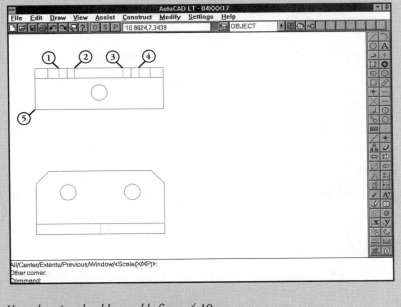

Figure 4.19

Completing the top view.

Your drawing should resemble figure 4.19

Completing the Right Side View of the Orthographic Drawing

In order to project the top view to the right side view, you will need a 45° projection line, originating from the upper right corner of the front view. This is a manual way of doing things, but it works well in aligning views and assuring accuracy.

In terms of AutoCAD LT features, you will use the LINE command, the Running Object Snap - Intersection, and the Ortho mode to draw lines. You will first draw lines from the top view to the 45° projection line, and then draw lines from the projection line down to the right side view. These lines will start to outline the right side view. Start the exercise to construct the next view.

Exercise

Starting the Right Side View

Continue drawing the bracket in figure 4.20.

Click on the arrow on the Current Layer name box, then choose the OBJECT *layer drop-down list*	Makes the OBJECT layer the current layer
You will first draw the 45° projection line.	
Command: *Choose the Line button*	Issues the LINE command
LINE From point: *Choose the Midpoint button*	Specifies the MIDpoint object snap
_MID of *Pick point at* ① *(see fig. 4.20)*	Places the starting point of the line
To point: @3,3 (Enter)	Places the ending point of the line
To point: (Enter)	Ends the LINE command

Now you will draw lines from the top view to the 45° projection line. You will use the running object snap of intersection. The Ortho button will help you draw horizontally and vertically (see fig. 4.20).

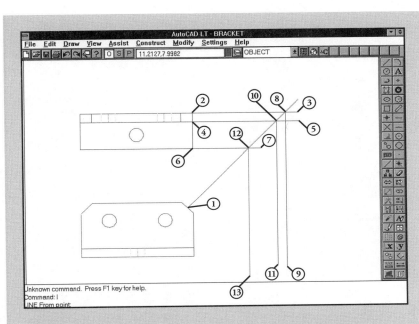

Figure 4.20

Lines start constructing the right side view.

Command: *Choose the Ortho button*	Activates Ortho mode
Command: *Choose the Object Snap button, and choose* **I**ntersection *to toggle on*	Sets the running object snap to INTersection (only the **I**ntersection running object snap should be on)
Command: *Choose Line button*	Issues the LINE command
LINE From point: *Pick point at* ②	Places the starting point of the line
To point: *Pick point at* ③	Places the ending point of the line
To point: (Enter)	Ends the LINE command
Command: (Enter)	Reissues the LINE command
LINE From point: *Pick point at* ④	Places the starting point of the line
To point: *Pick point at* ⑤	Places the ending point of the line
To point: (Enter)	Ends the LINE command
Command: (Enter)	Reissues the LINE command

continues

LINE From point: *Pick point at* ⑥	Places the starting point of the line
To point: *Pick point at* ⑦	Places the ending point of the line
To point: (Enter)	Ends the LINE command

Now you will draw lines from 45° projection line to the right side view (see fig. 4.20).

Command: *Choose the Line button*	Issues the LINE command
LINE From point: *Pick point at* ⑧	Places the starting point of the line
To point: *Pick point at* ⑨	Places the ending point of the line
To point: (Enter)	Ends the LINE command
Command: (Enter)	Reissues the LINE command
LINE From point: *Pick point at* ⑩	Places the starting point of the line
To point: *Pick point at* ⑪	Places the ending point of the line
To point: (Enter)	Ends the LINE command
Command: (Enter)	Reissues the LINE command
LINE From point: *Pick point at* ⑫	Places the starting point of the line
To point: *Pick point at* ⑬	Places the ending point of the line
To point: (Enter)	Ends the LINE command

Your drawing should resemble figure 4.20

Use the same LINE command, Running Object Snap - Intersection, and Ortho mode to complete the outline of the right side view.

Exercise

Completing Construction Lines for the Right Side View

Continue drawing the bracket in figure 4.21.

Command: *Choose the Line button*	Issues the LINE command

LINE From point: *Pick point at* ⑭   Places the starting point of the line
(see fig. 4.21)

To point: *Pick point at* ⑮   Places the ending point of the line

To point: (Enter)   Ends the LINE command

Command: (Enter)   Reissues the LINE command

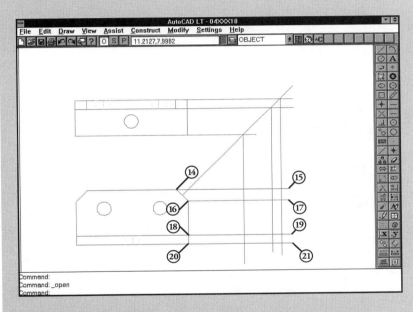

Figure 4.21

Completing construction lines for the right side view.

LINE From point: *Pick point at* ⑯   Places the starting point of the line

To point: *Pick point at* ⑰   Places the ending point of the line

To point: (Enter)   Ends the LINE command

Command: (Enter)   Reissues the LINE command

LINE From point: *Pick point at* ⑱   Places the starting point of the line

To point: *Pick point at* ⑲   Places the ending point of the line

To point: (Enter)   Ends the LINE command

Command: (Enter)   Reissues the LINE command

continues

LINE From point: *Pick point at* (20)	Places the starting point of the line
To point: *Pick point at* (21)	Places the ending point of the line
To point: (Enter)	Ends the LINE command
Your drawing should now resemble figure 4.21	

In this exercise, you project the rest of the lines using the LINE command. You then use the TRIM command again. This time when you are prompted with Select cutting edges...Select objects:, you will use C for Crossing window, window the entire right side view, and then press Enter. The next prompt is Select objects to Trim/UNDO:; you then pick the portions of the lines you want removed. You will select objects individually, and use F for Fence for selecting a number of objects at a time. Whatever the fence crosses will be selected and trimmed off. Press Enter, then select a few more objects and press Enter again. You have now completed the TRIM command.

Exercise

Trimming the Right Side View

Continue drawing the bracket in figure 4.22.

Click on the arrow on Current Layer name box, then choose the CENTER *layer*	Makes the CENTER layer the current layer
Command: *Choose the Line button*	Issues the LINE command
LINE From point: *Choose the Center button*	Specifies the CENter object snap
_CEN of *Pick circle at* (1) *(see fig. 4.22)*	Places the starting point of the line
To point: *Pick point at* (2)	Places the ending point of the line
To point: (Enter)	Ends the LINE command
Command: (Enter)	Reissues the LINE command

LINE From point: *Pick point at* ③	Places the starting point of the line
To point: *Pick point at* ④	Places the ending point of the line
To point: (Enter)	Ends the LINE command
Command: *Choose the Line button*	Issues the LINE command
LINE From point: *Choose the Center button*	Specifies the CENter object snap
_CEN of *Pick circle at* ⑤	Places the starting point of the line
To point: *Pick point at* ⑥	Places the ending point of the line
To point: (Enter)	Ends the LINE command
Command: *Choose the Object Snap button, then choose* **I***ntersection to toggle off*	Turns off **I**ntersection running object snap

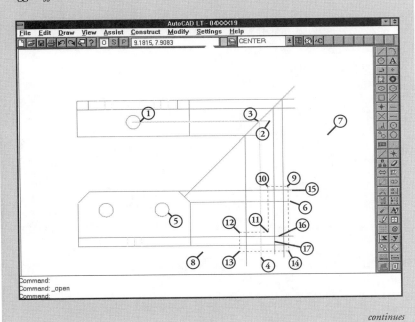

Figure 4.22

Trimming the right side view.

continues

`Command:` *Choose the Trim button*	Issues the TRIM command
`Select cutting edges` `Select objects:` **C** `Enter`	Activates the crossing window to select objects
`First Corner:` *Pick point at* ⑦ *(see fig. 4.22)*	Start a crossing window
`Other Corner:` *Pick point at* ⑧	Completes crossing window
`Select objects:` `Enter`	Ends selection process
`<Select Objects to Trim>` `/UNDO:` **F** `Enter`	Specifies the Fence selection method
`First fence point:` *Pick point at* ⑨	Picks first point of fence
`Undo/<Endpoint of Line>:` *Pick point at* ⑩	Picks next point of fence
`Undo/<Endpoint of Line>:` *Pick point at* ⑪	Picks next point of fence
`Undo/<Endpoint of Line>:` *Pick point at* ⑫	Picks next point of fence
`Undo/<Endpoint of Line>:` *Pick point at* ⑬	Picks next point of fence
`Undo/<Endpoint of Line>:` *Pick point at* ⑭	Picks next point of fence
`Undo/<Endpoint of Line>:` *Pick point at* ⑮	Picks next point of fence
`Undo/<Endpoint of Line>:` `Enter`	Completes fence selection process and trims items
`<Select Objects to Trim>/UNDO:` *Pick line at* ⑯	Removes line
`<Select Objects to Trim>/UNDO:` *Pick line at* ⑰	Removes line
`<Select Objects to Trim>` `/UNDO:` `Enter`	Ends the TRIM command

To complete the cleanup of the object, you can use the ERASE command (see fig. 4.23).

`Command:` *Choose the Erase button*	Issues the ERASE command
`Select objects:` **C** (Enter)	Activates the crossing window to select objects
`First Corner:` *Pick point at* ⑱ *(see fig. 4.23)*	Picks first corner of crossing window
`Other Corner:` *Pick point at* ⑲	Picks other corner at crossing window
`Select objects:` **C** (Enter)	Activates the crossing window to select objects
`First Corner:` *Pick point at* ⑳	Picks first corner of crossing window
`Other Corner:` *Pick point at* ㉑	Picks other corner at crossing window
`Select objects:` (Enter)	Ends the ERASE command

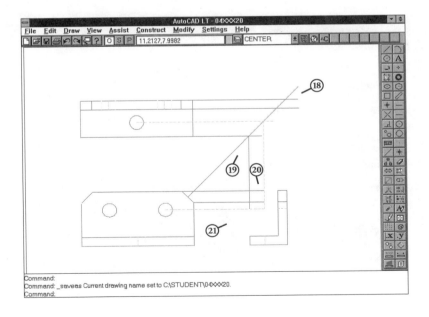

Figure 4.23

Erasing the remaining construction lines.

In this final exercise, you complete the right side view. You will use the OFFSET command and the DDCHPROP command to complete the drilled holes of the right side view. You will also use the DTEXT command to label each of the views. Chapter 5, "Text and Annotation," examines the DTEXT command in detail.

Exercise

Completing the Right Side View

Continue drawing the bracket in figure 4.24.

Command: *Choose* **C***onstruct, then*   Issues the OFFSET command
Offset

Offset distance or through   Sets offset distance to .1875
<0.1875>: **.1875** (Enter)

Select Object to offset: Selects line to offset
Select line at ① *(see fig. 4.24)*

Side to Offset: *Select point*   Offsets line
above line

Select Object to offset: Selects line to offset
Select line at ①

Side to Offset: *Select point*   Offsets line
below line

Select Object to offset: Selects line to offset
Select line at ②

Side to Offset: *Select to the right*   Offsets line
of line

Select Object to offset: Selects line to offset
Select line at ②

Side to Offset: *Select to the left*   Offsets line
of line

Side to Offset: (Enter)   Ends the OFFSET command

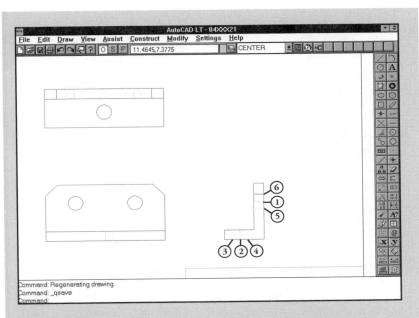

Figure 4.24

Offset lines and changed properties.

`Command:` *Choose the Change Properties button*	Issues the DDCHPROP command
`Select Objects:` *Select line at* ③	Selects the line to change properties
`Select Objects:` *Select line at* ④	Selects the line to change properties
`Select Objects:` *Select line at* ⑤	Selects the line to change properties
`Select Objects:` *Select line at* ⑥	Selects the line to change properties
`Select Objects:` (Enter)	Ends object selection
Choose the **L**ayer *button in the Change Properties dialog box*	Opens the Select Layer dialog box
Choose the HIDDEN *layer in the Select Layer dialog box and choose* OK	Selects the hidden layer as a new layer for the selected lines

Now to finish up the drawing, you will place text (see fig. 4.25).

Click on the arrow on Current Layer name box and choose the NOTES *layer*	Makes the NOTES layer the current layer

continues

Command: *Choose the Zoom button*   Issues the ZOOM command

All/Center/Extents/Previous /Window Scale(X/XP) **ALL** (Enter)   Zooms to the limits of the drawing

Command: *Choose the Text button*   Issues the DTEXT command

Justify/Style/<Start Point>: **6.25,1.375** (Enter)   Locates the start point for text

Height<0.1889>: **.2** (Enter)   Sets the height for text

Rotation Angle <0.0000>: (Enter)   Sets the rotation for text

Text: **Bracket Drawing** (Enter)   Places text in the drawing

Text: *Pick point at* ⑦

Figure 4.25

Placing text in the drawing.

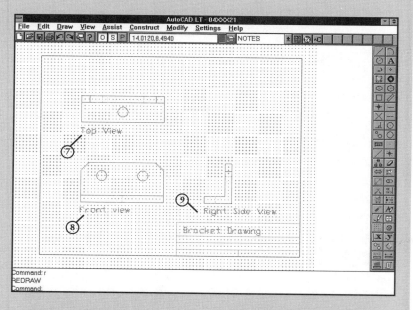

Top View (Enter)   Places text in the drawing

Text: *Pick point at* ⑧

Front View (Enter)   Places text in the drawing

Text: *Pick point at* ⑨	
Right Side View (Enter)	Places text in the drawing
Text: (Enter)	Ends the DTEXT command
Command: *Choose* **F**ile, E**x**it, *then choose* Yes *in the dialog box*	Saves the Bracket drawing and exits AutoCAD LT

Summary

You have learned about multiview drawings in this chapter, both mechanical orthographic and architectural plans and elevations. You have also learned how to create a prototype drawing that you can use as a template to create other drawings. The approach used to create the orthographic projection of the bracket is just one way of doing this type of drawing. You may want to look in some mechanical drafting textbooks to see other ways to approach an orthographic drawing. As far as text, there are many ways you can change and use text—you will see how in the next chapter.

Chapter
4

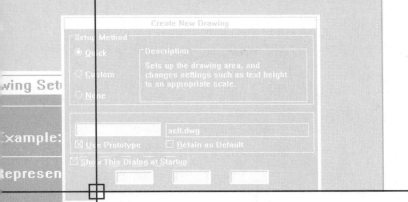

Part Two

Notes and Dimensions

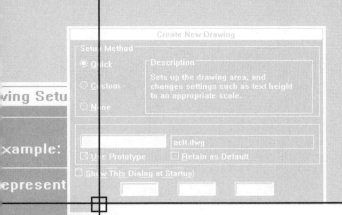

Text is a very important part of a technical drawing. You can use text for labels, title blocks, and descriptions of materials, and you can even bring text in from other drawings. You can vary the starting point, height, and rotation of text. You can also make text look different by choosing different text styles with different fonts. You can even input text from the Windows Clipboard. Topics included in this chapter are the following:

◆ Placing text

◆ Editing text

◆ Using text styles

◆ Bringing in text from other sources

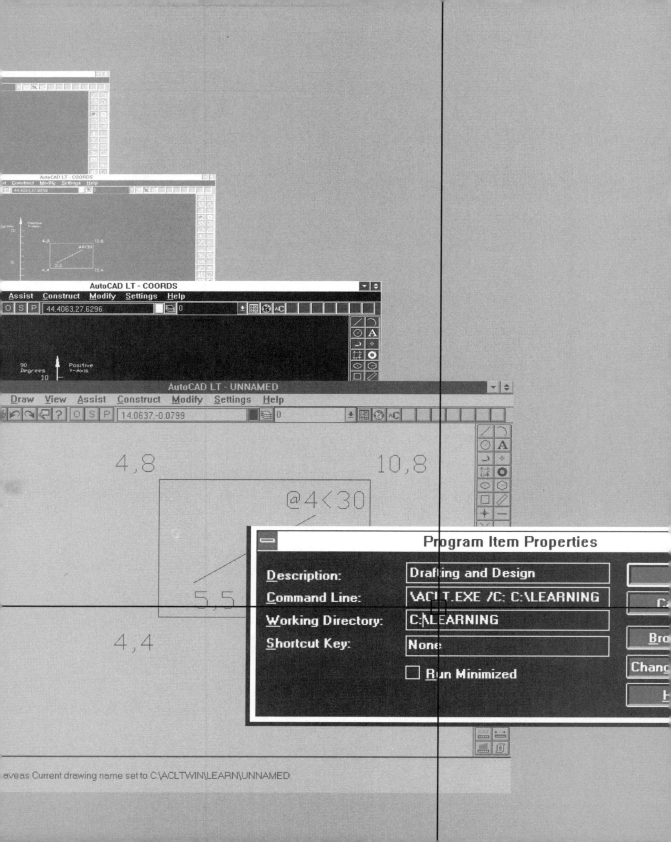

AutoCAD LT - COORDS

Assist   Construct   Modify   Settings   Help

O S P   44.4063,27.6296   0

90
Degrees   Positive
10   Y-Axis

AutoCAD LT - UNNAMED

Draw   View   Assist   Construct   Modify   Settings   Help

O S P   14.0637,-0.0799   0

4,8   10,8

@4<30

5,5

4,4

Program Item Properties

Description: Drafting and Design

Command Line: \ACLT.EXE /C: C:\LEARNING

Working Directory: C:\LEARNING

Shortcut Key: None

☐ Run Minimized

C...

Bro

Chang

H

aveas Current drawing name set to C:\ACLTWIN\LEARN\UNNAMED.

Text and Annotation

In this chapter, you explore the capabilities of AutoCAD LT by annotating a drawing. In terms of the appearance of text, you have a variety of text fonts to choose from—see figure 5.1 for just a sample. You learn more about text fonts later in this chapter.

Monotext looks like this

Scripts looks a little different

Gothicg looks real different

Placing Text

To place text in an AutoCAD LT drawing, you choose the button in the toolbox with the letter "A" on it. This activates the DTEXT command. DTEXT, which refers to dynamic text, dynamically shows text in the drawing as you place it. The command line prompts you for justification, style, and insertion point. The following are the typical responses you receive when placing text:

◆ **Start Point.** At this prompt, you decide where to pick the starting point of the text. From this point, text will be placed to the right (assuming that you are using the default settings in your dimension style). This is called *left-justified* text, which is the default text insertion method in AutoCAD LT. If you decide not to pick a point, and you press Enter instead, your text will automatically be placed under the last text that was placed. The height and rotation also match the last text entered.

◆ **Height.** The next prompt you receive is for the height of the text. The default is .2 inches. If you change the height, AutoCAD LT changes the default height to your new setting.

◆ **Rotation Angle.** The final prompt is for the rotation angle of the text. The default is 0, which is straight across. It can be any angle you want from 0° to 360°. If you change the rotation, AutoCAD LT changes the default rotation to your new setting.

TIP

Sometimes when you are using the DTEXT command, you may not receive the Height prompt. If you set up a text style and give it a text height, you will not receive the prompt for height when you use the DTEXT command. You'll learn about this in more detail when you create a text style later in the chapter.

To best learn how the DTEXT command works, use it first in an exercise. You will be using the Bracket drawing that you constructed in Chapter 4, "Multiview Drawings." To start this exercise, you use the ZOOM command to window in on the area, so you can better see what you are doing. You will also use the Drawing Aids dialog box to set up your grid and snap small enough to place text where it is shown in the figures.

Exercise

Placing Text

Command: *Choose the Open button*	Opens the drawing file BRACKET.DWG
Command: *Choose the arrow on the Current Layer drop-down list box*	
Choose the NOTES *layer*	Makes NOTES the current layer
Command: *Choose the Zoom button*	Issues the ZOOM command
All/Center/Dynamic/Extents/ Window/ <Scale(X/XP)>: *Select points at* ① *and* ② *(see fig. 5.2)*	Zooms window on title block
Command: *Choose* **S**ettings, *then* **D**rawing Aids	Opens the Drawing Aids dialog box
Double-click in the **X** Spacing *edit box under* **S**nap, *type* **.062** *and press Enter, then check the* On *box under* **S**nap	Sets snap to .062 and turns it on
Double-click in the X S**p**acing *box under* **G**rid, *type* **.125** *and press Enter, then check the* On *box under* **G**rid	Sets grid to .125 and turns it on
Choose OK	Closes dialog box and activates settings (see fig. 5.3)

continues

Figure 5.2

Picking points for a zoom window on the title block.

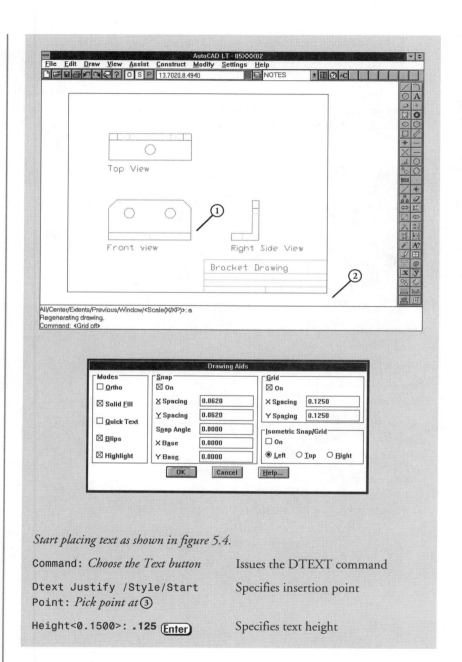

Figure 5.2

Picking points for a zoom window on the title block.

Figure 5.3

Choosing settings in the Drawing Aids dialog box.

Start placing text as shown in figure 5.4.

`Command:` *Choose the Text button*	Issues the DTEXT command
`Dtext Justify /Style/Start Point:` *Pick point at ③*	Specifies insertion point
`Height<0.1500>:` **.125** **(Enter)**	Specifies text height

Rotation Angle<0.0000>: (Enter)	Accepts default angle
Text: **Drawn by Joe Smith**(Enter)	Places text
Text: *Pick point at* ④	Specifies insertion point
Grand Rapids Community College (Enter)	Places text
Text: *Pick point at* ⑤	Specifies insertion point
Date 1/2/96 (Enter)	Places text
Text: *Pick point at* ⑥	Specifies insertion point
Full Scale (Enter)	Places text
Text: (Enter)	Ends DTEXT command

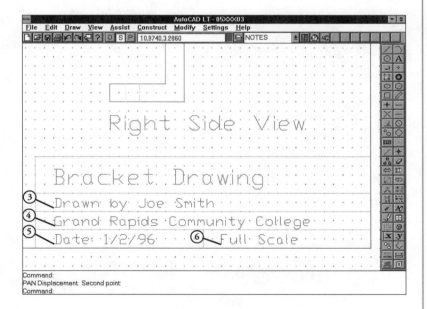

Figure 5.4

Placing text in the title block.

In the next exercise, you learn about the rotation angle prompt in the DTEXT command and about special characters. *Special characters* are a letter or character preceded by percent signs (%), appearing when you complete the DTEXT command. Some examples of special characters are the following:

◆ **%%o**—Overscores text on and off.

◆ **%%u**—Underscores text on and off.

◆ **%%d**—Places a degree symbol when the code occurs.

◆ **%%c**—Places a circle diameter symbol.

◆ **%%p**—Places a plus/minus sign when the code occurs.

◆ **%%%**—Places a single percent sign. Helps when a percent sign must be entered with another code.

See how it works in the next exercise.

Exercise

Placing Special Characters in Rotated Text

Continue working in the Bracket drawing.

Command: *Choose the Zoom button*	Issues the ZOOM command
All/Center/Dynamic/Extents/ Windows/<Scale(X/XP)>: **A** (Enter)	Zooms entire drawing
Command: *Choose the Zoom button*	Issues the ZOOM command
All/Center/Dynamic/Extents /Windows/ <Scale(X/XP)>: *Select points at* ⑦ *and* ⑧ *(see fig. 5.5)*	Zooms window on lower left of drawing
Place text as shown in figure 5.6.	
Command: *Choose the Text button*	Issues the DTEXT command
Dtext Justify /Style/Start Point: **1.625,1.625** (Enter)	Specifies insertion point
Height<0.1500>: **.2** (Enter)	Specifies text height
Rotation Angle<0.0000>: (Enter)	Accepts default angle
Text: **%%UGeneral Notes** (Enter)	Places text
Text: (Enter)	Ends the DTEXT command
Command: *Choose the Text button*	Issues the DTEXT command

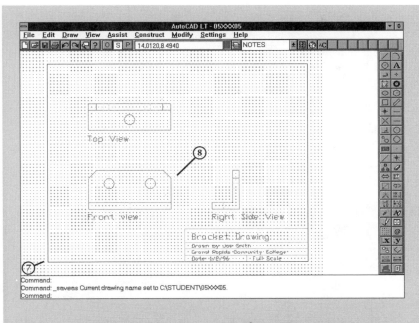

Figure 5.5

Picking points for a zoom window.

`Dtext Justify /Style/Start` `Point:` **`1.625,1.25`** `(Enter)`	Specifies insertion point
`Height<0.1500>:` **`.125`** `(Enter)`	Specifies text height
`Rotation Angle<0.0000>:` `(Enter)`	Accepts default angle
`Text:` **`Do not scale`** **`drawings`** `(Enter)`	Places text
`Text:` `(Enter)`	Ends the DTEXT command
`Command:` `(Enter)`	Reissues the DTEXT command
`Dtext Justify /Style/Start` `Point:` **`1.625,.875`** `(Enter)`	Specifies insertion point
`Height<0.1250>:` `(Enter)`	Accepts default height
`Rotation Angle<0.0000>:` `(Enter)`	Accepts default angle

continues

Text: **Refer to Specification** (Enter)	Places text
Text: **for Material Description** (Enter)	Places text
Text: (Enter)	Ends the DTEXT command
Command: (Enter)	Reissues the DTEXT command
Dtext Justify /Style/Start Point: **3.5,3.0** (Enter)	Specifies insertion point
Height<0.1500>: **.1** (Enter)	Specifies text height
Rotation Angle<0.0000>: **30** (Enter)	Specifies text angle
Text: **Part #6204** (Enter)	Places text
Text: (Enter)	Ends the DTEXT command
Command: (Enter)	Reissues the DTEXT command
Dtext Justify /Style/Start Point: **1.25,.625** (Enter)	Specifies insertion point
Height<0.1000>: **.125** (Enter)	Specifies text height
Rotation Angle<0.0000>: **90** (Enter)	Specifies text angle
Text: **Community College Property** (Enter)	Places text
Text: (Enter)	Ends the DTEXT command
Command: *Choose* File, Save	Saves the drawing in the BRACKET file

Your drawing should resemble figure 5.6

Justifying Text

One of the options you will be given when placing text involves the way you want to justify text. If you type **J** for justify after the first prompt for text, you receive the following options: Align, Fit, Center, Middle, and Right. One justify option you will not receive is the left justification option, which is the default. Left justification means that the lower left point is defined, and the text will be placed to the right. In the DTEXT command, left justification is a default that never changes. Samples of text formatted with the various options are shown in figure 5.7.

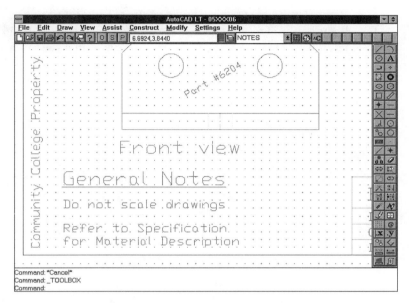

Figure 5.6

Placing text in the drawing.

◆ **Align.** The Align option enables you to place text at an inclined angle by picking two points, the first text line point and the second text line point. When you are picking the angle of the text, you are also picking the starting and ending points, which define the height of the text. The height of the text will vary depending on the number of characters between the first and second point.

◆ **Fit.** This option is similar to the Align option; it fits your text between the first text line point and the second text line point. Fit does differ from the Align option in the treatment of height, however. The height of the text remains the same and, instead, the width varies depending on the number of characters between two points. The text also is oriented horizontally and not at various angles.

◆ **Center.** The Center option centers a line of text on a picked point. The actual centering occurs after you complete the DTEXT command.

◆ **Middle.** The Middle option is similar to the Center option, but the line of text is centered both horizontally and vertically. Centering also occurs after you complete the DTEXT command.

◆ **Right.** The Right option works the reverse of left-justified text. Right-justified text means that the lower right point is defined, and text is placed to the left. Again, the text is justified after you finish the DTEXT command.

Figure 5.7

Different ways to justify text.

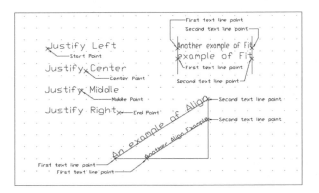

Editing Text

When you make a mistake, there is a simple way to edit text—use the DDEDIT command. This command provides a dialog box interface to edit text one line at a time; you can edit a letter or an entire line. To see how it is done, try editing a line of text. Start by changing a lowercase letter to an uppercase letter.

Exercise

Editing Text	
Start editing the text in figure 5.8.	
Command: *Choose the Open button, then choose the Bracket drawing*	Opens the BRACKET drawing
Command: *Choose the Edit_Text button*	Opens the Edit Text dialog box

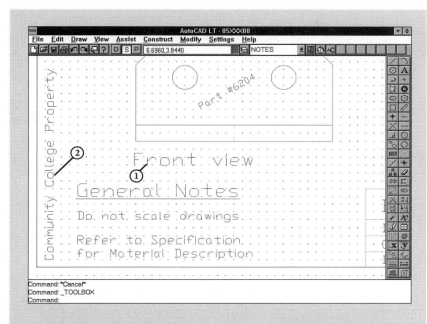

Figure 5.8

Editing text.

```
<Select a TEXT or ATTDEF      Selects text to be edited
object>/Undo: Select text at ①
```

Place your cursor just before   Changes the lowercase v to an
the letter v in the dialog box, click, uppercase V
then press Delete, type **V***, and press*
Enter (see fig. 5.9)

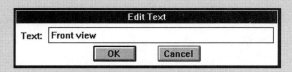

Figure 5.9

Editing text in the
Edit Text dialog box.

```
<Select a TEXT or ATTDEF      Selects text to be edited
object>/Undo: Select text at ②
(see fig. 5.8)
```

With all text highlighted, press   Edits the text within the dialog box
Delete, then type **Community**
College Drawing *and press Enter*

continues

```
<Select a TEXT or ATTDEF          Ends the DDEDIT command
object>/ Undo: (Enter)

Command: Choose File, Save        Saves the drawing to the
                                  BRACKET file
```

Your drawing should now resemble drawing 5.10

Figure 5.10

Final result of editing text.

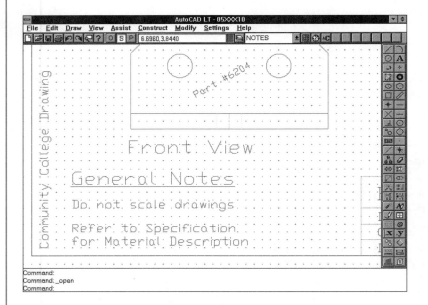

Using Text Styles

In this chapter, you have been using a STANDARD text style to label the drawing. You also have seen examples of different text fonts available in AutoCAD LT. What's the difference between text font and text style?

Text font is the way the text looks. It can be Monotext, Script Complex, or Gothic German, as shown earlier in figure 5.1. Text font is a property of *text style*; other properties of text style include height, width factor, obliquing angle, backwards, upside-down, and vertical. You can create, name, and recall different text styles for different types of text within AutoCAD LT.

You may want to create a bold text style for labeling views within a drawing, and a finer and smaller text style for labeling details within the views. To create a text style, there are a number of steps, as discussed in the following section.

Creating a Text Style

The simplest way to create a new text style is by using a dialog box. To activate this dialog box, choose **S**ettings, then **T**ext Style. The Select Text Font dialog box will be displayed (see fig. 5.11). You will see available choices of fonts in sample displays of each font in the dialog box.

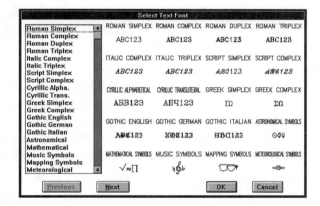

Figure 5.11

Font choices.

By choosing one of the fonts shown in the dialog box, you issue the STYLE command, which prompts you with a number of questions. After responding to these questions, you will have created a new text style. When a text style is created, the name of the text font chosen is used as the name of the text style. For example, a text style using the Roman Duplex text font, romand.shx font, is given the text style name of ROMAND upon completion of the command. This new text style is the default text style the next time the DTEXT command is issued.

After you select a text font from the Select Text Font dialog box, you issue the STYLE command and receive the following prompts:

◆ **Height.** The default value is 0. If a value other than 0 is entered, text using this style will all have the same height. Also, the height prompt will not appear in the TEXT and DTEXT commands when the style is used. The advantage of using the value of 0 is that a new height can be set each time DTEXT is used. The advantages to setting a height is that you never have to enter a text height again, as long as you are using that text style, and all of your text will be uniform.

◆ **Width factor.** The default value is 1, which makes the width equal to the height of the letter. If a value of .5 is entered, the text will be half as wide as it is tall, making for a narrower letter. If a value of 2 is entered, the text will be twice as wide as tall, for a wider letter (see fig. 5.12).

◆ **Obliquing angle.** The default value is 0, which means that text is straight up and down. A positive value slants text to the right; a negative value slants text to the left (see fig. 5.12).

◆ **Backwards.** The default value is NO, which means that text will be read in the normal manner from left to right. If you respond to the prompt with a Y, text will be produced that is backward. By using the Backwards option, you can print text on the back of a transparency, while the lines and geometry are on the front. Lines can be erased on the front, not affecting the text on the back (see fig. 5.12).

◆ **Upside-down.** The default value is NO, which means that text will read right-side up. If you respond with a Y, text will be upside-down (see fig. 5.12).

◆ **Vertical.** The default value is NO, which places text in a horizontal orientation. A Y response will place text from top to bottom, in a vertical orientation. You cannot apply vertical orientation to PostScript Type 1 fonts (see fig. 5.12).

After all these prompts are answered, the name of the current text style that you have created is displayed. The name for this new style of text comes from the font you selected.

Another important detail when selecting text fonts is when you press Next at the Select Text Font dialog box—you will see a dialog box resembling figure 5.13. The fonts from City Blueprint in the upper left corner to Technic Light in the lower right corner are PostScript Type 1 Fonts. They are outline fonts—you can expand, compress, or slant these fonts, but you cannot give them a vertical orientation. Some of them that appear as outlines in the

AutoCAD LT graphics window, such as Sans Serif Bold, can also be filled. To do this, choose **F**ile, then **I**mport/Export, then Po**s**tscript Out, and generate an EPS file. This process is described in detail in Chapter 10, "Reproducing Drawings."

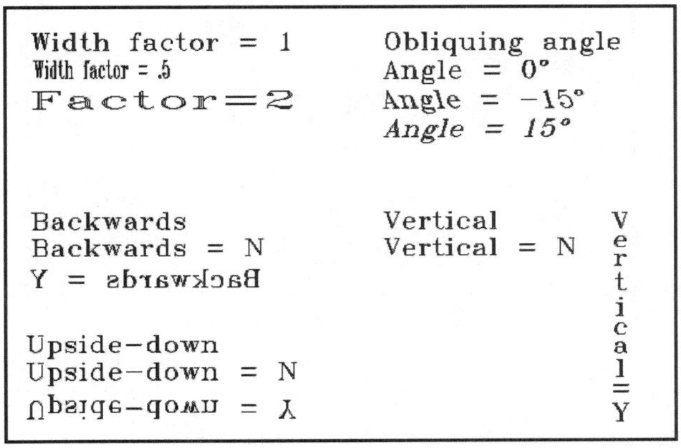

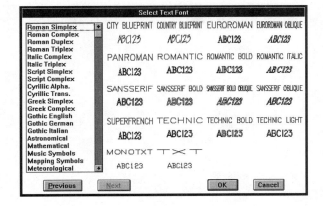

Figure 5.12

Variations on a text style.

Figure 5.13

PostScript Type 1 fonts.

Modifying a Text Style

Suppose you have created a new text style and want to change an existing text entity's style to this new text style. How would you do it? Choose **M**odify, then Modi**f**y Entity, from the pull-down menu. In the Modify Entity dialog box, click on the arrow on the **S**tyle box and then choose the name of the new text style (see fig. 5.14). In addition to changing the text style, you can also change text content, layer, location, height, rotation, and other properties. This flexible command is called the DDMODIFY command.

Figure 5.14

The Modify Text dialog box.

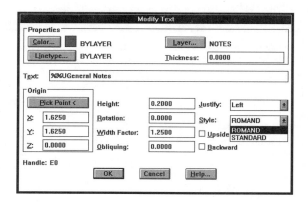

To understand how to create a text style and modify it, try the next exercise.

Exercise

Creating and Modifying a Text Style

Start creating a text style for figure 5.15.

Command: *Choose the Open button, then choose the Bracket drawing*	Opens the BRACKET drawing
Command: *Choose the Zoom button*	Issues the ZOOM command
All/Center/Extents/Previous/ Window/ Scale(X/XP)>: **ALL** (Enter)	Zooms to the limits of the drawing
Command: *Choose **S**ettings, **T**ext Style*	Opens the Select Text Font dialog box
Choose Roman Duplex *and choose* OK	Selects Roman Duplex as the text font for your text style; issues the STYLE command
Font file <txt>:romand Height <0.0000>: (Enter)	Accepts the default height of 0

`Width factor <1.0000>: `**`1.25`** `(Enter)`	Specifies a width factor
`Obliquing angle <0>: (Enter)`	Accepts the default
`Backwards?<N> (Enter)`	Accepts the default
`Upside-down?<N> (Enter)`	Accepts the default
`Vertical?<N> (Enter)`	Accepts the default
`ROMAND is now the current text style`	Informs you that the new text style is called ROMAND, which is now the default text style

TIP

If you want to change the name of the text style to a different name, you can choose **M**odify, then Re**n**ame. The Rename command enables you to rename a text style, from Romand to Boldone, for example.

`Command: `*`Choose the Text button`*	Issues the DTEXT command
`DTEXT Justify/Style/<Start point>:`**`6.2,7.5`** `(Enter)`	Determines start point for line of text
`Height <0.1250>: `**`.125`** `(Enter)`	Establishes height of the new text
`Rotation Angle<0.0000>: (Enter)`	Accepts default angle
`Text: `**`This drawing is just a sample`** `(Enter)`	Places text in the drawing
`Text: `**`of how AutoCAD LT can be`** `(Enter)`	Places text in the drawing
`Text: `**`used for mechanical drawing`** `(Enter)`	Places text in the drawing
`Text: (Enter)`	Ends the DTEXT command
`Command: `*`Choose `***M**`odify, Modi`**f**`y Entity`	Opens the Modify Entity dialog box
`Select object to modify:` *Pick text at* ① *(see fig. 5.15)*	Selects text to modify

continues

Figure 5.15

Creating and modifying a text style.

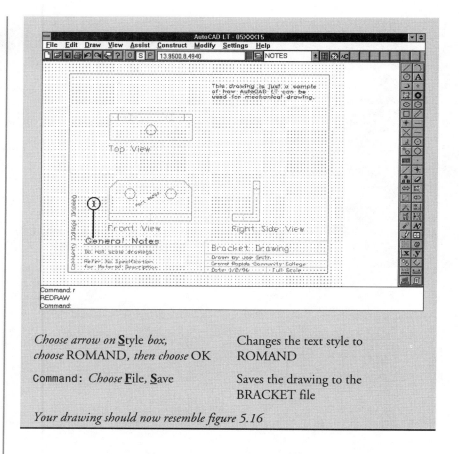

Choose arrow on **S**tyle *box, choose* ROMAND, *then choose* OK — Changes the text style to ROMAND

Command: *Choose* **F**ile, **S**ave — Saves the drawing to the BRACKET file

Your drawing should now resemble figure 5.16

TIP

Within the Bracket drawing, you now have two choices as far as text style—STANDARD and ROMAND. One text style comes with AutoCAD LT (STANDARD), and the other you created (ROMAND). If you would like to switch to the STANDARD style of text before issuing the next TEXT command, you can do it by using Entity Creation Modes. To use this dialog box, choose **S**ettings, **E**ntity Modes, then choose the Text **S**tyle button. Once you are in the Select Text Style dialog box, you can choose a text style name and see sample text, or you can choose Show All and see the entire alphabet. Choose OK to select a new text style.

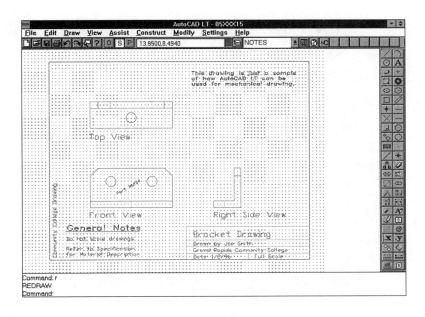

Figure 5.16

The final result of creating and modifying text style.

Bringing in Text from Other Sources

Just as you can create text within AutoCAD LT, you can also get text from other sources. One way is to get text from the Windows Clipboard. An easy way to show you how this works is by using the Notepad program under the Accessories program group within Windows. You can write a few lines of text in the Notepad program, copy it to the Clipboard, and then paste it into your AutoCAD LT drawing.

To show you how it works, try an exercise in AutoCAD LT.

Using Text from Other Sources

Exercise

Continue working on figure 5.16.

Command: *Choose the Text button*   Issues the DTEXT command

continues

`DTEXT Justify/Style/<Start point>:` **6.2,6.7** `Enter`	Specifies start point for line of text
`Height <0.1250>:` **.15** `Enter`	Establishes height of the new text
`Rotation Angle<0.0000>:` `Enter`	Accepts default angle
`Text:`	Places text in the drawing
Hold the Alt key while pressing the Tab key until the Program Manager icon appears, then release both keys	Makes Program Manager the current program
Double-click on Accessories Program Group icon	Opens up the program group to make selections
Double-click on Notepad (see fig. 5.17)	Activates the Notepad program

Figure 5.17

The Accessories Program Group with Notepad icon selected.

Type: **Here is an example of** `Enter`	Places text in the program
text from the notepad `Enter`	Places text in the program

Highlight the text with your cursor	Selects the text for the next command
Choose **E**dit, **C**opy *(see fig. 5.18)*	Copies selected text to the Clipboard
Hold the Alt key while pressing the Tab key until the AutoCAD LT icon appears, then release both keys	Makes AutoCAD LT the current program
Choose **E**dit, **P**aste	Pastes the text into the AutoCAD LT drawing
Text: **(Enter)**	Ends the DTEXT command
Command: *Choose* **F**ile, **S**ave	Saves the drawing to the BRACKET file

Your drawing should now resemble figure 5.19

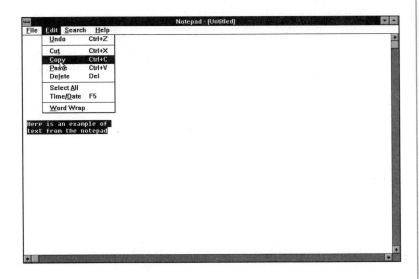

Figure 5.18

Highlighted text in Notepad.

Chapter

5

Placing text from Notepad into an AutoCAD LT drawing.

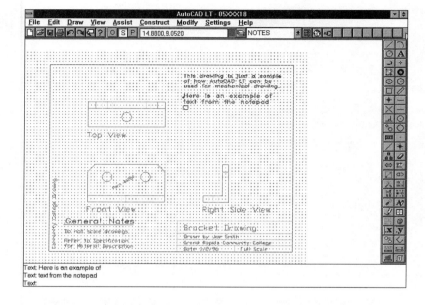

NOTE

Another way to bring in text from outside AutoCAD LT is to drag a file with a .TXT extension into an AutoCAD LT drawing. The drawing needs to be at the Text: prompt within the TEXT or DTEXT command in order for the .TXT file to be successfully inserted into the drawing, just like using the Clipboard.

Summary

In this chapter, you used the DTEXT command along with its different prompts to create text. You edited text with the DDEDIT command. You also created a new text style by using the STYLE command. In addition, you modified existing text with the DDMODIFY command, which can edit text, and change the text style, size, and location. Finally, you pasted text from the Notepad Program into AutoCAD LT.

Another important way to annotate drawings is with dimensions. You learn how to dimension drawings in the next two chapters.

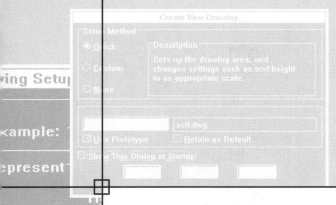

In addition to text, which was covered in Chapter 5, another important part of annotation is dimensioning. When you are constructing a part or building, dimensions are the critical measurements that ensure everything assembles correctly.

In this chapter, you explore the different dimensioning options available to you in AutoCAD LT for Windows. Basic dimensioning guidelines are also reviewed. This chapter examines the following items:

◆ Components of dimensioning

◆ Guidelines for dimensioning

◆ Choices of measuring units for a drawing

◆ Different types of dimension commands

◆ How architectural and mechanical dimensions differ

You need to approach dimensioning in an organized fashion to keep the drawing easy to read. Guidelines are thus presented in this chapter to help you stay organized. First, however, let's just talk about the actual dimensioning.

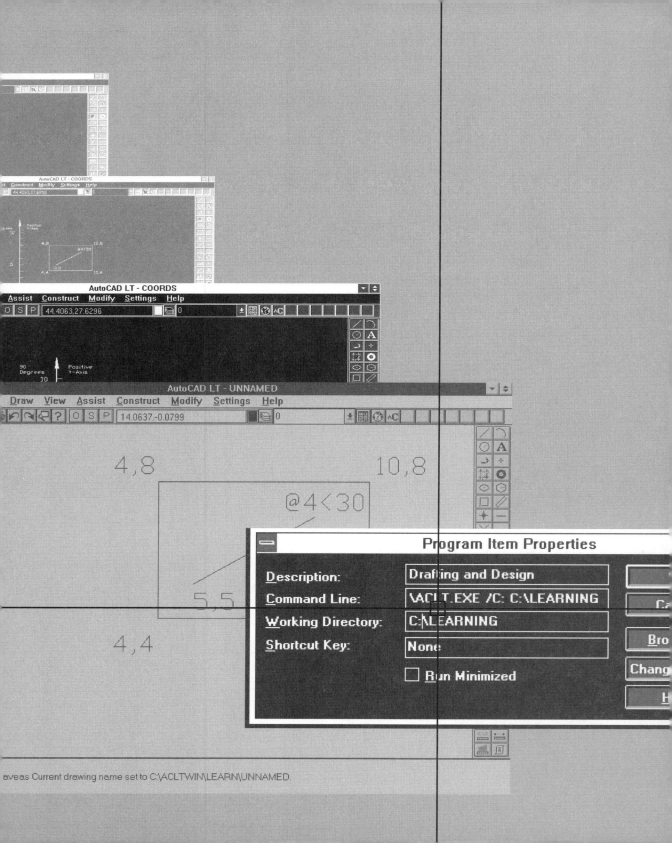

How to Start Dimensioning

In CAD, you draw precisely—when you start dimensioning, you just measure your drawing. To dimension in CAD, you pick two points on an object or the actual object you want to measure. CAD will give you the correct dimension measurement and the appropriate lines, arrows, and numbers to create the dimensions. To understand how dimensioning works in AutoCAD LT, however, you first need to understand the components of dimensioning.

Components of Dimensioning

In AutoCAD LT, there are a number of different types of dimensions: linear, radial, angular, ordinate, and leader. In terms of the components of dimensions, two types of dimensions are shown—linear and circular (see fig. 6.1).

Figure 6.1

*Components of
dimensioning.*

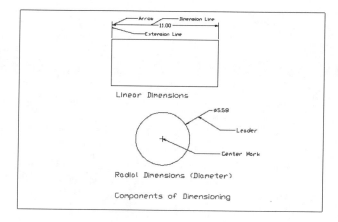

Some components for linear dimensions are the following:

◆ Dimension line

◆ Extension line

◆ Arrow

Components for radial dimensions are slightly different, as follows:

◆ Leader

◆ Center mark

These components are often referred to during the discussion of dimensions in this chapter.

Guidelines for Dimensioning

When you are constructing a drawing in CAD, it helps to have some guidelines. These guidelines are more conceptual—what to do and not to do when drawing in CAD—so your dimensions work well with your drawing. Some guidelines are as follows:

◆ **Draw precisely.** AutoCAD LT actually measures the drawing to determine the dimension. In order for this to work, you must draw precisely, as discussed in Chapter 3, "CAD Drawing Techniques."

◆ **Do not duplicate dimensions.** Only those dimensions necessary to produce a part or build a building as intended by the designer should be used. Too many dimensions end up crowding a drawing and, if they are duplicates of other dimensions, could be conflicting.

◆ **Give enough dimensions.** Enough dimensioning should be given to the manufacturer or contractor, so it will not be necessary to assume a size or scale a dimension.

◆ **Do not continue lines with dimension.** A dimension line should never line up with or form a continuation of a line of an object or building (see fig. 6.2).

◆ **Do not dimension to hidden lines.** Do not dimension to hidden lines; instead, dimension to continuous or solid lines (see fig. 6.2).

◆ **Do not dimension over extension lines.** When dimensioning a drawing, do not dimension over extension lines—it makes dimensions hard to read (see fig. 6.2).

◆ **Space dimensions uniformly.** Spacing of the dimension lines should be uniform throughout the drawing (see fig. 6.2).

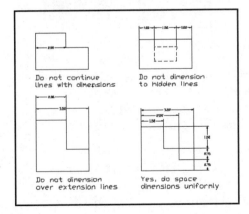

Figure 6.2

Guidelines for dimensioning.

Choice of Measuring Units for a Drawing

When you are drawing in CAD, you may need different types of units to create a drawing, depending on the work you are doing. If you are creating a mechanical drawing for a manufacturer, you might need decimal units. If you are creating a drawing of a building, you might need architectural units.

To set up units for a drawing, choose **S**ettings, **U**nits Style from the pull-down menu. You have a choice of five units. They range from **S**cientific (1.55E + 01) to **F**ractional (15 1/2"), as shown in the Units Control dialog box in figure 6.3.

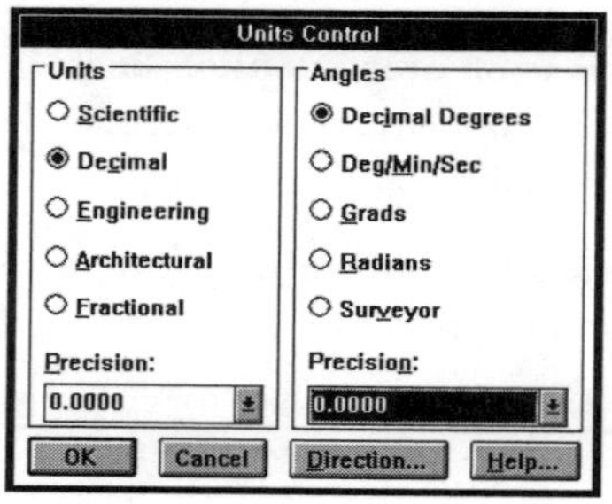

Figure 6.3

The Units Control dialog box.

Along with a choice of units, there also is a choice of precision. Precision affects to what accuracy distances are displayed in your dimensions (for example, .0000), or how coordinates are displayed in coordinate display. For the decimal units, for example, the precision can range from 0 to 0.00000000. For architectural units, precision can range from 0'–0" to 0'–1/256".

The default settings for decimal units is 0.0000; for architectural units, it is 1/16". If the default level is acceptable to you for dimensioning, and you draw with precision typed input and use object snaps to construct objects, then you can use the default settings. Remember, your drawing stores up to 16 decimal points of accuracy—if you use precise drawing techniques, your drawing should thus be precise.

An important point to keep in mind—decimal, fractional, or scientific units can represent any form of units you like: millimeter, feet, inches, or miles. Engineering and architectural units represent feet and inches; only these units accept precision input in feet and inches.

The various types of units are displayed in table 6.1.

Table 6.1
Types of Units

Units	Displays	Measurement
Scientific	1.35E + 01	units
Decimal	13.5	units
Engineering	1'-1.50"	inches and feet
Architectural	1'-1 1/2"	inches and feet
Fractional	13 1/2	units

Another type of measurement in the Units Control dialog box is Angles. Five different types of angle measurements exist, ranging from decimal degrees, 45.00°, to Surveyor, 45d0'0". The one used most often is Decimal Degrees (see fig. 6.3).

Along with a choice of angles, there is also a choice of precision. For decimal degrees, the precision can range from 0 to 0.00000000. The default setting for decimal degrees is 0. When you are doing mechanical drawings, you want angles more precise than the defaults. An accuracy of 0.0000 is recommended, so you can accurately measure and dimension angles after you have constructed the object. This should be done at the beginning of the drawing process, so you can check on the angles if necessary.

The different types of angles are shown in table 6.2.

Table 6.2
Types of Angles

Angles	Displays	Measurement
Decimal degrees	45.000°	degrees
Deg/Min/Sec	45°0'0"	degrees, minutes, seconds
Grads	50.000g	grads
Radians	0.785r	radians
Surveyor	45d0'0"	degrees, minutes, seconds

In the next exercise, you set the precision of the units for the Bracket drawing that you completed in Chapters 4 and 5. You will be dimensioning the drawing in this chapter, and also in Chapter 7.

Exercise

Setting the Precision for Units and Angles

Check the settings for precision units within the Bracket drawing.

Command: *Choose the Open button; choose the Bracket drawing*	Opens the Bracket drawing
Command: *Choose **S**ettings, then **U**nits Style*	Opens the Units Control dialog box
*The dialog box should have De**c**imal as Units, and De**c**imal Degrees as Angles.*	
Choose the arrow on the Precisio**n** *box of Angles, choose 0.0000, then choose OK (see fig. 6.4)*	Sets precision of angles to 0.0000
Choose the Save button	Saves drawing to BRACKET.DWG

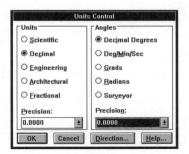

Figure 6.4

Setting the precision.

Different Types of Dimensioning Commands

In AutoCAD LT, there are different commands for different types of dimensioning. The dimensioning commands are organized into the following types:

◆ **Linear.** These are the most common dimensions found on a drawing, measuring straight distances. Included here are standard horizontal and vertical dimensions. The dimensions going off at different angles are called aligned and rotated. Two other options that dimension off existing linear dimensions are baseline and continue (see figures 6.5 and 6.6).

◆ **Radial.** These dimensions measure arcs, circles, and fillets. One option available is Diameter, used for measuring circles; another is the Radius option, available for measuring arcs and fillets. The Center Mark option is for putting a mark on the center rotation point of an arc, fillet, or circle (see fig. 6.5).

◆ **Angular.** This dimension type is used to place an angle measurement between two lines, or give the angle of a circle or arc. This type has only one option—Angular (see fig. 6.5).

◆ **Ordinate.** Ordinate datum dimensions are used to measure the distance to features or points in reference to an origin or datum point. You can either choose to dimension along the x axis using the x-datum option, the y axis using the y-datum option, or either the x or y axis using the automatic option (see fig. 6.5).

Chapter
6

◆ **Leader.** This dimension type enables you to draw an arrowhead, followed by one or two lines, followed by horizontal text (see fig. 6.5).

Figure 6.5

The different types of dimensions.

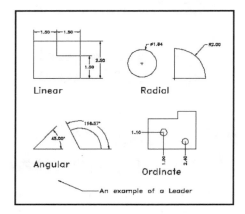

TIP

When dimensioning, make sure to use your object snaps for accurate dimensions. You can use temporary object snaps, such as **E**ndpoint or **C**enter, from your toolbox. The other option is to use running object snaps by choosing object snaps from the toolbox. You activate a dialog box, where you can choose object snaps like **E**ndpoint and **C**enter. Remember to toggle off your running object snaps when you are done dimensioning.

Linear Dimensions

The most often used of any of the AutoCAD LT dimensions are the linear dimensions; of these, horizontal and vertical dimensions are the most used. Figure 6.6 gives you a detailed look of the different types of linear dimensions.

Horizontal and Vertical Dimensions

To understand how to effectively use the dimension commands, let's look at how to use the horizontal dimensioning command. The prompts are essentially the same for both horizontal and vertical dimensions. To issue the horizontal dimensioning command, you Choose **D**raw, then L**i**near Dimensions, and then **H**orizontal from the pull-down menu.

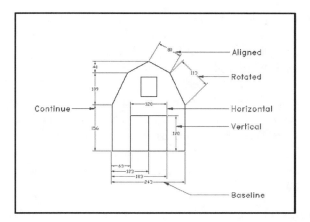

Figure 6.6

A detailed look at linear dimensions.

After performing these actions, you then receive a sequence of prompts at the command line—the following list shows how to respond to them. Remember to use object snap buttons for precision dimensioning:

- ◆ **First extension line origin or RETURN to select:** When you receive this prompt, you have two options. You can either select the first point from which you are dimensioning (using object snaps), or you can press Enter to select one complete entity to dimension. Usually, you pick the first point to dimension from.

- ◆ **Second extension line origin:** Select the point to dimension to. AutoCAD LT shows this prompt only if you pick a point in response to the preceding prompt. If you select an entity, AutoCAD LT uses the start and end points of the entity as the extension line origins. As a result, AutoCAD LT does not need to prompt you for a second point.

- ◆ **Select line, arc, or circle:** If you press Enter at the first prompt, this is the prompt you receive. You can select a complete line, a polyline, an arc, or a circle. There can be no breaks in the line or it will dimension just to the break.

- ◆ **Dimension line location (Text/Angle):** You can pick a point on the drawing where you want the dimension line located, or you can type in a distance from the entity to locate the dimension line. If you want a dimension line to be three units above the entity, for example, you type **@3<90** and then press Enter. The dimension is placed three units above the entity.

◆ **Dimension text<size>:** When you see this prompt, you usually just press Enter to accept the default dimension size. **Do not** type in the dimension—let AutoCAD LT do the work for you. One of the many benefits of accepting the dimension is that it will automatically update if you stretch the object. This is called *associative dimensioning*. If you want to add a prefix or suffix to the dimension, you need to include < > when you type. For example, if you type: < > **typ.⟨Enter⟩**, at the dimension text prompt, a 4.5000 unit dimension reads **4.5000 typ**. If you want a suffix or prefix for every dimension, set it up in the dimension style, as discussed in Chapter 7.

Now that you've become familiar with the terms for placing horizontal and vertical dimensions, let's try placing some in our Bracket drawing in the next exercise. You are going to have to modify some settings when placing dimensions so everything lines up. To do that, you will need to make some adjustments in the Dimension Style dialog box, which you can access from Settings in the pull-down menu. To make an adjustment in this dialog box, choose the appropriate button or other setting and then choose OK. The Dimension Style dialog box is discussed in more detail in the next chapter.

Exercise

Placing Horizontal and Vertical Dimensions

Continue the drawing from the previous exercise; begin by changing to the dimensioning layer.

Command: *Choose the arrow on the Current_Layer drop-down list box, then choose the DIM layer*
Makes DIM the current layer

Command: *Choose the Object_Snap button. Toggle off any other Running Object Snap. Toggle on* **E**ndpoint *and* **C**enter, *then choose* OK
Turns off all other object snaps; toggles on Endpoint and Center object snap

Command: *Choose* **S**ettings, Di**m**ension Style, Text **F**ormat. *Within Zero Suppression area, toggle on* L**e**ading *and* **T**railing, *choose* OK *(see figures 6.7 and 6.8)*
Removes leading and trailing 0's from dimensions; closes the dialog box

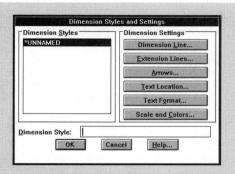

Figure 6.7

The Dimension Styles and Settings dialog box.

Figure 6.8

The Text Format dialog box.

*Within the Dimension Style dialog box, choose **T**ext Location. At the Horizontal pop-up list, choose the arrow to display the pop-up list. Choose Force Text Inside, choose OK, and then choose OK again (see fig. 6.9)*

Another setting within the Dimension Style dialog box, which forces text between the extension lines. The last OK closes the dialog box.

Now you are ready to start doing vertical and horizontal dimensioning (see fig. 6.10)

Command: *Choose the Snap button to highlight if it is not already on*

Turns drawing snap on

Command: *Choose **D**raw, then **L**inear Dimensions, then Horizontal*

Issues Dim: HORIZONTAL command

continues

Figure 6.9

The Text Location dialog box.

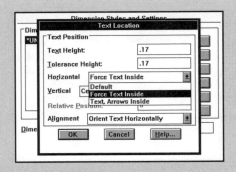

```
First extension line origin         Picks first point to dimension
or RETURN to select:                from
Pick the point at ①

Second extension line origin:       Picks center of circle to
Pick the circle at ②                dimension to

Dimension line location             Picks location for dimension
(Text/Angle): Pick ③                text

Dimension text <1.5>: (Enter)       Accept the numerical
                                    dimension
```

> **NOTE** When you are dimensioning, the numbers will be in the current text style. In the Bracket drawing, the current text style is ROMAND. If you used the STANDARD text style, the numbers in the dimension would appear different.

```
Command: Choose Draw, then Linear    Issues the Dim: VERTICAL
Dimensions, then Vertical            command

First extension line origin          Picks first point to dimension
or RETURN to select:                 from
Pick the point at ④

Second extension line origin:        Picks second point to
Pick the point at ⑤                  dimension to
```

```
Dimension line location          Picks location for dimension
(Text/Angle): Pick ⑥            text

Dimension text <1>: Enter        Accept the numerical
                                 dimension Ends the Dim:
                                 VERTICAL command

Command: Choose the Save button  Issues the SAVE command
```

Your drawing should now resemble figure 6.10

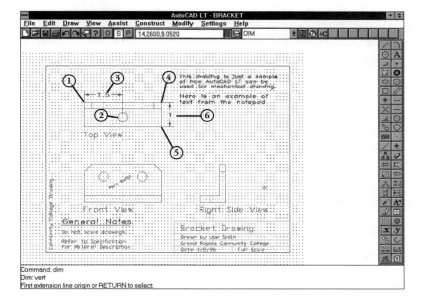

Figure 6.10

Horizontal and vertical dimensions.

When you are dimensioning in AutoCAD LT, notice that it reads Dim: vertical or Dim: horizontal at the command prompt. Dimensioning commands have their own mode. If you try to type in any commands at the dim: prompt—for example, type **line** Enter—AutoCAD LT will not recognize it. To exit dimensioning mode to get back to the command prompt, type **E** Enter at the dim prompt or choose the Cancel button on the toolbar. Any commands chosen from the pull-down menu, toolbar, or toolbox will automatically exit or enter the dimensioning mode.

NOTE

Baseline and Continue Dimensions

When you dimension objects or buildings, you may want to create a chain or stack of dimensions. That is why you use either the BASELINE or CONTINUE dimensioning commands. When you issue these commands, they both prompt you with: `Second extension line origin or RETURN to select`. The commands differ in how they respond to this prompt, however.

The BASELINE dimensioning command uses the *first* extension line of the previous dimension or a selected dimension as the starting point for the next set of dimensions. The prompt is requesting another second extension line to create the next baseline dimension. It creates a new dimension above the previous or selected dimension, creating stacked dimensions.

The CONTINUE dimensioning command uses the *second* extension line of the previous dimension or a selected dimension as the starting point for the next set of dimensions. The prompt is requesting another second extension line to create a continued dimension. It will create a new dimension aligned with the previous or selected dimension, creating a chain of dimensions.

To see how this all works, let's try using the CONTINUE and BASELINE dimensioning commands in the next exercise.

Exercise

Placing Continue and Baseline Dimensions

Continue the drawing from the previous exercise (see fig. 6.11).

`Command:` *Choose* **D**raw, *then* Li̲near Dimensions, *then* Continue	Issues Dim: CONTINUE command
`Second extension line origin or RETURN to select:` (Enter)	Press Enter for selection option
`Select continued dimension:` *Pick the point at* ⑦	Pick the dimension to continue
`Second extension line origin or RETURN to select:` *Pick the point at* ⑧	Picks second point to dimension to

`Dimension text <1.5>:` (Enter)	Accepts the number for the dimension
Command: *Choose* **D**raw, *then* L**i**near Dimensions, *then* Horizontal	Issues Dim: HORIZONTAL command
`First extension line origin or RETURN to select:` *Pick the point at* ⑨	Picks first point to dimension from
`Second extension line origin:` *Pick the circle at* ⑩	Picks center of circle to dimension to
`Dimension line location (Text/Angle):` Pick ⑪	Picks location for dimension text
`Dimension text <.75>:` (Enter)	Accept the numerical dimension
Command: *Choose* **D**raw, *then* L**i**near Dimensions, *then* Baseline	Issues Dim: BASELINE command
`Second extension line origin or RETURN to select:` *Pick the point at* ⑫	Picks the circle to dimension to
`Dimension text <2.25>:` (Enter)	Accepts the number for the dimension
Command: *Choose* **D**raw, *then* L**i**near Dimensions, *then* Baseline	Issues Dim: BASELINE command
`Second extension line origin or RETURN to select:` *Pick the point at* ⑬	Picks second point to dimension to
`Dimension text <3>:` (Enter)	Accepts the number for the dimension
Command: *Choose the Save button*	Issues the SAVE command

Your drawing should now resemble figure 6.11

Figure 6.11

*Continue and
baseline dimensions.*

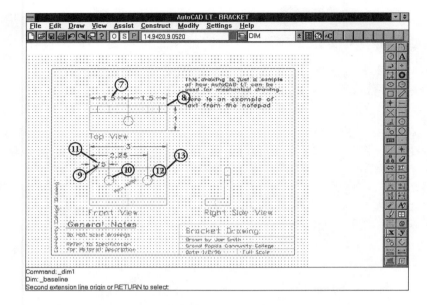

Aligned and Rotated Dimensions

When you are dimensioning, you sometimes may need to get a true dimension of a line that is neither horizontal or vertical. Generally, the ALIGNED dimensioning command accurately dimensions the true length of this line; the dimension line is parallel to the two points chosen. The ALIGNED command has the same prompts as the HORIZONTAL or VERTICAL commands.

The ROTATED command works a little differently. It will dimension the distance between two points at a specified angle. No matter what the actual angle is between these two points, the distance for the dimension and the dimension line will be at the specified angle. The ROTATED command has similar prompts as the HORIZONTAL and VERTICAL commands, and also uses the following prompts: `Dimension line angle <0>`.

The ROTATED dimensioning command can put an extension along with the dimension line at any desired angle, which can be helpful if you want the true measurement of a line at a specific angle. Of the two dimensioning commands (ALIGNED and ROTATED), however, ALIGNED is used the most and ROTATED is used seldom.

To see how this works, let's use the ALIGNED and ROTATED dimension commands on two lines of the same length in the Bracket drawing.

Exercise

Placing Aligned and Rotated Dimensions

You will continue the drawing from the previous exercise (see fig. 6.12).

Command: *Choose the Zoom button*	Issues the ZOOM command
`All/Center/Extents/Previous/Window/Scale(X/XP)>:` `1.1x` (Enter)	Zoom windows a little closer and ends the ZOOM command
Command: *Choose* **D**raw, *then* L**i**near Dimensions, *then* **A**ligned	Issues Dim: ALIGNED command
`First extension line origin or RETURN to select:` *Pick the point at* ⑭	Picks first point to dimension from
`Second extension line origin:` *Pick the circle at* ⑮	Picks second point to dimension to
`Dimension line location (Text/Angle):` Pick ⑯	Picks location for dimension text
`Dimension text <.4243>:` (Enter)	Accepts the numerical dimension and ends the Dim: Aligned command
Command: *Choose* **D**raw, *then* L**i**near Dimensions, *then* Rotated	Issues Dim: ROTATED command
`Dimension line angle <0>:` `60` (Enter)	Specifies an angle of 60°
`First extension line origin or RETURN to select:` *Pick the point at* ⑰	Picks first point to dimension from
`Second extension line origin:` *Pick the circle at* ⑱	Picks center of circle to dimension to
`Dimension line location (Text/Angle):` Pick ⑲	Picks location for dimension text

continues

Dimension text <.4098>: (Enter)   Accepts the numerical dimension

Command: *Choose the Save button*   Issues the SAVE command

Your drawing should now resemble figure 6.12

Figure 6.12

Aligned and rotated dimensioning.

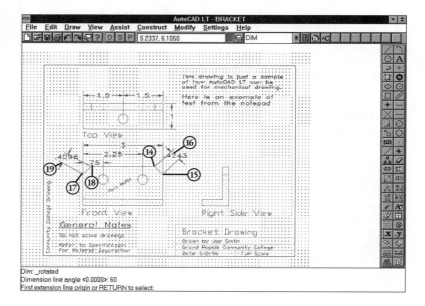

Radial Dimensions

Radial or circular dimensions often get used in a drawing with arcs, fillets, and circles in a drawing. To respond to these needs, AutoCAD LT has three dimensioning commands: DIAMETER, RADIUS, and CENTER.

The DIAMETER dimensioning command is used for dimensioning circles and includes the ANSI diameter symbol. RADIUS is used for dimensioning arcs and fillets and includes the radius symbol (R). CENTER is used for placing a center mark on either a circle or an arc. This center mark is placed automatically if text needs to be placed outside an arc or circle because of the size of text.

When you use Radius or Diameter dimensions, you receive a series of prompts. Here is how to respond to them:

◆ **Select arc or circle:** Pick the arc or circle you want to dimension. Wherever you pick the circle or arc is where the arrow will start for the dimensions.

◆ **Dimension text <degree>:** When you see this prompt, you usually press Enter. If you want to add a prefix or suffix to the dimension, you need to include <> in what you type; **<> typ.** (Enter) will create the result 3.0000 typ, for example.

◆ **Enter leader length for text:** You will see this prompt for the Diameter command when you are placing the numbers for the dimension outside the circle. You will always see this prompt for the RADIUS command.

To get more familiar with the DIAMETER and RADIUS dimensioning commands, you place these dimensions in the bracket drawing in the next exercise. You also need to modify some settings for dimensions within the Dimension Style dialog box. You will be covering the Dimension Style dialog box in more detail in the next chapter.

Exercise

Placing Diameter and Radius Dimensions

Continue the drawing from the previous exercise (see fig. 6.13).

Command: *Choose the Erase button*	Issues the ERASE command
Select objects: *Pick dimension at* ① *(see fig. 6.13)*	Selects the dimension
Select objects: *Pick line at* ②	Selects the angled line
Select objects: *Pick text at* ③	Selects the text
Select objects: (Enter)	Completes the selection set, erases the objects, and ends the ERASE command

continues

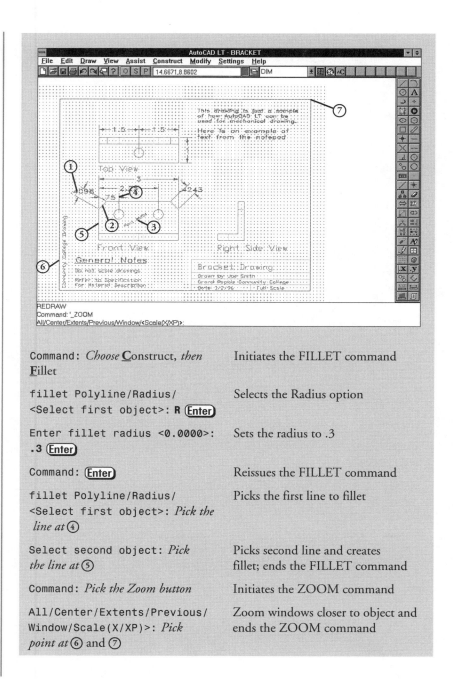

Command: *Choose* **C**onstruct, *then* **F**illet	Initiates the FILLET command
fillet Polyline/Radius/ <Select first object>: **R** (Enter)	Selects the Radius option
Enter fillet radius <0.0000>: **.3** (Enter)	Sets the radius to .3
Command: (Enter)	Reissues the FILLET command
fillet Polyline/Radius/ <Select first object>: *Pick the line at* ④	Picks the first line to fillet
Select second object: *Pick the line at* ⑤	Picks second line and creates fillet; ends the FILLET command
Command: *Pick the Zoom button*	Initiates the ZOOM command
All/Center/Extents/Previous/ Window/Scale(X/XP)>: *Pick point at* ⑥ *and* ⑦	Zoom windows closer to object and ends the ZOOM command

`Command:` *Choose* **S**ettings, **Di**mension Style, **T**ext Location. *At the horizontal pop-up list, choose the arrow to display the list. Choose* Default, *choose* OK, *then* OK *again (see fig. 6.14)*	Changes the settings for dimensions so that text will place outside the circle.

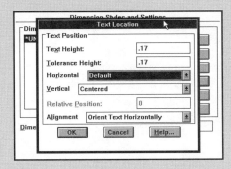

Figure 6.14

Changing settings in the Text Location dialog box.

To place dimensions, see figure 6.15

`Command:` *Choose* **D**raw, *then* Radial **Di**mensions, *then* **D**iameter	Issues Dim: DIAMETER command
`Select arc or circle:` *Pick the point at* ⑧	Picks the circle to dimension
`Dimension text <.375>:` `<> typ.` ⟨Enter⟩	Adds a Typ. suffix to the dimension
`Enter leader length for text:` *Pick point to place dimension, as shown in figure 6.15*	Places text on the drawing

When you are dimensioning with RADIUS or DIAMETER commands, the point you pick on the arc or circle is the starting point for the arrow of the dimension or leader line. Think carefully about where you want the leader line when you pick the point.

TIP

continues

Figure 6.15

The results of placing diameter and radius dimensions.

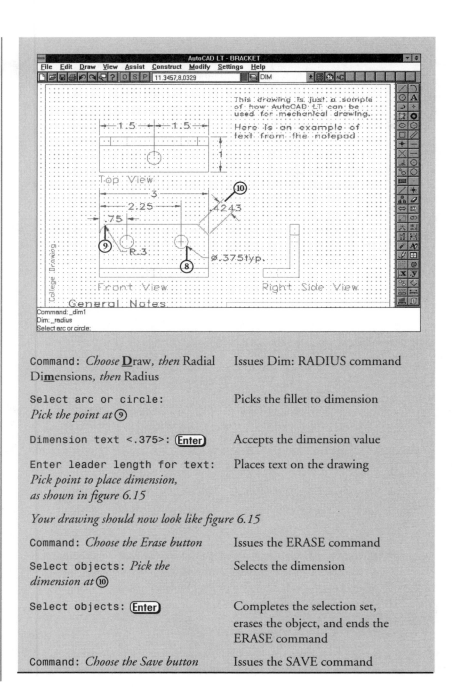

Command: *Choose* **D**raw, *then* Radial Di**m**ensions, *then* Radius	Issues Dim: RADIUS command
Select arc or circle: *Pick the point at* ⑨	Picks the fillet to dimension
Dimension text <.375>: [Enter]	Accepts the dimension value
Enter leader length for text: *Pick point to place dimension, as shown in figure 6.15*	Places text on the drawing
Your drawing should now look like figure 6.15	
Command: *Choose the Erase button*	Issues the ERASE command
Select objects: *Pick the dimension at* ⑩	Selects the dimension
Select objects: [Enter]	Completes the selection set, erases the object, and ends the ERASE command
Command: *Choose the Save button*	Issues the SAVE command

Angular Dimensioning

Angular dimensions are used in architectural and mechanical drawings. In an architectural drawing, you can have an angled wall in a building. You dimension the wall with an angle, so the contractor can properly align it to the rest of the walls in the building.

In a mechanical drawing, you can have a *chamfer*, an angled line between an intersection of a horizontal and vertical line of an object. On the bracket drawing, you created a chamfer in the upper right portion of the top view (see fig. 6.15). You can specify a chamfer in two ways—the distances from the intersection of the two line, or a single distance on one line and an angle. You will dimension the chamfer of the mechanical part with an angle and the distance from the top line.

First, review the prompts that you receive when you issue the command. To issue the command, you choose **D**raw, then An**g**ular Dimensions from the pull-down menu. You then receive a series of prompts at the command line, as follows:

◆ **Select arc, circle, line, or RETURN:** At this point, you usually select lines. The angle between two lines is usually what you dimension. If you select an arc, AutoCAD LT dimensions the total angle of the arc from endpoint to endpoint, using the center point of the arc as the center point of the angle. If you select a circle, you will be prompted for a second angle endpoint. Between the point you originally selected on the circle, the second endpoint you picked, and the center of the circle, an angle will be dimensioned. If you press Enter, you can specify an angle by three points: angle vertex, first angle endpoint, and second angle endpoint.

◆ **Dimension arc line location (Text/Angle):** At this point, you see some very dynamic choices in terms of different angle measurements with different dimension lines. The angle you measure and the location of the dimension and extension lines depends on where you pick your point.

◆ **Dimension text <*angle*>:** When you see this prompt, press Enter to accept the value.

◆ **Enter text location (or RETURN):** Pick a location for the text; if you press Enter, the text will appear in the center of the dimension arc.

To see how the angular dimension works, let's add an angular dimension to the bracket drawing in the next exercise.

Exercise

Placing Angular Dimensions

Continue the drawing from the previous exercise (see fig. 6.16).

Command: *Choose* **D**raw, *then* Ang**u**lar Dimension	Issues Dim: ANGULAR command
Select arc, circle, line, or RETURN: *Pick line at* ①	Picks the line to dimension
Second line: *Pick line at* ②	Picks the second line to dimension
Dimension arc line location (Text/Angle): *Pick point at* ③	Picks location for dimension line
Dimension text <45>: ⏎Enter	Accepts the numerical value
Enter text location(or RETURN): ⏎Enter	Place text on arc and ends the Dim: ANGULAR command
Command: *Choose* **D**raw, *then* **L**inear Dimensions, *then* **H**orizontal	Issues Dim: HORIZONTAL command
First extension line origin or RETURN to select: *Pick the point at* ④	Picks first point to dimension from
Second extension line origin: *Pick the point at* ⑤	Picks second point to dimension to
Dimension line location (Text/Angle): *Pick* ⑥	Picks location for dimension text
Dimension text <.3>: ⏎Enter	Accept the numerical dimension Ends the Dim: HORIZONTAL command
Command: *Choose the Save button*	Issues the SAVE command

Your drawing should now resemble figure 6.16

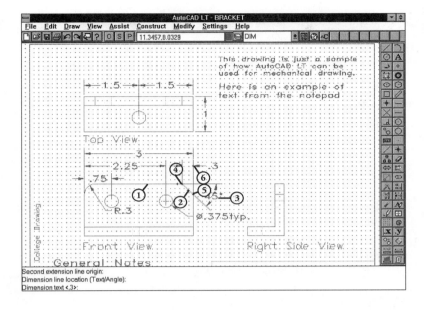

Figure 6.16

Placing angular and horizontal dimensions.

Ordinate Dimensioning

Ordinate dimensions are a set of dimensions that are measured from a common base point. These dimensions simply include a leader line and an aligned numerical value. Ordinate dimensions are useful if you are dimensioning a large number of features on an object. It is a less cluttered dimension—there are no dimension or extension lines to complicate the drawing.

To use the ORDINATE dimensioning command, you must first change the location of the *User Coordinate System* (UCS). The UCS marks the origin of the drawing, or the 0,0 point of the drawing, usually located in the lower left corner of the drawing. You need to change the origin point from the lower left corner of the drawing to the lower left corner of the object, so you can use ORDINATE dimensioning.

Three options are found within the Ordinate command: Automatic, X-Datum, Y-Datum. The Automatic option gives you a choice of either x or y datum as one of your prompts. The X-Datum option gives a distance on the x axis from the base point. The Y-Datum option gives a distance on the y axis from the base point.

When you use any one of these three options, you receive a series of prompts. Here is how to respond to them:

◆ **Select Feature:** Select the feature you want to dimension from the base point.

◆ **Leader endpoint:** Pick a point on which to place the end of the leader line. (Make sure the Ortho button is on, in order to get a nice, straight leader line.)

◆ **Dimension Text\<text\>:** When you see this prompt, press Enter to accept the value.

To see how the Ordinate dimension works, let's add some Ordinate dimensions to the bracket drawing.

Exercise

Placing Ordinate Dimensions

Continue the drawing from the previous exercise (see fig. 6.17).

Command: *Choose Object Snap button*	Issues the DDOSNAP command
Toggle on **E**ndpoint *and* **C**enter, *toggle off any other object snap, then choose* OK	Toggles on **E**ndpoint and **C**enter object snaps only
Command: *Choose* **A**ssist, *then Set ,* **U**CS *then choose* **O**rigin	Issues the UCS command
Origin point \<0,0,0\>: *Pick the point at* ① *(see fig. 6.17)*	Changes the basepoint of the drawing to the lower left corner of the front view

To see how the drawing basepoint has changed, move your cursor around and see the coordinates read about 0,0 at the lower left corner of the front view.

Command: *Choose the Ortho button and toggle it on*	Turns on ORTHO setting
Command: *Choose* **D**raw, *then* Ordi**n**ate Dimensions, *then* **A**utomatic	Issues the Dim: ORDINATE command

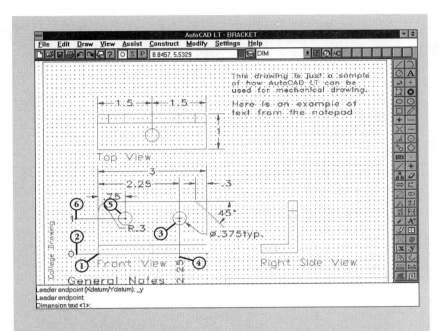

Figure 6.17

Placing ordinate dimensions.

`Select Feature:` *Pick the point at* ①	Picks the point to dimension
`Leader endpoint (Xdatum/ Ydatum):` `Y` **Enter**	Selects Y-datum
`Leader endpoint:` *Pick the point at* ②	Picks location for leader endpoint
`Dimension text <0>:` **Enter**	Accepts the numerical value Ends the ORDINATE command
`Command:` *Choose* **D**raw, *then* Ordi**n**ate Dimensions, *then* **X**-Datum	Issues the Dim: ORDINATE command
`Select Feature:` *Pick the point at* ③	Picks the point to dimension
`Leader endpoint:` *Pick the point at* ④	Picks location for leader endpoint
`Dimension text <2.25>:` **Enter**	Accepts the numerical value Ends the ORDINATE command

continues

Command: *Choose **D**raw, then* Ordi**n**ate Dimensions, *then* **Y**-Datum	Issues the Dim: ORDINATE command
Select Feature: *Pick the point at* ⑤	Picks the point to dimension
Leader endpoint: *Pick the point at* ⑥	Picks location for leader endpoint
Dimension text <1>: [Enter]	Accepts the numerical value Ends the ORDINATE command
Command: *Choose **A**ssist, then* Set U**C**S, *then choose **W**orld*	Issues the UCS command; sets UCS back to its original lower left corner of the drawing
Command: *Choose the Save button*	Issues the SAVE command

Your drawing should now resemble figure 6.17

Leader

Another important dimensioning command is the LEADER command. You will see a leader line incorporated in radial dimensions and also used as a way to just label parts of a drawing. The LEADER command draws an arrowhead with one or more leader lines and horizontal text.

To see how the LEADER dimensioning command works, let's try issuing the command and following the prompts. To issue the LEADER command, choose **D**raw, Leade**r** from the pull-down menu. You will receive the following prompts:

♦ **Leader start:** Pick the start point for the leader, whether it is an object you want to dimension or an object you want to label. The point you pick will be for the arrowhead.

♦ **To point:** Pick the points for the leader lines. Press Enter when you want to stop placing leader lines and place dimension text. If the angle of the last two lines is not horizontal, the LEADER command will automatically add a short horizontal line.

◆ **Dimension text<default>:** Press Enter to accept the default dimension text (from the most recent dimension text) or enter your own text.

To see how the LEADER dimensioning command works, let's add a leader to the bracket drawing in the next exercise.

Exercise

Placing Leaders and Text

Continue the drawing from the previous exercise (see fig. 6.18).

Command: *Choose the Object_Snap button. Toggle off any object snap, Choose* OK	Toggles off any running object snap; object snaps not needed for commands
Command: *Choose the Ortho button and toggle is off*	Turns Ortho setting off
Command: *Choose* **D**raw, *then* Leader	Issues the Dim: LEADER command
Leader Start: *Pick the point at* ①	Starts the leader line with the arrow
To point: *Pick the point at* ②	Completes the first leader line
Command: *Choose the Ortho button and toggle on*	Turns Ortho setting on
To point: *Pick the point at* ③	Draws a straight leader line
To point: ⏎Enter	Ends the leader lines
Dimension text <0>: **3/8"** **DRILL** ⏎Enter	Places a label on the drawing, replacing the default text

Your drawing should now resemble figure 6.18

continues

Chapter 6

Figure 6.18

Placing leaders and text.

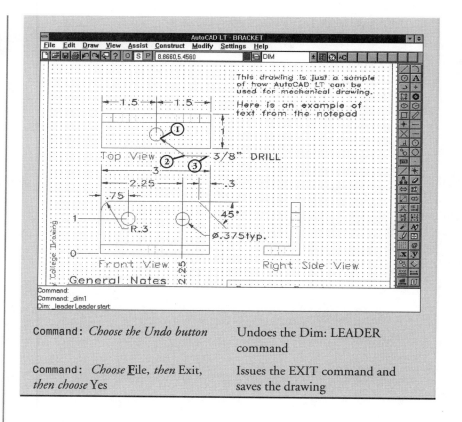

| Command: *Choose the Undo button* | Undoes the Dim: LEADER command |
| Command: *Choose* File, *then* Exit, *then choose* Yes | Issues the EXIT command and saves the drawing |

How Architectural and Mechanical Dimensions Differ

During this entire chapter, you have been doing mechanical dimensioning. Now let's make a brief comparison to architectural dimensioning with the floor drawing, found in the \acltwin directory of AutoCAD LT Release 2. In looking at figure 6.19, the following is evident:

◆ **Units are different.** Units used for the floor plan are architectural as compared to decimal, which were used in the bracket drawing. Floor plans are measured in feet and inches, 10'-2".

- ◆ **Ticks are used instead of arrows.** The dimension lines are terminated by slashes (tick marks) and not arrows.

- ◆ **Text does not break lines.** In an architectural plan, the text is printed above the line and does not break it.

- ◆ **Text for vertical dimensions reads from the right.** Text in mechanical drawings is read from the bottom of the sheet for vertical, horizontal, and aligned text.

You'll find out more details of dimensioning in architectural drawings when you do the architectural drawing project in Chapter 12.

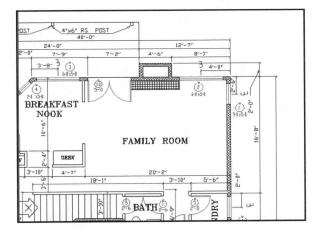

Figure 6.19

Dimensioning in an architectural drawing.

Summary

This chapter has familiarized you with dimensioning a mechanical drawing by having you dimension the Bracket drawing. Of all the dimensioning commands, the ones you will use the most are HORIZONTAL, VERTICAL, BASELINE, CONTINUE, ALIGNED, RADIUS, DIAMETER, and sometimes ANGULAR. You will use the LEADER command for placing notes in a drawing.

In the next chapter, you learn how to modify dimensions and create your own dimension style.

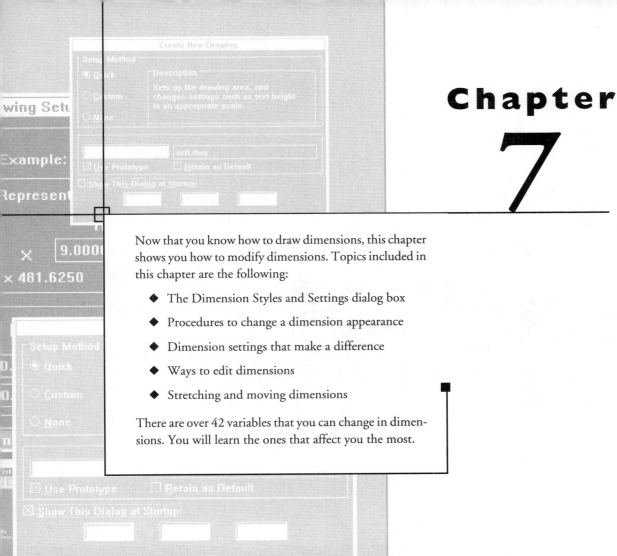

Now that you know how to draw dimensions, this chapter shows you how to modify dimensions. Topics included in this chapter are the following:

- ◆ The Dimension Styles and Settings dialog box
- ◆ Procedures to change a dimension appearance
- ◆ Dimension settings that make a difference
- ◆ Ways to edit dimensions
- ◆ Stretching and moving dimensions

There are over 42 variables that you can change in dimensions. You will learn the ones that affect you the most.

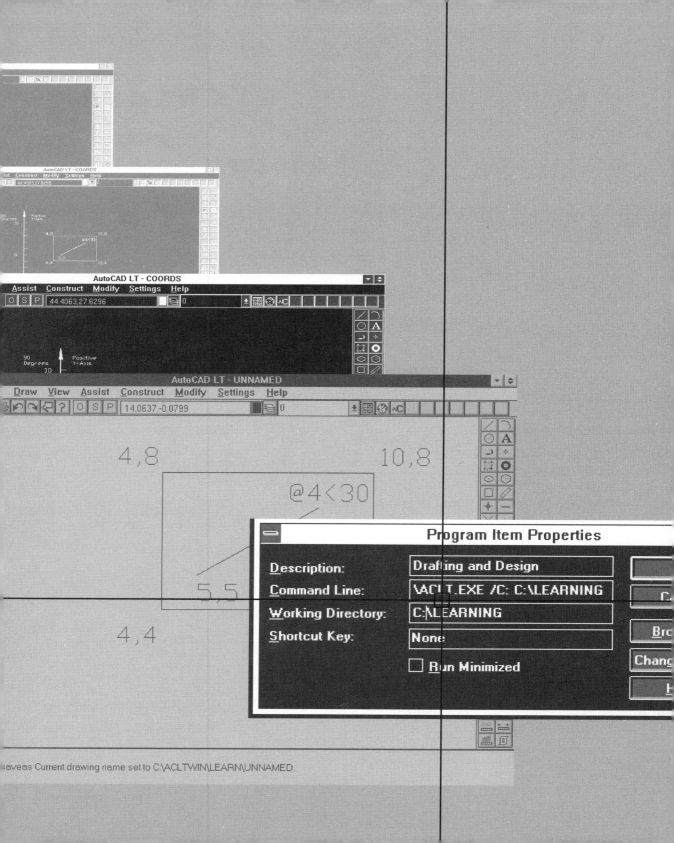

AutoCAD LT - COORDS

Assist   Construct   Modify   Settings   Help

O S P 44.4063,27.6296   0

90
Degrees   Positive
10   Y-Axis

AutoCAD LT - UNNAMED

Draw   View   Assist   Construct   Modify   Settings   Help

O S P 14.0637,-0.0799   0

4,8   10,8

@4<30

5,5

4,4

Program Item Properties

Description: Drafting and Design

Command Line: \ACLT.EXE /C: C:\LEARNING

Working Directory: C:\LEARNING

Shortcut Key: None

☐ Run Minimized

saveas Current drawing name set to C:\ACLTWIN\LEARN\UNNAMED.

How to Modify Dimensions

In the last chapter, you learned how to dimension the bracket drawing. To effectively control the appearance of your drawing, however, you need to understand dimension settings and dimension editing commands—this is the focus of this chapter. To begin, let's look at the Dimension Styles and Settings dialog box.

The Dimension Styles and Settings Dialog Box

The Dimension Styles and Settings dialog box is very powerful in AutoCAD LT for Windows—it can modify features of new or existing dimensions, and also can enable you to create unique dimension styles with their own names. To better understand how the dialog box works, you must first look at how it is organized (see fig. 7.1). The dialog box basically has three parts: on the left side is the Dimension **S**tyles list box, on the bottom is the **D**imension Style text box, and on the right side is a set of buttons listing the different dimension settings from which you can choose.

Figure 7.1

The Dimension Styles and Settings dialog box.

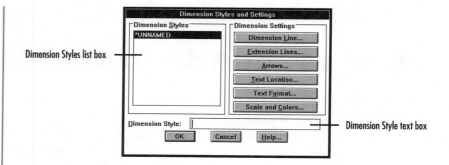

The Dimension Styles List Box

Let's first discuss the left half of the dialog box, the Dimension **S**tyles list box. A *dimension style* is a group of dimension settings that are given a name and have a particular appearance. If you are trying to get consistency of appearance of dimensions for a set of drawings, either for yourself or for an entire office, Dimension **S**tyles is a good way to do it. The list box shows the name of all dimension styles within the drawing (see fig. 7.2).

Figure 7.2

A list of dimension styles.

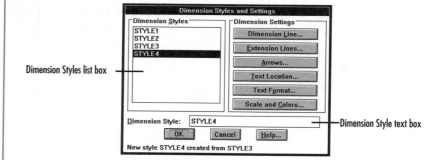

The Dimension Style Edit Box

The **D**imensions Style text box can be used two ways—either to select a dimension style from the list box, or to create a new dimension style based on an existing style. Once you create a new dimension style, change some dimension settings, and choose some OKs, and you now have a new dimension style with its own unique appearance. You'll get a chance to create your own dimension styles in detail in this chapter (see fig. 7.2).

Dimension Settings

Dimension settings buttons are on the right side of the dialog box. You can just change a particular dimension setting for an entire drawing and leave it unnamed as in figure 7.1, or you can give a particular set of dimension settings a dimension style name as in figure 7.2. Dimension settings buttons can be used either way.

There basic parts of a dimension are shown in figure 7.3. The six categories of dimension style settings are the following:

◆ **Dimension Line.** Settings in this area can force dimension lines between extension lines, and can create a dimension text with a box around it.

◆ **Extension Lines.** These settings can be used to eliminate extension lines, change the size of the center mark in circles and arcs, and change the size of the extension line above the dimension line.

◆ **Arrows.** The Arrows box can be used to select an arrow, tick, dot, or user-defined block to terminate the extension line. Arrow size can also be defined.

◆ **Text Location.** In this area, you can adjust the height of the text, the location of the horizontal and vertical dimensions above or below the extension line, and the orientation of the text, all horizontal or otherwise.

◆ **Text Format.** This box has a number of important settings, including length scaling to adjust dimensions to read accurately for larger scale details, and text prefix and suffix for adding lines automatically to a dimension. A choice of tolerances or limits can be shown in a dimension. Finally, an alternate set of dimensions, such as millimeters, can be shown in a dimension, if the Show Alternate Dimensions box is checked.

◆ **Scale and Colors.** The important settings in this area feature scaling and color for dimension lines, extension lines, and dimension text.

Figure 7.3

The parts of a dimension changed by dimension settings.

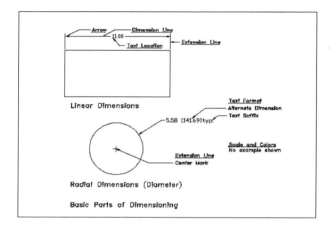

Procedures to Change a Dimension's Appearance

Before you learn about the details of different dimension settings, you might want to become familiar with the procedure to change the dimension's appearance with the different dimension settings. There are at least three ways to do this, as follows:

◆ **Change the dimension setting before creating the dimension.** If you want a dimension appearance to be affected by a change in the settings box, you can make a change in the Dimension Styles and Settings box before you create the dimension. When you start dimensioning, the settings change will be reflected in the dimension. You did this several times in Chapter 6, "How to Start Dimensioning."

◆ **Update the dimension.** Another way to change a dimension appearance is to make the settings change after you have created the dimension, and then use the dimension UPDATE command to change the appearance of the dimension. To do this, first make a change in the Dimension Styles and Settings box. Next, choose **M**odify, then Ed**i**t Dimension, and then **U**pdate Dimension from the pull-down menu. Select the dimension you want to update and press Enter. You will later try out this method in an exercise.

◆ **Use Dimension <u>S</u>tyles.** A third way to change a dimension appearance from the default is to create a dimension style. To do this, you first create a dimension style in the Dimension Styles and Settings box, make the changes in dimension settings, and then say OK to the appropriate dialog boxes. Whatever dimensioning you do in the drawing after that will be under the dimension style you just created with the appearance you have set. You will also try out this method in an exercise later in the chapter.

Updating a Dimension

To see how to change the appearance of a dimension, try the following exercise and update a dimension of the Bracket drawing.

Using the UPDATE command

Exercise

Open the Bracket drawing and learn how to use the UPDATE command.

Command: *Choose the Open button, and then choose the Bracket drawing under the directory where it was created*
Opens the Bracket drawing

Command: *Choose* <u>S</u>ettings, Di<u>me</u>nsion Style, *and choose* <u>T</u>ext Location. *At the Horizontal pop-up list, choose the arrow, then choose* Force Text Inside
Opens the Dimension Styles and Settings dialog box. Within the Text Location dialog box, choose the Force Text Inside setting

At the Alignment pop-up list, choose the arrow, choose Orient Text Horizontally, *choose* OK, *and then choose* OK *again (see fig. 7.4)*
Displays Alignment pop-up list, sets Orient Text Horizontally, and then closes the dialog box

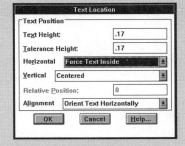

Figure 7.4

Chapter 7

The Text Location dialog box.

continues

See figure 7.5 to continue the exercise.

Command: *Choose* **M**odify, *then* Ed**i**t Dimension, *then* **U**pdate Dimension	Issues the Dim: UPDATE command
Select Objects: *Pick the point at* ①	Picks dimension to update
Select Objects: ⟨Enter⟩	Updates the dimension; ends the Dim: UPDATE command
Command: *Choose the Save button*	Issues the SAVE command

Your drawing should now resemble figure 7.5

Figure 7.5

An update that forces text inside the extension lines.

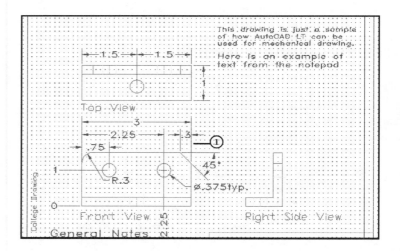

Creating a Dimension Style

Now let's try changing the appearance of dimensions by using a dimension style. To understand how dimension styles work, the first dimension style name you create replaces the "UNNAMED" style, and additional names are added to the Dimension Styles list box. After you create a new name, it is highlighted and becomes the current dimension style when you exit the dialog box.

You can also select an existing style from the Dimension **S**tyles listing simply by choosing it; as you choose OK to exit the dialog box, the selected style becomes the current style. Try an exercise with dimension style.

Using Dimension Styles

Exercise

You will continue the drawing from the previous exercise.

Command: *Choose* **S**ettings, Di**m**ension Style, *type* **Linear**, *then press* (Enter) *in the Dimension Style text box (see fig. 7.6)*

Opens the Dimension Styles dialog box; creates a new dimension style called Linear

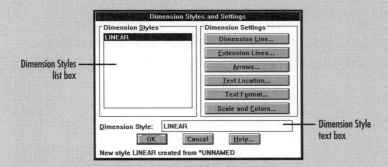

Dimension Styles list box

Dimension Style text box

Figure 7.6

Creating a linear dimension style.

Look at figure 7.7 to continue the exercise.

Command: *Choose Object Snap button*

Opens the DDOSNAP dialog box

Make sure the Endpoint and Center object snaps are checked on; choose OK

Sets the Endpoint and Center object snaps on for dimensioning

Command: *Choose* **D**raw, *then* **L**inear Dimensions, *then* Vertical

Issues Dim: VERTICAL command

First extension line origin or RETURN to select: *Pick the point at* ② *(see fig. 7.7)*

Picks first line to dimension from

Second extension line origin: *Pick the circle at* ③

Picks second line to dimension to

continues

`Dimension line location (Text/Angle):` Pick ④	Picks location for dimension text
`Dimension text <.25>:` **Enter**	Accepts the numerical dimension
`Command:` *Choose* **D**raw, *then* L**i**near Dimensions, *then* **B**aseline	Issues Dim: BASELINE command
`Second extension line origin or RETURN to select:` *Pick the point at* ⑤	Picks line to dimension to
`Dimension text <1.5>:` **Enter**	Accepts the number for the dimension and ends the Dim: BASELINE command

Your drawing should resemble figure 7.7

Figure 7.7

Vertical dimensions using linear dimension style.

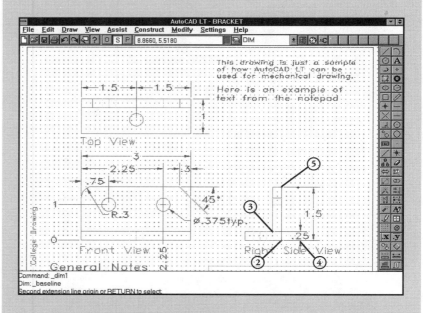

`Command:` *Choose* **S**ettings, Di**m**ension Style, *type* **Diameter**, *then press* **Enter** *in the Dimension Style edit box (see fig. 7.8)*	Opens the Dimension Styles dialog box; creates a new dimension style called Diameter

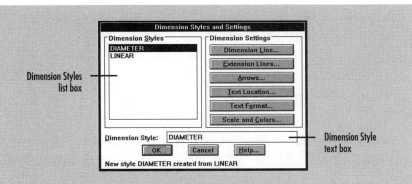

Figure 7.8

*Creating a Diameter
dimension style.*

Within the same dialog box, choose **T**ext
Location. *At the Horizontal pop-up list,
choose the arrow, choose* Default, *then
choose* OK.

Sets the Horizontal setting to
default; exits the Text Location
dialog box

*Within the Dimension Styles and Settings
dialog box, choose* Text F**o**rmat. *In
Text Suffix: type* **typ.** (Enter) *and in
Text Prefix: type* **%%C** (Enter), *then choose*
OK *(see fig. 7.9)*

Creates a default suffix of typ.
and default prefix of the
diameter sign. Exits the two
dialog boxes, Text Format and
Dimensions Styles. Diameter
dimension style is current

Figure 7.9

*The Text Format
dialog box.*

To place the dimension, see figure 7.10.

Command: *Choose* **D**raw, *then* Radial
Di**m**ensions, *then* **D**iameter

Issues Dim: DIAMETER
command

continues

7

Chapter

`Select arc or circle:` *Pick the circle at ⑥ (see fig. 7.10)*	Picks the circle to dimension
`Dimension text <.375>:` `(Enter)`	Accepts the value
`Enter leader length for text:` *Pick point to place dimension, as shown in figure 7.10*	Places text on drawing

Figure 7.10

Dimension using Diameter dimension style.

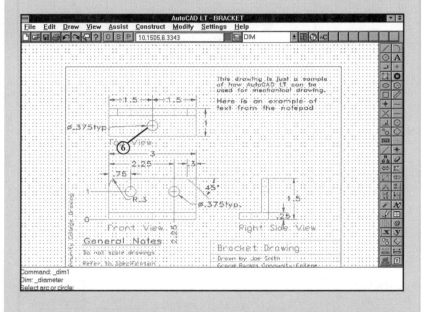

`Command:` *Choose Object Snap button*	Opens the DDOSNAP dialog box
Make sure the **E**ndpoint *and* **C**enter *object snaps are checked off; choose* OK	Endpoint and Center object snaps are toggled off; dialog box is closed
`Command:` *Choose the Save button*	Issues the SAVE command

TIP

It may be a little challenging to get the dimension diameter just right. Within the Dimension Styles and Settings dialog box, make sure the settings are as requested. When prompted for an arc or circle for the dimension diameter, make sure to pick the left edge of the circle in this exercise.

Dimension Settings That Make a Difference

In this chapter, you are not going to become an expert with dimension settings, but you will learn about the dimension settings that can affect the appearance of what you are dimensioning. You'll learn about dimension settings that make a difference. Let's look at the details of the different dialog boxes of the Dimension Styles and Settings dialog box.

Dimension Lines

Figure 7.11 displays the default settings for the Dimension Line dialog box. The two settings you'll use the most in this dialog box are Force **I**nterior Lines and Basic **D**imensions.

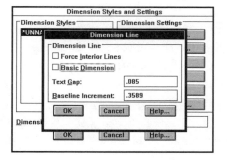

Figure 7.11

The Dimension Line dialog box.

When toggled on, Basic **D**imensions puts a box around the dimension for use with geometric tolerancing symbols. When Force **I**nterior Lines is toggled on, dimension lines are forced between extension lines, even when text doesn't fit. See figure 7.12 for examples of this.

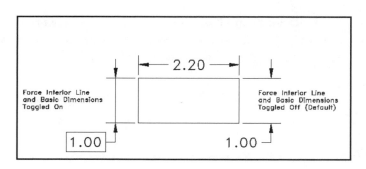

Figure 7.12

The result when Dimension Line settings are changed.

Chapter

7

Extension Lines

The important settings in the Extension Lines dialog box are the Visibility of the extension lines, the value for the Center Mark Size, and whether the Mark with Center Lines box is checked or not. Figure 7.13 shows the default settings in this dialog box.

Figure 7.13

The Extension Lines dialog box.

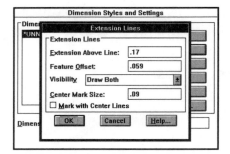

These settings for extension lines are as follows:

◆ **Visibility.** The choices in terms of drawing extension lines are draw both, suppress first, suppress second, and suppress both (see fig. 7.14).

◆ **Center Mark Size.** Sets the size for the center mark for the CENTER, DIAMETER, and RADIUS dimensioning commands. The following values affect the center mark in different ways:
   ◆ 0—No center mark
   ◆ Positive value—A cross is drawn in the center of a circle or arc when the CENTER dimension command is issued or when DIAMETER or RADIUS have the text outside a circle or arc. Figure 7.14 shows a value of .15.

◆ **Mark with Center Lines.** When it is checked, this setting draws a cross in the center of a circle or arc; after a space, it also extends the cross beyond the diameter of the circle (see fig. 7.14).

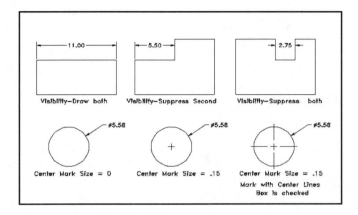

Figure 7.14

*The result when
Extension Lines
settings are changed.*

Arrows

The settings for the Arrows dialog box consists of line terminators. You have a choice of size and a choice of type: arrow, tick, dot, and user. To choose a different dimension line terminator, just pick the one you want in the dialog box (see fig. 7.15).

You can even create your own line terminator or arrow, and then make a block out of it. After making a block of the arrow, choose **U**ser in the arrows dialog box, type in the arrow's block name in the User Arro**w** entry box, and then choose OK. You now have your own custom arrow. Figure 7.15 shows the default settings for the Arrows dialog box.

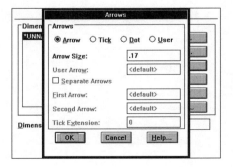

Figure 7.15

*The Arrows
dialog box.*

To get an idea of the appearance of the different arrows, see figure 7.16.

Figure 7.16

Arrows, ticks, dots, and a user-defined arrow1.

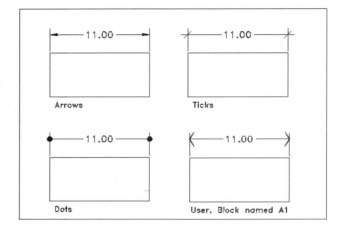

Text Location

In previous exercises, you changed the settings in the Text Location dialog box. At the Horizontal pop-up list, you chose the arrow and selected Force Text Inside. This setting forces the dimension text inside the extension lines in linear dimensions. In the **V**ertical pop-up list, you can adjust the setting so the dimension text does not break the dimension line. In the A**l**ignment pop-up list, you can adjust the settings so the dimension text aligns with the dimension line for use in architectural drawings (see fig. 7.17).

Figure 7.17

The Text Location dialog box.

You can do all preceding adjustments of dimensions, plus adjust text height and text height for tolerances. Some of the options available in the Text Location dialog box are as follows:

◆ **Horizontal.** This setting controls the placement of text in relation to the extension line (see fig. 7.18). There are three options:

> **Default**—Places text inside the extension lines unless there is insufficient space; in that case, the text is forced outside. Used for radial dimensions so that text is not jammed inside a circle or arc.
>
> **Force Text Inside**—Forces the text inside the extension lines. Used for linear dimensions, so the text isn't always jumping outside the extension lines.
>
> **Text, Arrows Inside**—Forces both text and arrows inside extension lines. The arrows are often deleted if the space gets too tight.

◆ **Vertical.** This controls the placement of text in relation to the dimension line. The options are the following:

> Centered and breaking the dimension line.
>
> Above and not breaking the dimension line.
>
> Relative position can be above or below the dimension line, depending on the value in the Relative **P**osition box (a negative value puts it below the dimension line). See figure 7.18 for some examples.

◆ **Alignment.** This option controls the alignment of text with the dimension line. The two most important settings in this box are Orient Text Horizontally, which is used for most mechanical drawings, and Align With Dimension Line, which is used more for architectural drawings.

The other two settings are Aligned When Inside Only and Aligned When Outside Only; both of these refer to the relation between the text and whether it is inside or outside the extension lines (see fig. 7.18).

Figure 7.18

Some different settings for the Text Location dialog box.

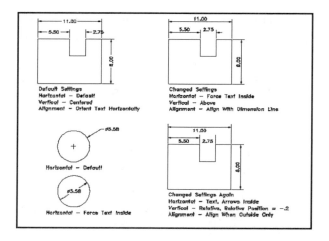

Text Format

In the Text Format dialog box, you have already used the Text Suffi̲x portion of Basic Units to put Typ. in dimension text. You have also checked the Le̲ading and T̲railing boxes, so extraneous Os do not appear in dimension text. In this dialog box, there are basically four sections: Basic Units, Tolerances, Zero Suppression, and Alternate Units. Default settings of the dialog box are displayed in figure 7.19.

Figure 7.19

The Text Format dialog box.

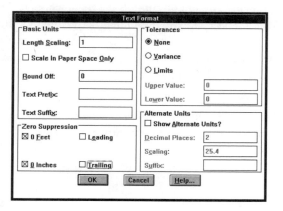

The important sections of the Text Format dialog box are as follows:

◆ **Basic Units.** Key settings in this section include:

Length Scaling—A factor that is entered to adjust the dimensioning for a larger scale detail. If you are dimensioning a detail that is four times larger then the full-scale drawing, you would key in a factor of 25 so the dimension would measure properly.

Round Off—To round off dimensions to a smaller number of units then what is in Units control, type in the appropriate number in this box. Type **.5** ⁅Enter⁆ if you want to round off to the nearest .5.

Text Prefix and Suffix—If you want to automatically add text to a dimension, enter it into these boxes (see fig. 7.20).

◆ **Zero Suppression.** If these items are checked, here is what happens:

0 Feet—Suppresses the 0 in a zero foot dimension, 4 3/4" instead of 0'–4 3/4". Default setting has this box checked.

0 Inches—Suppresses the 0 for inches when not needed, 21' instead of 21'–0". Default setting has this box checked.

Trailing and Leading Zeros—Applies to decimal degrees. Default is not to have this checked. If checked, results are shown in figure 7.20.

◆ **Tolerances.** Variance or Limits can be toggled on. If Units are set to Decimal with a precision of .00, an upper value of .02, and a lower value of .01, the results are shown in figure 7.20.

◆ **Alternate Units.** If the Show Alternate Units box is checked, the default scaling value of 25.4 is activated. This allows an alternate unit of millimeters to be shown, along with inches being the standard unit. This scaling value can also be set to other values. One setting would be .03937 if the alternate unit is inches, with millimeters being the standard unit. Of course, other scaling values could be used with other units.

Figure 7.20

Some different settings from the Text Format dialog box.

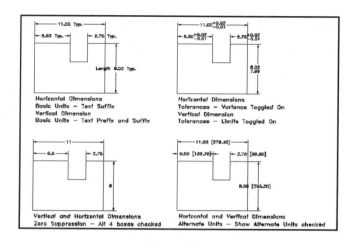

Scale and Colors

The Scale and Colors dialog box, as shown in figure 7.21, has one setting that gets adjusted for different scale plots—1:4, 1:12, 1:48—and that is the Feature Scaling setting. This setting enables all the dimension components to be adjusted at once. The adjustments in the Scale and Colors dialog box are the following:

◆ **Feature Scaling.** This is primarily used to adjust the size of the dimension when it is plotted to a much smaller scale than the full size scale it is drawn. This Feature Scaling affects numerous dimension features all at once, such as text size, arrow or tick size, extension line offsets, and other variables. For an architectural plan that is plotted 1 = 48, Plotted Inches = Drawing Units (a plot of a 1/4" = 1'), the feature scaling should be 48.

◆ **Colors.** The dimension line color is by block, unless a specific color is typed in.

Feature scaling is used in architectural drawings so that dimension text and arrows can be read and seen in a plot. Setting Feature Scaling once for the entire drawing avoids having to make settings for each individual dimension and for text height and arrow size. If you would like the text height and arrow size to be 3/16" high in the actual plot, feature scaling should be as follows:

◆ 48 for a 1/4" = 1' plot

◆ 96 for a 1/8" = 1' plot
◆ 192 for a 1/16" = 1' plot

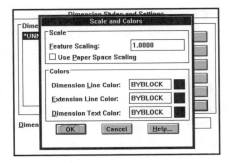

Figure 7.21

The Scale and Colors dialog box.

Ways to Edit Dimensions

When you were dimensioning the Bracket drawing, you were dimensioning with associative dimensions. You can tell if you are dimensioning with an associative dimension because when you pick on the dimension, dimension text and lines, extension lines, and arrows are all one item. A non-associative dimension would be broken into separate entities, with the arrows, text, and dimensions being individual parts.

To make sure you are dimensioning with associative dimensions, you can see whether Associative Dimensions under the **S**ettings pull-down menu is checked, which means it is on. This is a default setting in AutoCAD LT. Whenever this is checked, you are creating associative dimensions. This setting affects linear, ordinate, diameter, radius, and angular dimensions. It does not affect center marks and leaders, however; they are separate parts, no matter how Associative Dimensions is set.

There are some big advantages to dimensioning with Associative Dimensions checked on. If you stretch an object that has an associative dimension, AutoCAD LT automatically updates the dimension and its text. There are also six **M**odify, Ed**i**t Dimension commands, shown in figure 7.22. These commands can only be used with associative dimensions, as follows:

◆ **Change Text.** This menu item issues the NEWTEXT dimension command. NEWTEXT changes the dimension text to any alphanumeric characters you want. It prompts with: `Enter new dimension text:`. You can respond with text to replace the existing dimension text, or Press Enter with no response if you want to restore the actual dimension measurement in the dimension. The second prompt is: `Select objects:`. With this, you select the dimension(s) you want to change.

◆ **Home Position.** This menu item issues the HOMETEXT dimension command. The command restores dimension text back to its original default location after being moved by the TEDIT command. It prompts with: `Select Objects.`

◆ **Move Text.** This menu item issues the TEDIT dimension command. TEDIT moves, rotates, and restores the dimension text back to its original location. It prompts with `Select dimension:`, and you select the dimension you want to move or edit. The second prompt is `Enter text location (Left/Right/Home/Angle):`, and you can respond with the appropriate choice. Left/Right changes the location of the dimension text to the right or left of the dimension line for linear, radius, or diameter dimensions. Home restores dimension text back to its original default location. Angle rotates dimension text to the angle you specify. You can also just pick a point in the drawing for the new location of the text.

◆ **Rotate Text.** This menu item issues the TROTATE dimension command. TROTATE rotates dimension text to the angle you determine. It prompts you with `Enter text angle:`, and you can respond with the appropriate angle. The second prompt is `Select objects:` —you select the dimension(s) you want to change.

◆ **Oblique Dimension.** This menu item issues the OBLIQUE dimension command. OBLIQUE changes the angle of the dimension lines for isometric drawings or for other applications. It prompts you with `Select objects:`; you select the dimension(s) you want to change. The second prompt is `Enter obliquing angle (RETURN for none):`, after which you type in the appropriate angle.

◆ **Update Dimension.** This menu item issues the UPDATE dimension command. UPDATE updates dimensions to the current dimension settings in the Dimension Styles and Settings dialog box. It prompts with `Select Objects.`

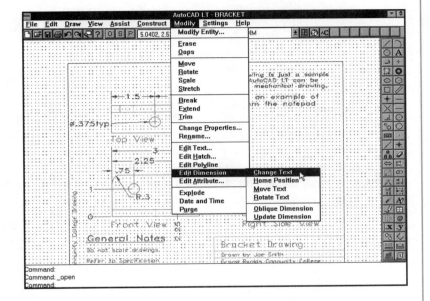

Figure 7.22

Edit Dimension commands.

Be careful when using the UPDATE dimension command, STRETCH command, or some of the other Edit Dimension commands. The dimension will be updated to the current settings in the Dimension Styles and Settings dialog box. If you changed the adjustment in the Horizontal pop-up list back to Default in the Text Location dialog box, all your linear dimension text may suddenly jump outside the extension lines (see fig. 7.17). If you use the UPDATE dimension command and you select ALL, the appearance of many dimensions could be changed.

CAUTION

To better understand how these Edit Dimension commands work, let's try an exercise with the Bracket drawing.

Chapter

7

Exercise

Using the Edit Dimension Commands

Use the Edit Dimension commands to modify dimensions in figure 7.23.

Command: *Choose the Open button;*
choose the Bracket drawing

Opens the bracket drawing

Command: *Choose **S**ettings, Di**m**ension*
Style, choose Linear *in the Dimension*
Styles list box, and then choose OK

Opens the dialog box, and
makes Linear the current
dimension style

Command: *Choose **M**odify, Ed**i**t*
*Dimension, then **M**ove Text*

Issues the Dim: TEDIT
command

Select Dimension: *Pick dimension*
at ① *(see fig. 7.23)*

Picks the dimension to change

Figure 7.23

Using Edit
Dimension
commands to
modify the
Bracket
drawing.

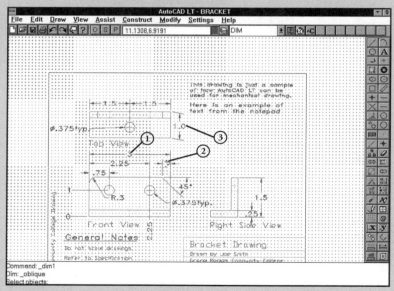

Enter text location (Left/
Right/Home/Angle): **L** Enter

Moves dimension text to the
left on the dimension line,
then ends the command

Command: *Choose **M**odify, Ed**i**t*
*Dimension, then **R**otate Text*

Issues the Dim: TROTATE
command

`Enter text angle: 45` (Enter)	Specifies a 45 degree rotation
`Select objects:` *Pick dimension at* ②	Picks the dimension to rotate
`Select objects:` (Enter)	Ends the Dim: TROTATE command
Command: *Choose* **M**odify, Ed**i**t Dimension, *then* **H**ome Position	Issues the Dim: HOMETEXT command
`Select objects:` *Pick dimension at* ①	Picks the dimension to restore to original default position
`Select objects:` (Enter)	Ends the Dim: HOMETEXT command
Command: *Choose* **M**odify, Ed**i**t Dimension, *then* **C**hange Text	Issues the Dim: NEWTEXT command
`Enter new dimension text:` `1.0` (Enter)	Sets value of dimension text
`Select objects:` *Pick dimension at* ③	Picks the dimension to change
`Select objects:` (Enter)	Ends the Dim: NEWTEXT command
Command: *Choose* **M**odify, Ed**i**t Dimension, *then* **O**blique Dimension	Issues the Dim: OBLIQUE command
`Select objects:` *Pick dimension at* ③	Picks the dimension to change
`Select objects:` (Enter)	Ends the selection set
`Enter obliquing angle(RETURN for none):-10` (Enter)	Specifies -10 angle; changes the extension lines
Command: *Choose the Save button*	Issues the SAVE command

Your drawing should resemble figure 7.23

7
Chapter

Stretching and Moving Dimensions

There are several ways you can move and stretch dimensions. You can use the dimension command TEDIT to move dimension text around. The STRETCH command can be used to stretch the dimension, or the dimension and the object. You can also change the location of your dimension or your dimension and object by using the MOVE command. Finally, you can utilize GRIPS to move dimension text, stretch the dimension, or move the dimension and the object. Details of the ways you can stretch and move dimensions are as follows:

◆ **TEDIT.** This command can be used to move dimension text around, and it will maintain its association with the rest of the dimension. It can be found on the pull-down menu. Choose **M**odify, Ed**i**t Dimension, then **M**ove Text. You next select a dimension and, when prompted with `Enter text location (Left/Right/Home/Angle):`, pick a point on the drawing for the new location of the dimension text.

◆ **STRETCH.** This command can be used to stretch the object and the dimension combined, or just the dimension alone. When the dimension and object are stretched at the same time, the dimension text will automatically be updated, one of the advantages of associative dimensions. The STRETCH command can be found in the toolbox. The prompts will ask you for the `First corner:` and `Other corner:` of a crossing window to select the entities you want to stretch; pick two points to select the window. At the prompt `Base point or displacement:`, pick a point. At the prompt `Second point of displacement:`, type in or pick the distance you want to stretch the entities.

◆ **MOVE.** You can move the entire associative dimension with this command. The MOVE command can be found in the toolbox. At the prompt `Select Objects:`, select the dimension you want to move. At the prompt `Base point or displacement:`, pick the point where you want to move it from (using your object snap modes, if needed). At the prompt `Second point of` displacement:, pick the point where you want to move the dimension to (again, use your object snap modes).

◆ Grips are created when you select an object and you aren't in a command—they appear as little blue squares. After a grip is created and selected a second time, it turns from a blue outline to a solid red square, and the following commands are available from which to choose: STRETCH, MOVE, ROTATE, SCALE, and MIRROR. Press the spacebar to look at the different available commands. Start picking points or typing in coordinates to choose the command you want to use. You can cancel them at any time by choosing the Cancel button on your toolbar. Grips will be discussed in more detail in Chapter 11, "Creating a Mechanical Drawing."

To understand how the STRETCH and TEDIT command work in AutoCAD LT, try these commands in an exercise.

Using the STRETCH and TEDIT commands

Exercise

Continue the drawing from the previous exercise (see fig. 7.24).

Command: *Choose the Stretch button*	Issues the STRETCH command
Select objects to stretch by window or polygon.	
Select objects: *Pick the point at* ④ *(see fig. 7.24)*	Picks the first point of the crossing window
Other corner: *Pick the point at* ⑤	Completes the crossing window selects the dimension
Select objects: (Enter)	Completes the selection set
Base point or displacement: *Pick a point near the dimension*	
Second point of displacement: *Pick a point at* ⑥	Picks the new location of the dimension line

continues

7

Chapter

Figure 7.24

Using the STRETCH command.

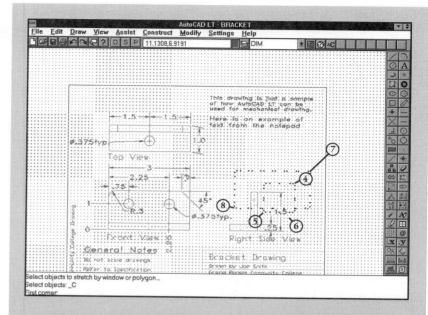

Command: *Choose the STRETCH command*	Issues the STRETCH command
Select objects to stretch by window or polygon.	
Select objects: *Pick the point at ⑦ (see fig. 7.24)*	Picks the first point of the crossing window
Other corner: *Pick the point at ⑧*	Completes the crossing window selects the dimension
Select objects: (Enter)	Completes the selection set
Base point or displacement: @ (Enter)	Establishes the base point
Second point of displacement: @0,.5 (Enter)	Stretches the objects
Command: *Choose* **M**odify, Ed**i**t Dimension, *then* **M**ove Text	Issues the Dim: TEDIT command
Select Dimension: *Pick the 2 dimension in the Right Side View (see fig. 7.25)*	Picks the dimension to change

Enter text location (Left/ Right/Home/Angle): *Pick a point at* ⑨ *(see fig. 7.25)*	Moves dimension text to a new location that is picked.
Command: *Choose the Save button*	Issues the SAVE command
Your drawing should now resemble figure 7.25	
Command: *Choose the Undo button*	Undoes the last command

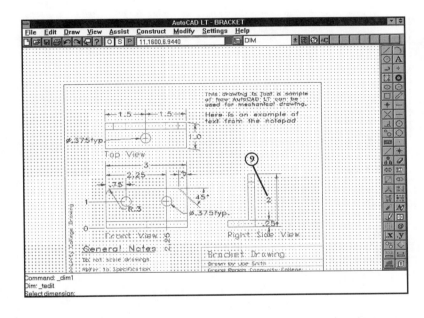

Figure 7.25

The results of stretching dimensions.

Using Grips to Stretch a Dimension

Continue the drawing from the previous exercise (see fig. 7.25).

Command: *Create a crossing window by choosing* ① *and* ② *(see fig. 7.26)*	Creates grips on the dimension
All crossed entities display small blue boxes	

continues

Figure 7.26

Using grips to stretch a dimension.

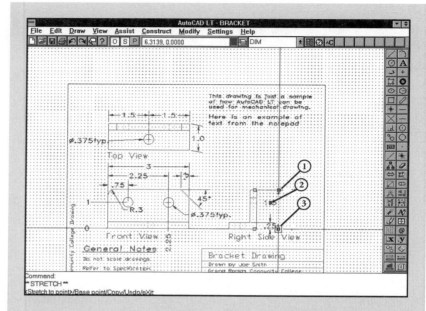

Command: *Hold down the Shift key and pick ③, ④, and ⑤*	Turns the control points red
Command: *Pick any of the red points*	Issues the GRIP option command; starts with STRETCH command
`<Stretch to point>/Base Point/Copy/Undo/eXit:` *Pick the point at ⑥*	Stretches the dimension to the new location
Your drawing should look like figure 7.27	
Command: *Choose* File, Exit, *then choose* Yes	Issues the EXIT command and saves the drawing

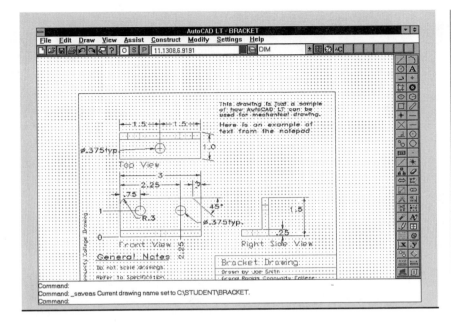

Figure 7.27

The results of stretching the dimension with grips.

NOTE

One other command you can use to edit your dimension is the EXPLODE command found in your toolbox. When you explode your dimension, it is no longer a single entity, just a bunch of arrows, text, and extension lines— a non-associative dimension. You can edit these parts individually, but it will no longer update automatically if stretched, and the Edit Dimension commands can no longer be used. The EXPLODE command is not recommended for editing dimensions.

Summary

In the last two chapters, you have learned how to create dimensions and how to modify dimensions. In this chapter, you learned how different dimension settings affect the appearance of the dimension. You can add a suffix or a prefix to a dimension, force text inside extension lines, or use ticks instead of arrows with your dimension lines. You also examined the procedure for changing the appearance of the dimension. You can use Dimension Styles to change them, or use the UPDATE dimensioning command.

The Edit Dimension commands can be used to modify the appearance of dimensions. For example, you can use the TEDIT dimensioning command to move around dimensioning text. In addition, the STRETCH command can be used to stretch both the object and the dimension, and the dimension text automatically updates to the new measurement of the object. Finally, grips can also be used to move or stretch the dimension.

In the next chapter, you learn about sectional and auxiliary views, which are the different ways to cut an object.

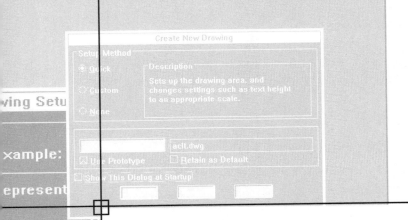

Part Three

Viewing and Reproducing Drawings

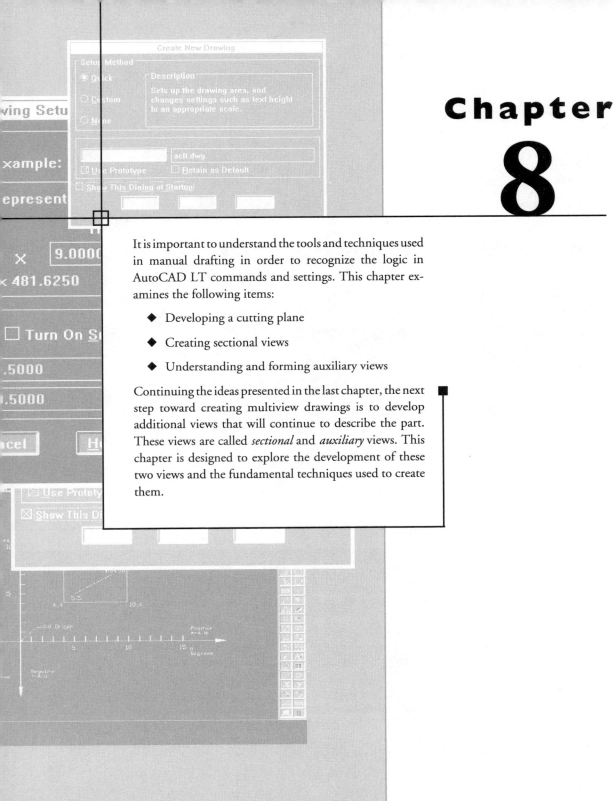

It is important to understand the tools and techniques used in manual drafting in order to recognize the logic in AutoCAD LT commands and settings. This chapter examines the following items:

◆ Developing a cutting plane

◆ Creating sectional views

◆ Understanding and forming auxiliary views

Continuing the ideas presented in the last chapter, the next step toward creating multiview drawings is to develop additional views that will continue to describe the part. These views are called *sectional* and *auxiliary* views. This chapter is designed to explore the development of these two views and the fundamental techniques used to create them.

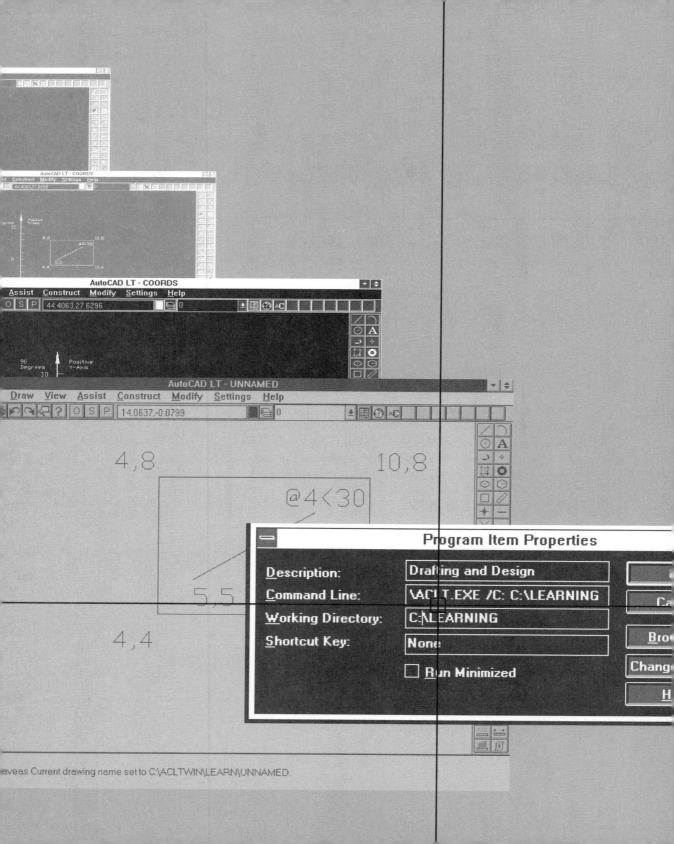

Developing Sectional and Auxiliary Views

The exterior surface of a part, a machine, or a building is best illustrated with orthographic views or elevations. Describing the inside of these parts requires the use of additional drawings. These drawings are necessary in all disciplines of drafting. No matter which discipline you are drafting under, each of these additional drawings emanates from the same fundamental techniques introduced in this chapter. In order to properly explore the logic of these drawings, you will use mechanical parts for examples.

Creating a Sectional View

In the previous chapters, all of the parts that were created could be adequately described with two or three views. Most parts can be developed by a machinist given these typical views. At times, however, you will be forced to describe complex internal shapes or voids that cannot be delineated with hidden lines. In order to accomplish this, you must utilize a *sectional* drawing.

Sectional drawings are also referred to as cross sections, cut-away views, or simply just sections. In the first half of this chapter, you will create two sectional drawings.

The first will be a bushing. A *bushing* is a metal lining or part that reduces the friction on other moving parts (see fig. 8.1).

Figure 8.1

The completed bushing.

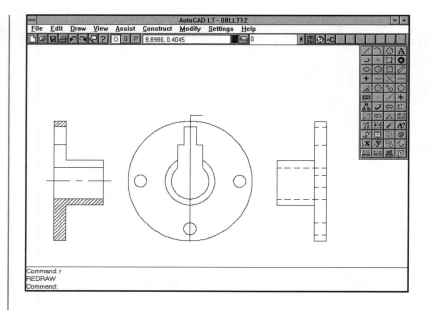

In order to create these drawings, a part that will be used to demonstrate sectioning techniques must be created, as described in the following exercises.

NOTE

To create these drawings quickly, reference points are used. You should create the drawings with the skills developed in previous chapters.

Starting the Front View

Command: *Choose* File, *then* New	Opens the Create New Drawing dialog box
Choose OK *in the dialog box*	Opens the Quick Drawing Setup dialog box
Choose OK *in the dialog box*	Accepts the default settings
Command: *Choose the Circle button*	Issues the CIRCLE command
_CIRCLE 3P/TTR/<Center point>: 4,4 (Enter)	Sets circle center point
Diameter/<Radius>: D (Enter)	Selects Diameter option

Exercise

`Diameter: 5` (Enter)	Creates the circle
`Command:` *Right click on the mouse*	Reissues the CIRCLE command
`CIRCLE 3P/TTR/<Center point>:` *Choose the Center button*	Sets the object snap to CENter
`_CEN of` *Pick on* ① *(see fig. 8.2)*	Selects circle center point
`Diameter/<Radius>: 2` (Enter)	Creates the circle
`Command:` *Right click on the mouse*	Reissues the CIRCLE command
`CIRCLE 3P/TTR/<Center point>:` *Choose the Center button*	Sets the object snap to CENter
`_CEN of` *Pick on* ①	Selects circle center point
`Diameter/<Radius>: .75` (Enter)	Creates the circle
`Command:` *Right click on the mouse*	Reissues the CIRCLE command
`CIRCLE 3P/TTR/<Center point>:` *Choose the Center button*	Sets the object snap to CENter
`_CEN of` *Pick on* ①	Selects circle center point
`Diameter/<Radius>: 1` (Enter)	Creates the circle
`Command:` *Right click on the mouse*	Reissues the CIRCLE command
`CIRCLE 3P/TTR/<Center point>:` *Choose the Quadrant button*	Sets the object snap to QUAdrant
`_QUA of Pick on` ②	Selects circle center point
`Diameter/<Radius>: .25` (Enter)	Creates circle
`Command:` *Choose the Copy button*	Issues the COPY command
`Select objects: L` (Enter)	Selects last circle drawn
`1 found`	
`Select objects:` *Right click on the mouse*	Accepts the selection set
`<Base point or displacement>/` `Multiple: M` (Enter)	Selects the multiple option

continues

`Base point:` *Choose the Center button*	Sets the object snap to CENter
`_CEN of` *Pick on* ③	
`Second point of displacement:` *Choose the Quadrant button*	Sets the object snap to QUAdrant
`_QUA of` *Pick on* ④	
`Second point of displacement:` *Choose the Quadrant button*	Sets the object snap to QUAdrant
`_QUA of` *Pick on* ⑤	
`Second point of displacement:` *Right click on the mouse*	Ends the COPY command
`Command:` *Choose the Save button on the toolbar*	Opens the Save As dialog box
Save the drawing as BUSHING.DWG in the ACLTWIN directory	

Figure 8.2

Developing the front view.

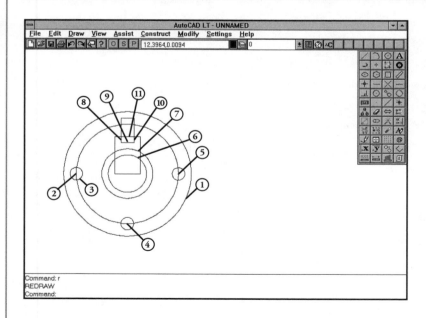

Now that the basic front view is constructed, continue with the exercise by forming the key way and cleaning up the drawing.

Exercise

Continuing with the Front View

Continue with the BUSHING.DWG.

Command: *Choose the Rectangle button*	Issues the RECTANGLE command
_RECTANG First corner: **3.5, 4** (Enter)	Sets the first corner
Other corner: **@1,1.5** (Enter)	Sets the opposite diagonal corner
Command: *Right click on the mouse*	Reissues the RECTANGLE command
_RECTANG First corner: **3.75, 5.25** (Enter)	Sets the first corner
Other corner: **@.5,1** (Enter)	Sets the opposite diagonal corner
Command: *Choose the Erase button*	Issues the ERASE command
_ERASE Select objects: *Pick on ② (see fig. 8.2)*	Highlights the object
1 found	
Select objects: (Enter)	
Command: *Choose the Trim button*	Issues the TRIM command
_TRIM Select cutting edges Select objects: *Pick the two rectangles*	Highlights the objects
Select objects: *Right click on the mouse*	Accepts the selection set
<Select object to trim>/Undo: *Pick on ⑥⑦⑧⑨⑩⑪ (see fig. 8.2)*	Trims the objects
<Select object to trim>/Undo: *Right click on the mouse*	Ends TRIM command
Command: *Right click on the mouse*	Reissues the TRIM command

continues

TRIM Select cutting edges Select objects: *Pick on* ① *(see fig. 8.3)*	Highlights the object
Select objects: *Right click on the mouse*	Accepts the selection set
<Select object to trim>/Undo: *Pick on* ②③④ *(see fig. 8.3)*	Trims the objects
<Select object to trim>/Undo: *Right click on the mouse*	Ends the TRIM command
Command: *Choose the Rectangle button*	Issues the RECTANGLE command
_RECTANG First corner: **7.5, 3** (Enter)	Sets first corner
Other corner: **@1.5,1.86** (Enter)	
Command: *Right click on the mouse*	Issues the RECTANGLE command
RECTANG First corner: **9, 1.5** (Enter)	Sets first corner
Other corner: **@.5,5** (Enter)	
Command: *Choose the Line button*	Issues the LINE command
_LINE From point: **7.5, 3.25** (Enter)	
To point: **@2<0** (Enter)	
To point: *Choose* **C**onstruct, *then* **O**ffset	Issues the OFFSET command
_OFFSET Offset distance or Through <.5000>: **1.309** (Enter)	Sets the offset distance
Select object to offset: *Pick the line you just created*	Selects last object drawn
Side to offset? *Pick above the line*	Copies the line
Side to offset? (Enter)	Completes the command
Save the drawing	

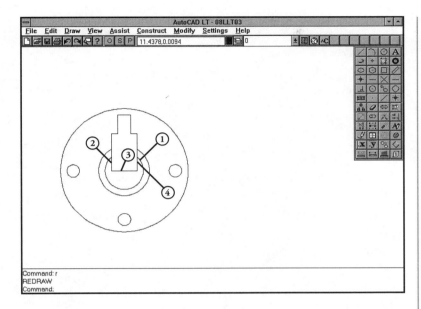

You are still in the midst of completing the BUSHING.DWG. In the next
exercise, you finish the front view that you have already started, and then add
finishing touches to complete the bushing.

Continuing with the Front View

Exercise

Continue with the BUSHING.DWG.

Command: *Choose the Line button*	Issues the LINE command
_LINE From point: **9,1.75** (Enter)	Starts the LINE
To point: **@.5<0** (Enter)	
To point: *Choose* **C**onstruct, *then* **O**ffset	Issues the OFFSET command
_OFFSET Offset distance or Through <1.309>: **.5** (Enter)	Sets the offset distance
Select object to offset: *Pick on* ① *(see fig. 8.4)*	Highlights the object
Side to offset? *Pick above the line*	Copies the line

continues

8
Chapter

`Select object to offset:` *Right click on the mouse two times*	Reissues the OFFSET command
`_OFFSET Offset distance or Through <.5000>:` **2** (Enter)	Sets the offset distance
`Select object to offset:` *Pick on* ①	Selects line
`Side to offset?` *Pick above the line*	Copies the line
`Select object to offset:` *Right click on the mouse two times*	Reissues the OFFSET command
`_OFFSET Offset distance or Through <2.000>:` **2.5** (Enter)	Sets the offset distance
`Select object to offset:` *Pick on* ①	Selects line
`Side to offset?` *Pick above the line*	Copies the line
`Select object to offset:` *Right click on the mouse two times*	Reissues the OFFSET command
`_OFFSET Offset distance or Through <2.500>:` **3.75** (Enter)	Sets the offset distance
`Select object to offset:` *Pick on* ①	Selects line
`Side to offset?` *Pick above the line*	Copies the line
`Select object to offset:` *Right click on the mouse two times*	Reissues the OFFSET command
`_OFFSET Offset distance or Through <3.750>:` **4.5** (Enter)	Sets the offset distance
`Select object to offset:` *Pick on* ①	Selects line
`Side to offset?` *Pick above the line*	Copies the line
`Select object to offset:` *Right click on the mouse*	Ends the OFFSET command
`Command:` **CH** (Enter)	Issues CHANGE command

`Select objects:` *Pick at* ② *then* ③	Creates implied crossing window
`Select objects:` *Right click on the mouse*	Accepts selection set
`Properties/<Change point>:` **P** (Enter)	Select change properties
`Change what property (Color /Elev/LAyer/LType/Thickness):` **LT** (Enter)	Selects the Line Type option
`New linetype <CONTINUOUS>:` **HIDDEN** (Enter)	Selects the Hidden linetype
`Change what property (Color /Elev/LAyer/LType/Thickness):` (Enter)	Ends CHANGE command
`Command:` *Choose the Save button on the toolbar*	Saves the drawing

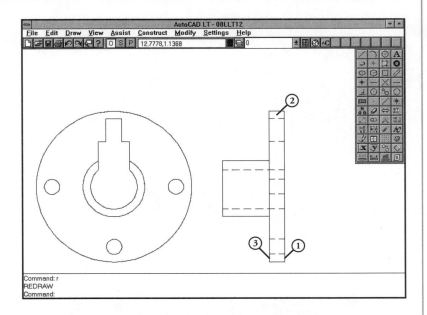

Figure 8.4

The finished part.

NOTE

In the last exercise, the line type was changed by typing the CHANGE command from the keyboard. This eliminated the need to load the line type before using it within a drawing. If you were to try changing the line type via DDMODIFY or some other means, you would have had to load the line type prior to specifying it.

Now that the part is complete, you will begin to explore the logic of section views.

Understanding the Cutting Plane

The interior of a part can be shown by passing an imaginary plane through the object. This imaginary plane is called the *cutting plane*, as shown in figure 8.5.

Figure 8.5

The cutting plane.

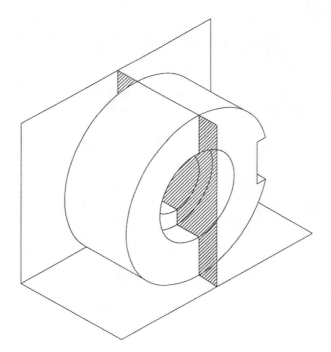

The cutting plane typically divides the part into two equal halves. The two halves are pulled apart, which exposes the interior (see fig. 8.6). The right half is used to create the section view and the left half is discarded.

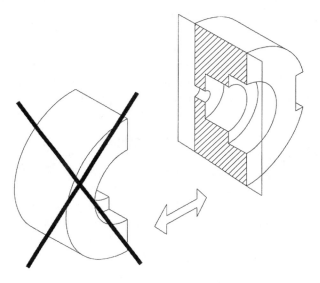

Figure 8.6

Separating the part.

For the purposes of this particular drawing, the right half of the section is illustrated and will be placed on the drawing, where a left side view is shown.

In figure 8.7, the correct cutting plane in the front view and the correct section view are shown. The other section drawings represent improper techniques.

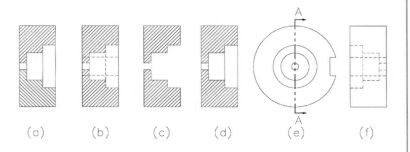

(a)   (b)   (c)   (d)   (e)   (f)

Figure 8.7

Proper drawing placement.

Understanding the Lines in a Section

The lines of the section should show a profile of the sectional areas. The section drawing should show all of the visible lines that extend beyond the cutting plane. These lines should have a lighter line weight in order to emphasize the section. Drawing just the areas coinciding with the plane in figure 8.7 creates a drawing that has no reference and appears disjointed. Do not show hidden lines in the section 8.7(b). Hidden lines in the section do not provide additional information and often confuse the viewer. Also, do not show the lines of the section as hidden lines 8.7(b); this too is the result of improper technique.

The most important line within a section drawing is the *reference* line, which depicts the cutting plane. This line is typically drawn with a *phantom* line type, consisting of a long line with two short dashes (see fig. 8.7). This cutting plane line should be the same thickness as the object lines in order to remain visible in a complex orthographic drawing. Multiple cutting plane lines can exist in a drawing depending on necessity. In addition, the cutting plane line does not have to remain a straight line (see fig. 8.8); at times it may be necessary to jog a section line in order to include additional information. This type of section is called an *offset* section.

Figure 8.8

Section line techniques.

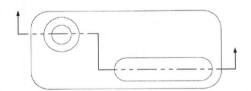 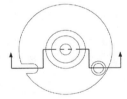

The arrowheads point to the direction at which the section is viewed. This is typically the opposite direction of the sections location in the drawing. Large capital letters denote the section—for example, the sectional view in figure 8.7 would be labeled, Section A-A.

Utilizing Section Lining Techniques

Section lining or poché (pronounced *po-shay*) is used to depict the materials of an object in section. In the past, different types of section lining was used to depict the sectioned material that was used to create the part. Although this technique is still used in architectural detailing, mechanical drafting (which involves parts that are typically created out of metal) has had to adjust. So many types of metal and plastic exist—for example, there are at least one hundred types of steel—that keeping a hatch for each material has became cumbersome. It is now accepted that for objects that have been sectioned, the ANSI31 hatch or 45° diagonal lines can be used. Refer back to figure 8.8 for examples of this section lining technique.

In AutoCAD LT, the command used to add the section lining is Boundary Hatch or BHATCH. The Boundary_Hatch button in the toolbox will access the Boundary Hatch dialog box. There are several preloaded hatches that can be used to identify the material in the section lining.

Many firms have established the type of hatch used to depict a material. Do not make these decisions on your own. Ask a supervisor for the correct material scheme.

While creating a section of one solid part, the section lining should not change direction. Part (a) in figure 8.7 shows an improper section lining technique. The only time the section lining direction should change is when there are adjacent parts within a section that are made of different materials.

Creating Different Section Types

There are a variety of different types of section drawings. Each type will be dependent on the part. It could take years of experience to fully understand which method should be used at any given time, but the fundamentals of each type are examined in the next few sections.

Creating a Full Section

The most common sectional drawing that drafters generate is the *full section*, which is created when the cutting plane line passes through the whole object. As illustrated in figure 8.8, the full section is typically a mirrored image of the right side view; this is the case for the part you have created. In the next exercise, begin by creating the cutting plane line.

CAUTION

Make sure that all areas to be hatched are closed (no gaps). If gaps exist in the area, an error message will be returned or the hatch will leak out into the adjacent areas. Also, if a boundary definition error message is displayed and you know the area is closed, try again by picking a new point within the area—BHATCH is occasionally temperamental.

Exercise

Creating the Full Section

Continue working on the BUSHING.DWG.

Command: *Choose the Line button*	Issues the LINE command
_LINE From point: **4,1.25** (Enter)	
To point: **4,6.75** (Enter)	
To point: **@.5<0** (Enter)	Points to the direction that the section will be viewed
To point: *Choose* **C**onstruct, *then* **M**irror	Issues the MIRROR command
Select objects: *Pick on* ① *then* ② *(see fig. 8.9)*	Selects objects with an implied window

Figure 8.9

Creating the full section.

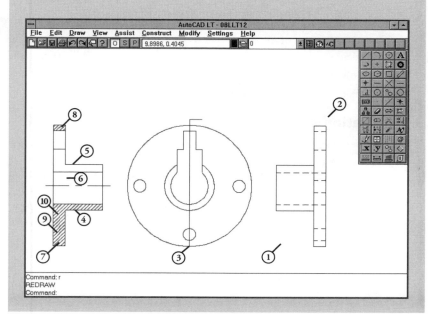

```
10 found
```

`Select objects:` *Right click on the mouse*	Accepts the selection set
`First point of the mirror line:` *Pick on* ③	Sets the first point
Press F8 *to turn on Ortho*	
`Second point:` *Drag the cursor straight up and click*	Sets the second point directly above the first
`Delete old objects?<N>:` *Right click on the mouse*	Accepts the default of not erasing the selection set
`Command:` *Choose the Trim button*	Issues the TRIM command
`_TRIM Select cutting edge(s)` `Select objects:` *Pick on* ④ *and* ⑤	Highlights the objects
`Select objects:` *Right click the mouse*	Accepts the edges
`<Select object to trim>/Undo:` *Pick on* ⑥	Trims the objects
`<Select object to trim>/Undo:` *Choose the Change_Properties button*	Issues the CHANGE command
`Select objects:` *Pick* ⑦ *then* ⑧	Selects lines with an implied crossing window
`Select objects:` *Right click on the mouse*	Accepts the selection set and opens the Change Properties dialog box
Pick on the Linetype *button*	Opens the Select Linetype dialog box
Pick on CONTINUOUS, *then* OK	Sets the selection set to continuous linetype
Pick OK *in the Change Properties dialog box*	Changes the lines to continuous and returns to the drawing area
`Command:` *Choose the Erase button*	Issues the ERASE command
`_ERASE Select objects:` *Pick* ⑨⑩	Highlights the objects
`Select objects:` (Enter)	Erases the objects

continues

Command: *Pick on the Boundary*   Opens the Boundary Hatch
_Hatch tool   dialog box

Set Pattern T**y**pe *to* User-Defined,
then **A**ngle: *to* **45**, *then* **S**pacing: *to* **.1**,
then click on the Pick **P**oints < *button*

Select internal point: Pick ⑦⑧   Highlights the sectioned areas

Select internal point: *Right click*
on the mouse

Pick on the Preview Ha**t**ch < *button,*
then pick the **C**ontinue *button in the*
Boundary Hatch dialog box

Pick on the Apply button

Command: *Choose the Save button on*   Saves the drawing
the toolbar

Creating a Half Section

If the part is symmetrical, it may be necessary to only pass the section line
halfway through the object, exposing half of the interior (see fig. 8.10). This
is referred to as a *half section.*

Figure 8.10

*Developing the
half section.*

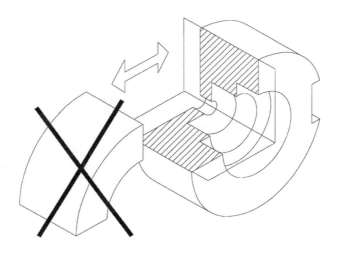

This method of sectioning is often used in assembly drawings where it is necessary to show both the interior and exterior construction. The drawbacks to this method include the fact that most parts are not symmetrical and dimensioning the interior of half an object can be confusing to the viewer. Hidden lines are typically not shown in either half unless absolutely necessary for dimensioning. The other unique feature of this drawing is that a center line divides the sectioned half from the unsectioned half.

The results of a half section in the current part are shown in figure 8.11.

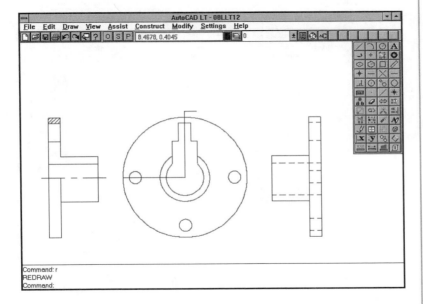

Figure 8.11

The half section.

The only portion of the part shown with section lining is at the top of the key way.

The only area of a half section that contains section lining is the area above or below the center line. The drawing generally dictates which side receives the hatching.

TIP

Creating a Broken-Out Section

Depending on the geometry of the part, it might be necessary to show a small portion in section. This type of drawing is called a *broken-out section.* This partial section is denoted with an irregular line called a *break line*. Break lines are discussed in more detail later in this chapter, but are used when full or half sections might obscure important aspects of the part. Unlike the other sections, it is common to see these sections included in side views because the rules regarding section lines are implied. If a machinist sees a broken-out section in a drawing, they will understand why this drawing is required and what it actually means.

Using a splined polyline (similar to those created in Chapter 3), a broken-out section of the current part would appear as illustrated in figure 8.12.

Figure 8.12

A broken-out section.

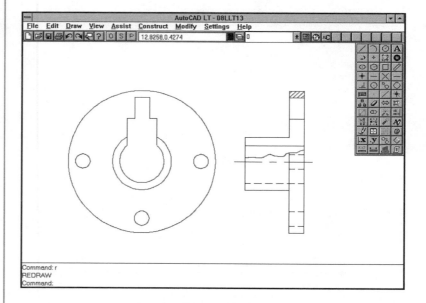

Creating a Revolved Section

At times, a part may require several sections in order to explain its geometry. This happens frequently with spokes, bars, arms, and other extrusions. These are drawn with a sectioning technique called *revolved sections*. The cutting line planes are created perpendicular to the center line of the element, and then are rotated in line with the view.

The revolved section can be incorporated into the part or broken-out. Both methods are displayed in figure 8.13.

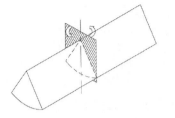
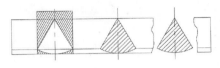

Figure 8.13

Developing a revolved section.

The next major drawing will be a metal punch. A *punch* is a tool used for driving bolts, nails, or pins in or out of a hole (see fig. 8.14). Punches can also be used to perforate or emboss a piece, as well as to recess finish nails for a smooth finish in cabinet making.

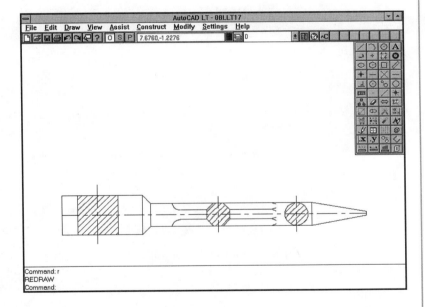

Figure 8.14

The completed punch.

In the next exercise, you create a punch to experiment with revolved sections. Because of its symmetry, the top half will be created and mirrored down to form the whole punch.

Creating the Top Half of the Punch

Start a new drawing with all of the default settings.

Command: *Choose the Line button*	Issues the LINE command
_LINE From point: **1,1** (Enter)	Establishes the first point
To point: **@.5<90** (Enter)	
To point: **@2<0** (Enter)	
To point: **@.3<-45** (Enter)	
To point: **@4<0** (Enter)	
To point: **@1.38<-10** (Enter)	
To point: **@.0482<270** (Enter)	
To point: **C** (Enter)	Closes the line
Command: *Choose the Zoom button on the toolbar*	Issues the ZOOM command
All/Center/Extents/Previous /Window/<Scale(X/XP)>: **E** (Enter)	Selects the Extents option
Command: *Right click on the mouse*	Reissues the ZOOM command
All/Center/Extents/Previous /Window/<Scale(X/XP)>: **.9x** (Enter)	Zooms out to 90 percent
Command: *Choose the Osnap button*	Opens the Object Snap dialog box
Click on **E**ndpoint *and* **P**erpendicular, *then pick* OK	Sets the running object snap
Command: *Choose the Line button*	Issues the LINE command
From point: *Pick on* ① (*see fig. 8.15*)	
To point: *Pick on* ②	Runs line perpendicular to center line
To point: *Right click on the mouse twice*	Ends and reissues the LINE command
From point: *Pick on* ③	

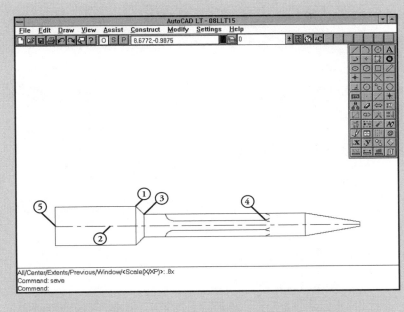

Figure 8.15

The punch drawing.

To point: *Pick on* ②	
To point: *Choose* **C**onstruct, *then* **O**ffset	Issues the OFFSET command
Offset distance or Through <Through.5000>: **4** (Enter)	Sets distance
Select object to offset: *Pick the last line drawn under* ③	Highlights the line
Side to offset? *Pick to the right*	Copies the line to the right
Select object to offset: *Choose the Line button*	Issues the LINE command
From point: **3.962,1.119** (Enter)	Establishes the first point
To point: **@2.25<0** (Enter)	
To point: **@.125<18** (Enter)	
To point: *Right click on the mouse twice*	Ends and reissues the LINE command
From point: *Pick on* ④	

continues

Chapter

8

```
To point: @.125<-18 (Enter)

To point: Choose the Copy button          Issues the COPY command

Select objects: L (Enter)                 Selects the last object drawn

Select objects: Right click on the        Accepts the selection set
mouse

<Base point or displacement>              Sets the base point
/Multiple: Pick on ④

Second point of displacement:             Copies the line up
Drag the object straight up and pick the
top line of the punch

Command: Choose the Arc button;           Issues the ARC command
make sure running osnaps are off

Center/<Start point>: 3.75,               Sets the first point
1.2879 (Enter)

Center/End/<Second point>:
3.8235,1.1626 (Enter)

End point: 3.962,1.119 (Enter)

Command: Choose the Save                  Opens the Save Drawing As
button on the toolbar                     dialog box

Save the drawing as PUNCH.DWG
```

Now that the top half of the punch is drawn, the next step is to mirror the top half down to form the whole punch.

Exercise

Completing the Punch

Continue with the PUNCH.DWG.

Command: *Choose* **C***onstruct, then* **M***irror*

Select objects: **ALL** (Enter)   Select all of the visible objects

15 found

Turn osnaps back on

Select objects: *Right click on the mouse*	Accepts the selection set
First point of the mirror line: *Pick on* ⑤	
Press F8 to turn Ortho mode on	
Second point: *Drag cursor to the right and pick*	Sets the second point
Delete old objects? <N>: *Right click on the mouse*	Accepts the default of not erasing the selection set
Command: *Choose the Zoom button on the toolbar*	Issues the ZOOM command
All/Center/Extents/Previous /Window/ <Scale(X/XP)>:.**8x** (Enter)	Zooms out to 80 percent
Command: *Choose the Erase Button*	Issues the ERASE command
Select objects: *Pick on the center line*	Highlights the line
Select objects: *Right click on the mouse*	Removes one of the two centerlines
Command: **R** (Enter)	Redraws the view
Command: **CH** (Enter)	Issues the CHANGE command
Select objects: *Pick on the center line*	
Select objects: *Right click on the mouse*	Accepts the line
Properties/<Change point>: **P** (Enter)	Selects Properties option
Change what property (Color /Elev/LAyer/LType/Thickness)? **LT** (Enter)	Selects linetype option
New linetype <BYLAYER>: **CENTER** (Enter)	Sets the linetype

continues

```
Change what property (Color        Ends CHANGE command
/Elev/LAyer/LType/Thickness)?
Right click on the mouse

Command: LTSCALE (Enter)           Issues the LTSCALE command

New scale factor <1.000>:          Sets LTSCALE
.4 (Enter)

Command: Choose the Save button on  Saves the drawing
the toolbar
```

CAUTION

Remember to change the LTSCALE when using different line types. If the setting is too large or too small, you will not see the line displayed correctly.

In order to create the cross sections, you must know the shape of the part. The large end has a square profile, the middle has an octagon, and the point is a cylinder that eventually tapers to a point. You use the POLYGON command to create the shapes.

The POLYGON command prompts you for the number of sides in the shape. Next, you are prompted for the center point or, alternately, the length of one edge, which will drag the shape size. If the center point option is selected, you then create a circle into which the polygon will fit. The command makes sense when you observe the results. As you will see, this command will prove quite useful.

Exercise

Creating the Revolved Sections

Continue working on the PUNCH.DWG.

```
Command: Choose the Polygon button   Issues the POLYGON command

POLYGON Number of sides <4>:         Sets polygon to octagon
8 (Enter)

Edge/<Center of polygon>:            Sets the object snap to NEArest
Choose the Nearest button

of Pick on ① (see fig. 8.16)
```

`Radius of circle:` *Pick on top line*	Perpendicular object snap is still running
`Command:` *Right click the mouse*	Reissues the POLYGON command
`POLYGON Number of sides <8>:` **4** (Enter)	Sets polygon to octagon
`Edge/<Center of polygon>:` *Choose the Nearest button*	Sets the object snap to NEArest
to *Pick on* ②	
`Radius of circle:` *Pick on top line*	Perpendicular osnap is still running
`Command:` *Choose the Circle button*	Issues the CIRCLE command
`_CIRCLE 3P/TTR/<Center point>:` *Choose the Nearest button*	Sets the object snap to NEArest
to *Pick on* ③	Sets the center point
`Diameter/<Radius>:` *Pick on the top line*	Sets the radius
`Command:` *Choose the Trim button*	Issues the TRIM command
`Select cutting edge(s)`	
`Select objects:` *Pick the octagon*	Highlights the octagon
`Select objects:` *Right click on the mouse*	Accepts the selection set
`Select objects to trim:` *Pick all of the lines in the octagon, except the center line*	
`Select objects:` *Pick on the Boundary_Hatch tool*	Opens the Boundary Hatch dialog box
Set Pattern Type *to* User-Defined, *then* **A**ngle: *to* **45**, *then* **S**pacing: *to* **.1**, *then click on the* Pick **P**oints < *button*	
`Select internal point:` *Pick inside the top and bottom half of each polygon*	Highlights the sectioned areas

continues

Select internal point: *Right click on the mouse*

Pick on the Preview Ha<u>t</u>ch < *button, then pick the* <u>C</u>ontinue *button in the Boundary Hatch dialog box*

Pick on the Apply button

Add vertical center lines to the sections, as shown in figure 8.16

Command: *Choose the Save button on the toolbar*   Saves the drawing

Figure 8.16

The revolved sections drawing.

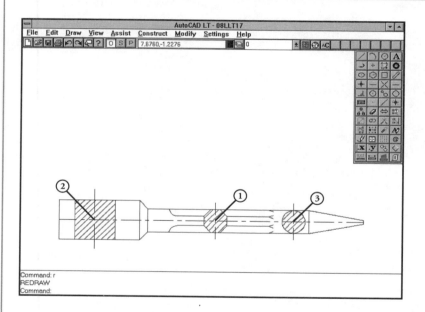

Revolved sections can be more complicated than those illustrated here. Make sure you understand the underlying principle because all revolved sections have a fundamental premise.

Creating Removed Sections

Revolved sections do not always stay on top of the drawing of the element they represent. There may be circumstances in which the part is too complex or there are too many sections to be represented. When cross sections are located on another part of the drawing or slightly pulled away from the part, they are called removed sections. Figure 8.17 shows both methods as they would appear in a drawing.

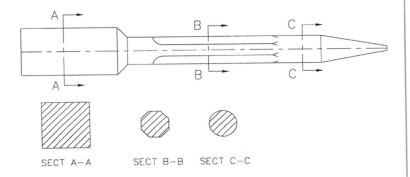

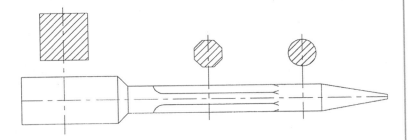

Figure 8.17

Examples of removed sections.

Removed sections should be labeled with letters. Typically, letters I and O are not used—they are often confused with the numbers one and zero.

TIP

Removed sections can be scaled up in order to explain more detail and to provide ample space for proper dimensioning. They can also be located on a completely different drawing sheet, as long as a cross-reference is included on both sheets—this technique is frequently used by architects. Sheets of details are usually cross-referenced back to their location in the building section drawings or on the elevation drawings.

Chapter

8

Creating Aligned Sections

Aligned sections are used when a part cannot be sectioned using typical techniques due to its geometry. Offset cutting plane lines were demonstrated earlier in the chapter and go hand in hand with aligned sections. An example of where an aligned section should be developed is shown in figure 8.18.

Figure 8.18

An example of an aligned section.

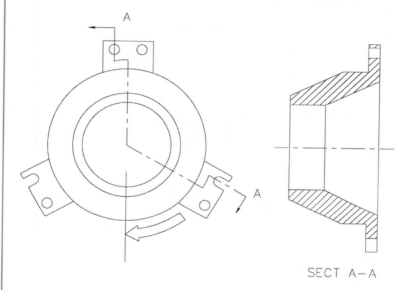

SECT A–A

CAUTION

The aligned part should never rotate beyond 90°, and caution should always be taken because aligned sections can cause confusion. Only use this technique if clarity cannot be gained by other means.

Utilizing Breaks

Break lines are used to conserve space in a drawing. In figure 8.13, an example of a break was illustrated as an option when using revolved sections. Breaks are also used when a part is longer than the space provided for the drawing. If a break is to be used, the part must have the same section along its length or it must taper uniformly. If a machinist notes a break on a drawing, he or she will realize that the dimension given will not scale on the drawing because the part is longer than what has been drawn. Figure 8.19 illustrates an example of a break, along with the typical methods used to illustrate breaks.

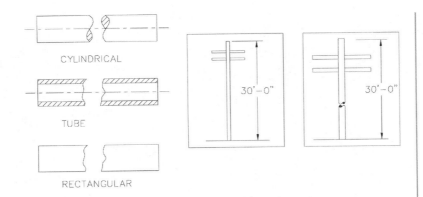

Figure 8.19

Using breaks on long parts.

CYLINDRICAL

TUBE

RECTANGULAR

The following exercise demonstrates how cylindrical break lines are constructed. Using the punch drawing, you will stretch out the piece and perform a break.

Exercise

Creating the Cylindrical Break Lines

Continue working on the PUNCH.DWG.

Command: *Choose the Stretch button*	Issues the STRETCH command
First corner: *Pick on* ① *(see fig. 8.20)*	Selects first corner
Other corner: *Pick on* ②	Highlights the objects
Select objects: *Right click on the mouse*	Accepts the selection set
Base point or displacement: *Pick near* ①	
Second point of displacement: @1<0 (Enter)	
Command: *Choose the Polyline button*	Issues the PLINE command
First point: 6.75,1.21 (Enter)	
Arc/.../Width/<Endpoint of line>: .42<30 (Enter)	

continues

```
Arc/.../Width/<Endpoint of
line>: .266<135 Enter
```

```
Arc/.../Width/<Endpoint of                    Ends the PLINE command
line>: Right click the mouse
```

```
Command: Choose Modify, then                  Issues the PEDIT command
Edit Polyline
```

```
Select polyline: L Enter                       Selects the last entity drawn
```

```
Close/.../eXit <X>: S Enter                    Selects the spline option
```

```
Close/.../eXit <X>: Choose                     Issues the MIRROR command
Construct, then Mirror
```

```
Select objects: L Enter                        Selects the polyline
```

```
1 found
```

```
Select objects: Right click on                 Accepts the selection set
the mouse
```

```
First point of mirror line:
Pick on ①
```

```
Press F8 to turn on Ortho
```

```
Second point: Drag the cursor
straight up and pick
```

```
Delete old objects? <N>:                       Mirrors the polyline
Right click on the mouse
```

```
Command: Choose the Copy button                Issues the COPY command
```

```
Select objects: Pick both polylines            Highlights the polylines
```

```
Select objects: Right click on the             Accepts the selection set
mouse
```

```
<Base point or displacement>                   Sets the base point
/Multiple: Pick anywhere that is
not on an object
```

```
Second point of displacement:
@.2<0 Enter
```

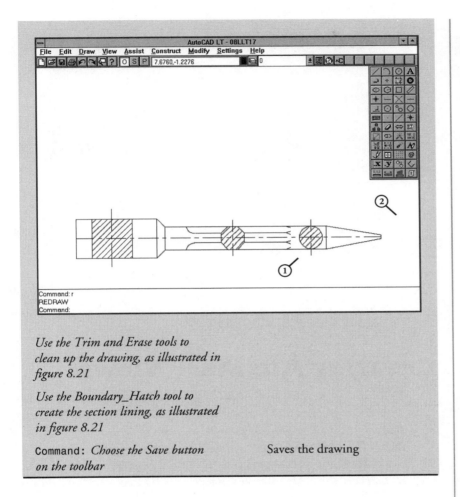

Figure 8.20

Developing the cylindrical break lines.

Use the Trim and Erase tools to clean up the drawing, as illustrated in figure 8.21

Use the Boundary_Hatch tool to create the section lining, as illustrated in figure 8.21

Command: *Choose the Save button on the toolbar*   Saves the drawing

Figure 8.21

*The cylindrical
break lines.*

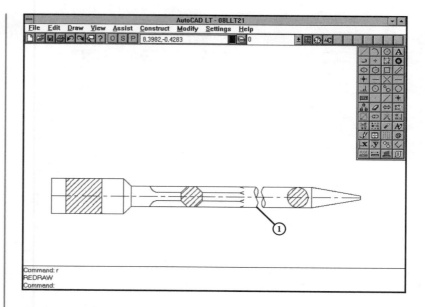

Creating Auxiliary Views

There are some parts whose shapes or surfaces will not be parallel to the normal viewing planes. These surfaces are known as *oblique planes.* The true shape of these oblique planes is never shown in any of the standard multiview drawings. Therefore, it is necessary to create a new picture plane that is parallel to the oblique surface. The object lines of the part are projected perpendicular to this plane to create the oblique surface. These drawings are called *auxiliary views.*

The subject of auxiliary views can be rather confusing if you are not up to speed on the terminology. Most beginning drafters develop an understanding of the concept after working with the drawings and studying examples.

In order to construct the true shape, the oblique surface must be oriented perpendicular to one of the standard views. Using the example from earlier in the chapter, you will see that the oblique surface is currently perpendicular to the front view. This makes the construction of the auxiliary view quite simple. Study the illustration shown in figure 8.22.

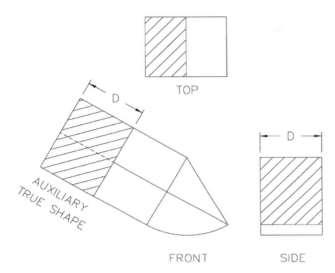

Figure 8.22

Developing the auxiliary view.

Using the OFFSET and LINE commands with a running osnap of ENDpoint and PERpendicular, the auxiliary views are easier to construct. The goal of this type of drawing is to find the true shape of an oblique plane.

Prior to the next exercise, you will need to develop the front, top, and right side orthographic views for the part illustrated in figure 8.23. Use this illustration as a reference while creating the part.

Take a few minutes and recreate the orthographic drawings shown in figure 8.24. Start a new drawing using the defaults. Save the drawing as BLOCK.DWG.

Now that the orthographic views are created, you will find the true shape of the triangular area. The main idea is to find at least two sides of the triangle that are displayed at true length. The true length of a line will occur on one of the orthographic views. For example, the top view depicts the true length of the line BC, whereas the right view displays line AC at its true length. If you view either of these lines in the front view, you will see a distorted or foreshortened view of each line. This identification of true length will take practice and can be much more complicated than the example here. Begin the exercise by creating an auxiliary plane parallel to line BC.

Figure 8.23

The reference part.

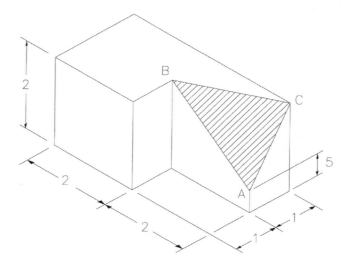

Exercise

Creating a True Shape

Command: *Choose* **C**onstruct, *then* **O**ffset	Issues the OFFSET command
Offset distance or Through <Through>: **1** (Enter)	
Select object to offset: *Pick line BC in the Top view*	Highlights the object
Side to offset?: *Pick anywhere down and to the right of BC*	Copies the line parallel
Select object to offset: *Right click on the mouse twice*	Reissues the OFFSET command
Offset distance or Through <1.0000>: **2** (Enter)	
Select object to offset: *Pick the new line just created*	

Side to offset?: *Pick anywhere*   Copies the line parallel
down and to the right of the new line

Zoom the view to Extents and set the
running osnap to ENDpoint

Command: *Choose the Line button*   Issues the LINE command

From point: *Pick on A in the Top view*

To point: *Pick the Perpendicular tool*

PER to *Pick on* ① *(see fig. 8.24)*

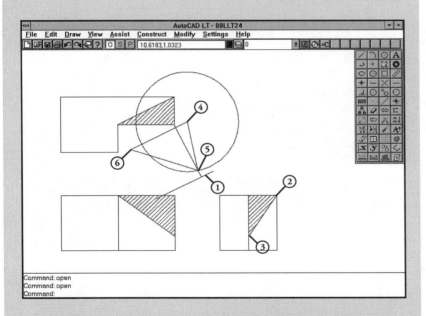

Figure 8.24

Developing the
auxiliary view.

To point: *Choose the Distance button* Issues the DISTANCE command

DIST First point: *Pick on C or*
② *in the Right view*

Second point: *Pick on A or* ③ *in*
the Right view

Distance = 1.8028, Angle in the XY Plane
= 2356, Angle from the XY Plane = 0

continues

X = 1.0000, Y = 1.5000, Z = 0.0000

Command: *Choose the Circle button*   Issues the CIRCLE command

CENTER 3P/TTR/<Center of   Sets the center point
circle>: *Pick on* ④

Diameter/<Radius>: **1.8028** (Enter)

Command: *Choose the Line button*   Issues the LINE command

From point: *Pick on* ④

To point: *Choose the Intersection*   Sets the object snap to INTersection
button

INT of *Pick on* ⑤

To point: *Pick on* ⑥

Zoom to Extents

*Clean up the drawing and add
hatching to the triangle to show the
true shape*

Save the drawing (see fig. 8.25)

Figure 8.25

*The completed
auxiliary view.*

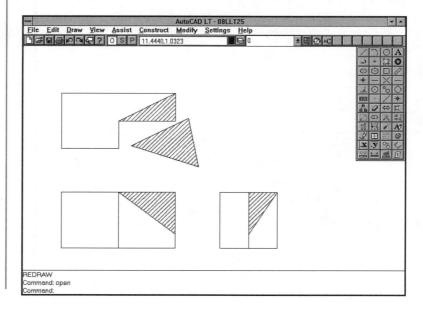

Again, there are a variety of methods used to establish an auxiliary view. The previous exercise will help you to identify and understand the concept behind such a view.

Identifying the Types of Auxiliary Views

When looking at the two-dimensional front view of any part, the distance along the x axis is referred to as the *width*. The distance along the y axis is referred to as the *height*, and the distance you cannot see in the front view is the *depth* of the part or the z axis.

The top view will show the x distance (width) and the z distance (depth). Likewise, the right view will show height and depth, or y and z.

NOTE Do not become confused by the X, Y, and Z coordinates that are discussed here. They are not AutoCAD LT 3D coordinates; they are the 2D equivalents. Although they reference the 3D construction, they are still two-dimensional views.

Auxiliary views are labeled based on the principal dimension of the part, used in the view. The view developed in the last example would be described as a height auxiliary because it was projected from the top view. Any auxiliary view projected from an orthographic view will display the dimension not shown in the view. The top view does not show height; therefore, the auxiliary view will show the true height of the part.

Understanding the Auxiliary Types

There are three types of auxiliary drawings; an infinite number of drawings can be generated from these types. The types are established by the dimension, which stays constant in any of the auxiliary views. The three types and their descriptions are as follows:

◆ **Height auxiliary views.** If you were to continue drawing the whole part from the last exercise, you would quickly see that the height is the only consistent dimension in the drawing. The only dimension missing from the Top view is the height. Any auxiliary view that is projected off the Top view will show the height dimension. These drawings are titled *height auxiliary views*.

◆ **Depth auxiliary views.** The only dimension missing from the Front view is the depth. Any auxiliary views projected from the Front view will show the depth as a true dimension.

◆ **Width auxiliary views.** The only dimension missing from the Right view is the width. Any auxiliary views projected from the Right view will show the width as a true dimension.

All auxiliary views can be placed into one of these categories. In the next section, you explore the method of drawing a curved surface in an auxiliary view.

NOTE

Although an infinite number of views can be generated from the orthographic views, the purpose of an auxiliary view is to fully describe the part. Establishing the true shape of an element does not require lots of auxiliary views.

Creating Curves in Auxiliary Views

A cylinder that has been cut by an inclined plane produces another set of problems. An auxiliary view is required to find its true shape. A cylinder that has been cut this way will be an ellipse in true shape. The next two exercises take you through the process of creating the auxiliary view of a cylinder sliced by an inclined plane.

Exercise

Creating an Auxiliary View of a Cylinder

Start a new drawing and accept all the defaults.

Command: *Choose the Circle button*	Issues the CIRCLE tool
CIRCLE 3P/TTR/<Center of circle>: **7,2** (Enter)	Sets the center point
Diameter/<Radius>: **1** (Enter)	Creates the Right view
Command: *Choose the Line button*	Issues the LINE command
LINE From point: **.5,1** (Enter)	Sets the first point
To point: **@2<90** (Enter)	
To point: **@1<0** (Enter)	

`To point: @3.2148<321` (Enter)	
`To point: C` (Enter)	Creates the Front view
`Command:` *Choose* **C**onstruct, *then* **O**ffset	Issues the OFFSET command
`Offset distance or Through <Through.5000>: 2` (Enter)	Sets the offset distance
`Select object to offset:` *Pick on* ① *(see fig. 8.26)*	
`Side to offset?` *Pick on* ②	Sets the center line of the auxiliary view
`Select object to offset:` *Right click on the mouse twice*	Reissues the OFFSET command
`Offset distance or Through <2.000>: 1` (Enter)	Sets the offset distance to 1
`Select object to offset:` *Pick on* ②	
`Side to offset?` *Pick on* ①	Sets half of the depth
`Select object to offset:` *Pick on* ②	
`Side to offset?` *Pick on* ③	Sets other half of the depth
`Select object to offset:` *Choose the Ellipse button*	Issues the ELLIPSE command
`<Axis endpoint1>/Center: C` (Enter)	Specifies the center option
`Center of ellipse:` *Choose the Midpoint button*	Sets the center
`MID of` *Pick on* ②	
`Axis endpoint:` *Choose the Endpoint button*	
`END of` *Pick on* ④	

continues

<Other axis distance>/Rotation: Issues the MIDpoint object snap
Choose the Midpoint button

MID of *Pick on* ③   Creates the true shape

Save the drawing as CIRC_AUX.DWG
in the ACLTWIN directory

Now that the true shape of the inclined plane is illustrated, proceed with the exercise by constructing the rest of the cylinder and cleaning up the drawing.

Exercise

Completing the Auxiliary View of the Cylinder

Continue working on the CIRC_AUX.DWG. Set ENDpoint as a running object snap.

Command: *Choose the Line button*   Issues the LINE command

LINE From point: *Pick on* ⑤
(see fig. 8.26)

To point: *Choose the Perpendicular*   Issues the PERpendicular object
button   snap

Figure 8.26

Developing the auxiliary view.

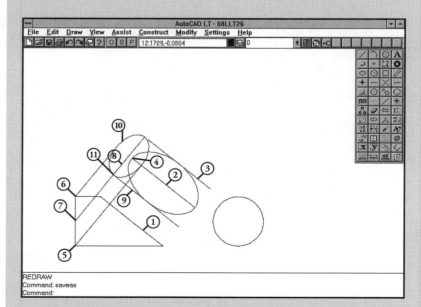

`PER to` *Pick on* ②	
`To point:` *Right click on the mouse two times*	Reissues the LINE command
`LINE From point:` *Pick on* ⑥	
`To point:` *Choose the Perpendicular button*	Issues the PERpendicular object snap
`PER to` *Pick on* ②	Sets half the depth
`To point:` *Right click on the mouse two times*	Reissues the LINE command
`LINE From point:` *Choose the Endpoint button*	
`END of` *Pick on* ⑦	
`To point:` *Choose the Perpendicular button*	Issues the PERpendicular object snap
`PER to Pick on` ②	Sets the other half of the depth
`To point:` *Choose* **M**odify, *then* **E**xtend	Issues the EXTEND command
`Select boundary edge(s)` `Select objects:` *Pick on* ⑧	
`Select objects:` *Right click on the mouse*	Accepts the selection set
`<Select object to extend>` `/Undo:` *Pick on* ⑨	Creates an axis point
`<Select object to extend>` `/Undo:` *Choose the Ellipse button*	Issues the ELLIPSE command
`<Axis endpoint1>/Center:` **C** (Enter)	
`Center of ellipse:` *Pick on* ⑧	Sets the center
`Axis endpoint:` *Pick on* ⑩	
`<Other axis distance>` `/Rotation:` *Pick on* ⑪	

continues

Use the TRIM, CHANGE, and
ERASE commands to create the
drawing illustrated in figure 8.27

Save the drawing

Figure 8.27

*The completed
cylinder auxiliary
view.*

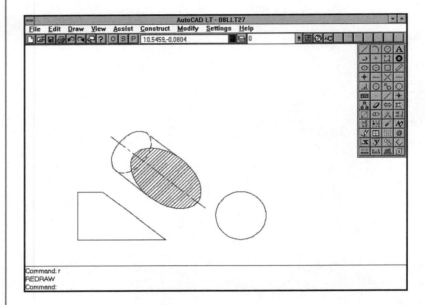

Auxiliary views are quite complicated, and any seasoned draftsperson will
verify this fact or will wax eloquently about creating these drawings in a
descriptive geometry class. CAD techniques make most of the more difficult
drawing constructions easy and more accurate.

Summary

In this chapter, you were introduced to the concept of section and auxiliary views. Within these two types of drawings, you explored the following concepts:

- ◆ The cutting plane and where it should be placed within a drawing

- ◆ Proper line types and their weights in section drawings

- ◆ Section lining and how to use the BHATCH command to create poché within the section drawing

- ◆ The difference between a full and half section

- ◆ Why you would have to develop a broken-out section

- ◆ The ability to use a revolved section to illustrate the shape of the part at the section line

- ◆ How to create and define a removed section

- ◆ The concept of an aligned section and how to use break lines

- ◆ The definition of an auxiliary view and what it can illustrate to the machinist

- ◆ The three types of auxiliary drawings: height, width, and depth

- ◆ How to create a cylindrical auxiliary view

Be prepared to repeat this chapter at a later date after you have had an opportunity to utilize some of the skills learned in this book. Many of these concepts were presented to drafters and machinists during their education or apprenticeship, only to be forgotten or neglected.

In the next chapter, you will learn about the benefits of creating pictorial drawings, which will help to convey your ideas to everyone.

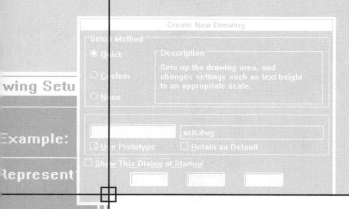

In this chapter, you explore the techniques used to represent objects with pictorial drawings. These drawings have a distinct characteristic of creating a three-dimensional quality. The following topics are explored:

- ◆ Developing axonometric projections
- ◆ Understanding isometric drawings
- ◆ Developing dimetric drawings
- ◆ Creating trimetric drawings
- ◆ Creating assembly drawings
- ◆ Understanding oblique drawings

These drawings are a natural extension to the ideas formulated by orthographic drawing techniques. AutoCAD LT will offer some invaluable tools to create these drawings with ease. Although these types of drawings are considered advanced, you can develop the ability to mentally translate orthographic drawings to the third dimension quickly by honing this drawing technique.

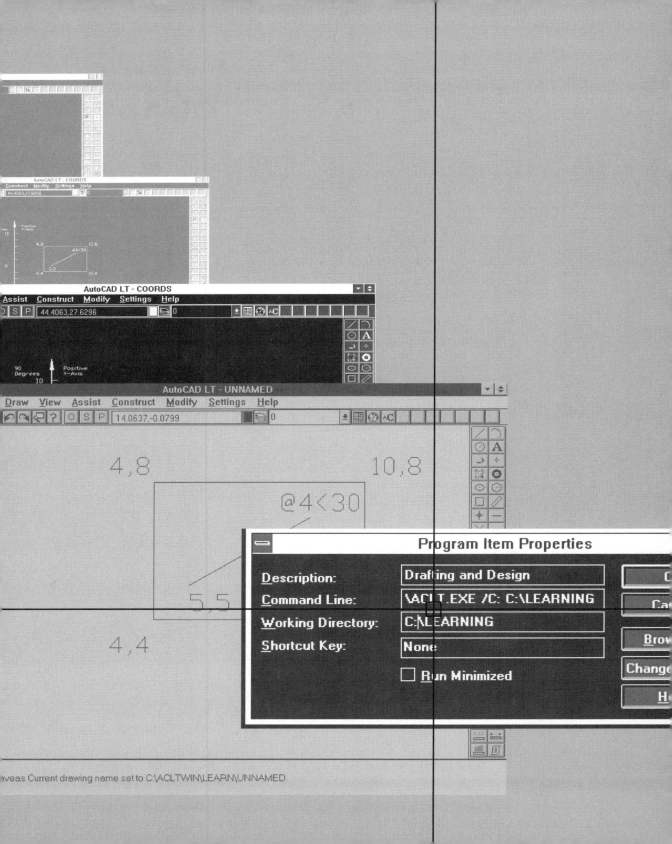

Exploring Paraline Projection

The ability to draw an object with three-dimensional quality will add a great deal of enjoyment and satisfaction to your CAD experience. Many people who lack technical training have a hard time interpreting drawings that are drawn orthographically. As soon as the drawing is projected to the third dimension, you will immediately see the impact of this method.

Paraline projection can be accomplished in a variety of ways, depending on the geometry and the type of drawing needed. Paraline projection should not be confused with perspective drawing. Perspective drawings taper off to vanishing points, whereas paraline drawings maintain true height. Also, do not confuse this method of drawing with 3D. The drawing is considered pictorial in nature because you can see all three dimensions (height, width, and depth) at the same time. A true 3D AutoCAD LT drawing gives each entity's defining points (like an endpoint or vertex) a third coordinate (dimension) in the Cartesian grid. This third, or Z, coordinate describes the point's height above or below the XY grid. Although AutoCAD LT is capable of creating 3D drawings, this book is designed to be an introductory text and will not cover the complexity of 3D drawing.

Pictorial drawings are not typically used to show complex sections or parts, but can be quite useful to describe the overall geometry of the object and help a machinist or builder visualize the part as they interpret the orthographic drawings. Pictorial drawings are also used in catalogs, manuals, and other marketing literature.

Developing Axonometric Projections

The most recognizable difference between pictorial or axonometric drawings and multiview drawings is the rotated position of the object in reference to the picture plane. Figure 9.1 illustrates that orthographic drawings are parallel to the picture plane, whereas axonometric drawings are created at an angle to this plane.

Figure 9.1

Defining the axonometric projection.

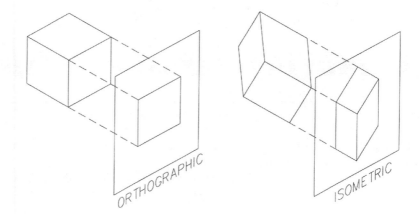

Axonometric drawings have three popular viewing angles: isometric, dimetric, and trimetric. The following sections describe the unique qualities of each type.

Understanding Isometric Projection

The term "isometric" actually means equal measure. When constructing the view, the angles created by the main axes are equal angles to the plane of projection. Interestingly enough, if you look at figure 9.1, the true length of the axes is never shown. All of the sides have lengths that are foreshortened. Figure 9.2 displays the axes angles in relation to the picture plane.

Figure 9.2

An isometric projection diagram.

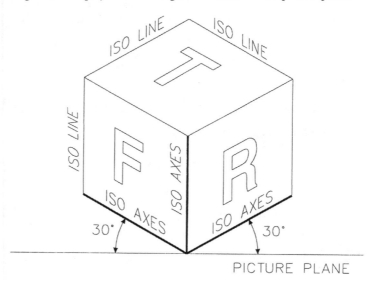

Each angle is conventionally drawn at 30° to the picture plane. These lines are referred to as the *isometric axes*. Any line that is parallel is described as an *isometric line*. Any lines created that are non-parallel are *non-isometric lines*.

Working with Axes

The ability to manipulate the three major drawing axes is a necessary step to understanding any 3D representation. Isometric drawings display these three axes at the same time. Although they are rotated 30° to the picture plane, the F (Front) and R (Right side) are used to describe the geometry along the width and depth of the part, whereas the height is determined along the vertical axis, as displayed in figure 9.2.

The orthographic drawings that will be used to create the isometric part are shown in figure 9.3.

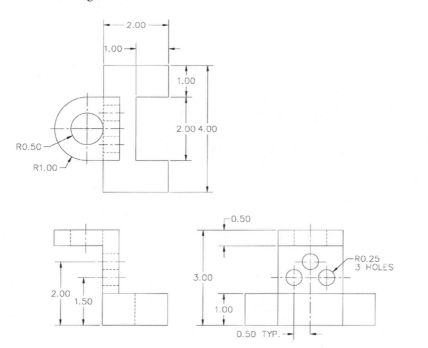

Figure 9.3

A table support bracket.

All of the lines must be drawn or duplicated to an angle of 30, 150, 210, or 330. In order to create an isometric drawing, AutoCAD LT has some built-in controls to ease the drawing setup.

In the Drawing Aids dialog box, you can set the Snap as Isometric. This check box will toggle the crosshairs to an isometric orientation. There are also buttons to determine the isometric face in which you would like to work. Although you can set the crosshairs orientation within the dialog box, the preferred method is Ctrl+E.

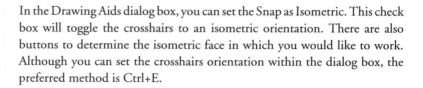

TIP

You can toggle between the three isometric settings by pressing Ctrl+E. This is much quicker than opening the Drawing Aids dialog box.

There are three radio buttons, which describe the three faces. Figure 9.2 displays the right and top views; the left view is the same as the front, which is shown as F in the figure. These three views can be accessed by pressing Ctrl+E. If you continue pressing Ctrl+E, you will cycle through the three faces.

As you gain experience with this drawing method, it comes more naturally. Begin this exercise by setting up the isometric drawing of the table bracket.

Exercise

Setting Up the Isometric Drawing

Start a new drawing and accept the defaults.

Command: *Choose* **S***ettings, then* **D***rawing Aids*	Opens the Drawing Aids dialog box
Choose the On *box under* **I***sometric Snap/ Grid, then* OK	Closes the dialog box and returns to the drawing area
The crosshairs are now in isometric	
Press Ctrl+E to toggle between the Top, Left, and Right planes	
Set the running osnap to ENDPoint	
Command: *Choose the Line button*	Issues the LINE command
LINE From point: *Pick anywhere near the bottom of the drawing area*	
To point: @4<30 (Enter)	Sets the z axis

`To point:` *Right click on the mouse twice*	Reissues the LINE command
`LINE From point:` *Pick on* ① *(see fig. 9.4)*	
`To point:` `@3.5<150` `Enter`	Sets the x axis
`To point:` *Right click on the mouse twice*	Reissues the LINE command
`LINE From point:` *Pick on* ①	
`To point:` `@3<90` `Enter`	Sets the y axis
`To point:` *Choose the Copy button*	Issues the COPY command
`COPY Select objects:` `L` `Enter`	Selects the last object drawn
`Select objects:` *Right click on the mouse*	Reissues the COPY command
`<Base point or displacement> /Multiple:` `M` `Enter`	
`Base point:` *Pick on* ①	
`Second point of displacement:` *Pick on* ②	Creates a vertical boundary box edge
`Second point of displacement:` *Pick on* ③	Creates a vertical boundary box edge
`Second point of displacement:` *Right click on the mouse twice*	Reissues the COPY command
`COPY Select objects:` *Pick on* ①③	Highlights the objects
`Select objects:` *Right click on the mouse*	Accepts the selection set
`<Base point or displacement>/ Multiple:` *Pick on* ①	
`Second point of displacement:` *Pick on* ④	Creates the top of the boundary box

Use the Copy button to finish the boundary box, as illustrated in figure 9.4

Zoom the view using the All option and save the drawing

Figure 9.4

The boundary box.

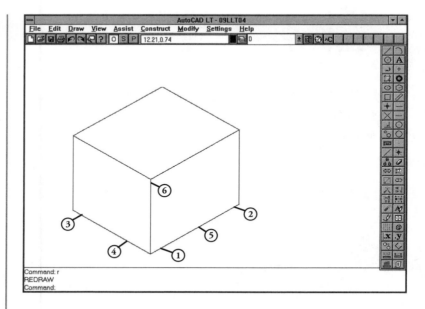

Working with the Scale

The scale of this drawing will be unique. Each of the sides is measured on a true scale. As mentioned earlier, this is not quite correct because the isometric rotation reduces the true lengths by *foreshortening* them. Foreshortening occurs when lines are not drawn to their true lengths because they are at an angle to the picture plane. To illustrate this point, hold a pencil between your thumb and first finger in front of you. Hold the pencil parallel to your shoulders (right and left). Your face is the picture plane. Slowly spin the tip toward your nose. As you turn it, the length appears to become shorter. If you were to draw the pencil relative to the picture plane, it would be drawn shorter than its true length.

If you were to draw a part with true isometric projection, all of the dimensions would be 82 percent of their actual size to accommodate this foreshortening. Manual drafters use an isometric scale in order to draw a true isometric projection. The benefits of creating an isometric that has been reduced to 82 percent of its true size do not outweigh the problems it creates while trying to be productive. It is therefore acceptable to use a true scale measurement when creating these drawings. An isometric that has been scaled to 82 percent can be described as an isometric *projection,* while the drawings created with a true scale are called isometric *drawings.*

The next exercise will continue to create the part using construction lines. Keep in mind that the techniques used are shortcuts toward completing the drawing.

Developing the Base of the Isometric Drawing

Exercise

Continue working with the previous drawing.

Command: *Choose the Copy button*	Issues the COPY command
Turn the running ENDpoint osnap off	
COPY Select objects: *Pick on* ①② *(see fig. 9.5)*	Highlights the objects
Select objects: *Right click on the mouse*	
<Base point or displacement> /Multiple: *Choose the Endpoint button*	
END of *Pick on* ①	Selects the ENDpoint
Second point of displacement: @1<90 (Enter)	
Second point of displacement: *Right click the mouse twice*	Reissues the COPY command
COPY Select objects: *Pick on* ③	Highlights the object
Select objects: *Right click on the mouse*	
<Base point or displacement>/ Multiple: M (Enter)	
Base point: *Choose the Endpoint button*	Uses the ENDPoint osnap to select a basepoint
END of *Pick on* ①	
Second point of displacement: @1<30 (Enter)	Copies the object

continues

```
Second point of displacement:          Copies the object
@3<30 (Enter)

Second point of displacement:          Copies the object
@2<150 (Enter)

Second point of displacement:          Reissues the COPY command
Right click on the mouse twice

COPY Select objects: Pick on ④        Highlights the object

Select objects: Right click on
the mouse

<Base point or displacement>
/Multiple: Choose the Endpoint button

END of Pick on ④                       Uses the ENDPoint osnap as
                                       a basepoint

Second point of displacement:          Copies the object
@2<150 (Enter)

Second point of displacement:          Ends the COPY command
(Enter)

Command: Choose the Save button on the  Opens the Save Drawing As
toolbar                                 dialog box
```

Save as BRACKET.DWG

Exercise

Developing More of the Base

Continue working with the previous drawing.

```
Command: Choose the Copy button        Issues the COPY command

COPY Select objects: Pick on ⑤        Highlights the object
(see fig. 9.5)

Select objects: Right click on the mouse

<Base point or displacement>
/Multiple: M (Enter)

Base point: Choose the Endpoint button
```

END of *Pick on* ⑤	Uses the ENDPoint osnap as a basepoint
Second point of displacement: @1<30 (Enter)	Copies the object
Second point of displacement: @3<30 (Enter)	Copies the object
Second point of displacement: @4<30 (Enter)	Copies the object
Second point of displacement: *Right click on the mouse twice*	Reissues the COPY command
COPY Select objects: *Pick on* ⑥	Highlights the object
Select objects: *Right click on the mouse*	
<Base point or displacement> /Multiple: M (Enter)	
Base point: *Choose the Endpoint button*	
END of *Pick on* ⑥	Uses the ENDPoint osnap as a basepoint
Second point of displacement: @1<150 (Enter)	Copies the object
Second point of displacement: @2<150 (Enter)	Copies the object
Second point of displacement: *Right click on the mouse*	
Command: *Choose the Save button on the toolbar*	Saves the drawing

Figure 9.5

Adding construction lines.

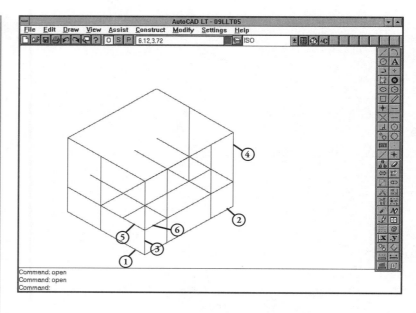

By this point, you have drawn several construction lines. The next step in the exercise is to create some forms to solidify the base of the object.

Exercise

Solidifying the Base

Continue with the BRACKET.DWG.

Command: *Choose the Layers button*	Opens the Layer Control dialog box
Type **ISO** *and pick the New button*	
Highlight the new layer and set the color to RED, then make this layer current	
Set the running osnap to INTersection	
Command: *Choose the Line button*	Issues the LINE command
LINE From point: *Pick on* ① (*see fig. 9.6*)	Sets the first point
To point: *Pick on* ②	Sets the second point and so on

To point: *Pick on* ③

To point: *Pick on* ④

To point: *Pick on* ⑤

To point: *Pick on* ⑥

To point: *Pick on* ⑦

To point: *Pick on* ⑧

To point: **C** (Enter)   Forms the top of the base

Return the running osnap to NONE

Command: *Choose the Copy button*   Issues the COPY command

Select objects: *Choose the*   Issues the DDLMODE
Layers button   command

Pick Select **A**ll, *then* Free**z**e, *then* OK   Freezes all but the current
                                                                  layer

Select objects: *Pick on* ⑨⑩⑪
⑫⑬

Select objects: *Right click on*   Accepts the selection set
the mouse

<Base point or displacement>
/Multiple: *Choose the Endpoint button*

END of *Pick on* ⑫

Second point of displacement: Copies the object
@1<270 (Enter)

*Set the osnap to ENDpoint and use the Line button
to continue the drawing, as shown in figure 9.6*

Thaw all of the layers and save the drawing

Figure 9.6

The base drawing.

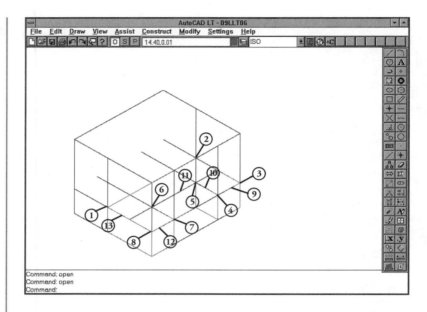

The base of the table support bracket is essentially complete. The construction lines will be used throughout the parts development. You may need to toggle this layer on and off as the drawing gets more complex. In the next exercise, you begin to lay out the arm of the bracket.

Exercise

Laying Out the Arm

Continue with the BRACKET.DWG.

Command: *Choose the Copy button*   Issues the COPY command

Select objects: *Pick on* ①   Highlights the object
(see fig. 9.7)

Select objects: *Right click on the mouse*

<Base point or displacement>
/Multiple: *Pick anywhere in the drawing
area that is not on the object*

Second point of displacement:
@.5<150 (Enter)

Second point of displacement: *Choose the Copy button*	Reissues the COPY command
Select objects: *Pick on* ②③④⑤	
Select objects: *Right click on the mouse*	Accepts the selection set
<Base point or displacement> /Multiple: *Pick anywhere that is not on the object*	Issues the ENDPoint osnap
Second point of displacement: **@2<90** (Enter) *Endpoint button*	Issues the ENDPoint osnap
Second point of displacement: *Choose the Line button*	Issues the LINE command
Set a running osnap of INTersection	
LINE From point: *Pick on* ⑥	Sets first point
To point: *Pick on* ⑦	Sets second point
To point: *Choose the Copy button*	Issues the COPY command
Select objects: **L** (Enter)	Uses the Last selection option
Select objects: *Right click on the mouse*	Accepts the selection set
<Base point or displacement> /Multiple: **M** (Enter)	Uses the Multiple option
Base point: *Pick on* ⑧	
Second point of displacement: *Pick on* ⑨⑩	
Freeze all the layers	
Set the osnap to ENDpoint and use the Line button to continue the drawing, as shown in figure 9.7	
Save the drawing	

Figure 9.7

Developing the arm.

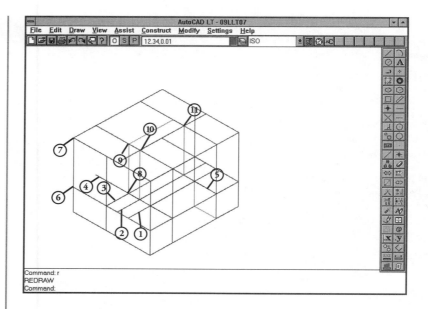

Creating Circles in Isometric Drawings

Constructing curves and circles in isometric drawings was once a laborious task. In order to construct a circle in any of the three planes, you had to follow these steps (see fig. 9.8):

1. Draw the circle as a square whose sides represent the diameter of the circle.

2. Draw lines from the two largest angles perpendicular to the two opposite sides.

3. Draw two arcs from the largest angles with a radius equal to the length of the line just drawn.

4. Draw two circles whose center points are at the intersection of the perpendicular lines that were drawn.

5. Clean up the drawing.

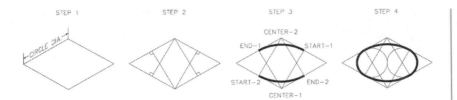

Figure 9.8

Steps for creating an isometric ellipse in manual drafting.

Fortunately, AutoCAD LT has a tool that creates an ellipse in any of the three planes. This type of ellipse is called an *isocircle*. It is available as an option within the ELLIPSE command and will be used to create the arc on the arm of the table bracket. The ELLIPSE command can be accessed in the following ways:

◆ Choose the Ellipse button in the toolbox

◆ Choose **D**raw, then **E**llipse

◆ Type **ELLIPSE** or **EL** at the Command: prompt

All of these methods will issue the ELLIPSE command. The most accessible method is from the toolbox.

After issuing the ELLIPSE command from the menu or the toolbox, the Isocircle option will not be available unless it has been marked in the Drawing Aids dialog box.

In the next exercise, use an isocircle to create the circles and arcs in the table support bracket arm.

Drawing Isometric Circles

Exercise

Continue with the BRACKET.DWG.

Thaw all layers

Command: *Choose the Copy button*	Issues the COPY command
Select objects: *Pick on* ① (*see fig. 9.9*)	
Select objects: *Right click on the mouse*	Accepts the selection set

continues

Figure 9.9

Developing the isometric circles.

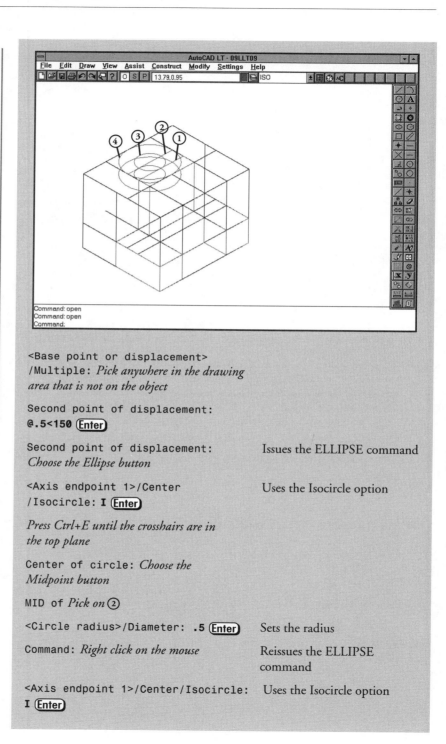

```
<Base point or displacement>
/Multiple: Pick anywhere in the drawing
area that is not on the object

Second point of displacement:
@.5<150 Enter

Second point of displacement:          Issues the ELLIPSE command
Choose the Ellipse button

<Axis endpoint 1>/Center               Uses the Isocircle option
/Isocircle: I Enter

Press Ctrl+E until the crosshairs are in
the top plane

Center of circle: Choose the
Midpoint button

MID of Pick on ②

<Circle radius>/Diameter: .5 Enter     Sets the radius

Command: Right click on the mouse      Reissues the ELLIPSE
                                       command

<Axis endpoint 1>/Center/Isocircle:    Uses the Isocircle option
I Enter
```

```
Center of circle: Choose the
Midpoint button

MID of Pick on ②

<Circle radius>/Diameter: 1 (Enter)          Sets the radius

Command: Choose the Copy button            Issues the COPY command

Select objects: Pick on ③④               Highlights the objects

Select objects: Right click on the mouse   Accepts the selection set

<Base point or displacement>
/Multiple: Pick anywhere in the drawing
area that is not on the object

Second point of displacement:
@.5<270 (Enter)

Second point of displacement:             Ends the COPY command
Right click on the mouse
```

Freeze all layers and clean up the drawing, as illustrated in figure 9.10

Save the drawing

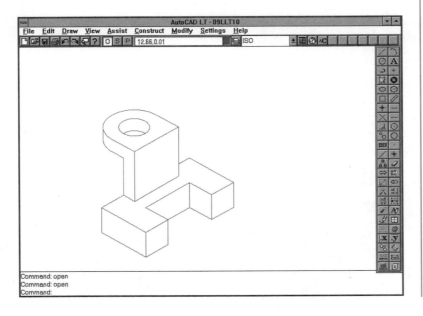

Figure 9.10

Cleaning up the isometric circles.

Continue the exercise by adding the three holes in the arm using the right isometric plane. It is important to know the center point location of the circles in order to create construction lines.

TIP

It is essential to keep construction lines for ellipses within the drawing—unlike a circle, AutoCAD LT cannot locate a center for an ellipse.

Exercise

Creating More Isometric Circles

Continue with the BRACKET.DWG.

Command: *Choose the Copy button*	Issues the COPY command
COPY Select objects: *Pick on* ① *(see fig. 9.11)*	Highlights the object

Figure 9.11

Developing the screw holes.

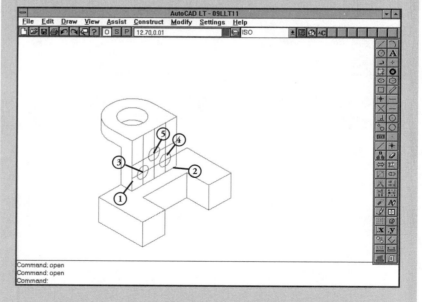

Select objects: *Right click on the mouse*	Accepts the selection set
<Base point or displacement> /Multiple: **M** (Enter)	Uses the Multiple option

`Base point:` *Pick anywhere in the drawing area that is not on the object*	
`Second point of displacement:` `@.5<30` (Enter)	Copies the object
`Second point of displacement:` `@1<30` (Enter)	Copies the object and so on
`Second point of displacement:` `@1.5<30` (Enter)	
`Second point of displacement:` *Right click on the mouse twice*	Reissues the COPY command
`COPY Select objects:` *Pick on* ②	Highlights the object
`Select objects:` *Right click the mouse*	Accepts the selection set
`<Base point or displacement>` `/Multiple:` `M` (Enter)	Uses the Multiple option
`Base point:` *Pick anywhere in the drawing area that is not on the object*	
`Second point of displacement:` `@.5<90` (Enter)	Copies the object
`Second point of displacement:` `@1<90` (Enter)	Copies the object
`Second point of displacement:` *Pick on the Ellipse tool*	Issues the ELLIPSE tool
`<Axis endpoint 1>/Center` `/Isocircle:` `I` (Enter)	
Press Ctrl+E until the crosshairs are in the right plane	
`Center of circle:` *Choose the Intersection button*	Uses the INTersection osnap
`INT of` *Pick on* ③	
`<Circle radius>/Diameter:` `.25` (Enter)	Sets the radius
`Command:` *Right click on the mouse*	Reissues the ELLIPSE command

continues

```
<Axis endpoint 1>/Center
/Isocircle: I (Enter)

Center of circle: Choose the          Uses the INTersection osnap
Intersection button

INT of Pick on ④

<Circle radius>/Diameter:            Sets the radius
.25 (Enter)

Command: Right click on the mouse

<Axis endpoint 1>/Center
/Isocircle: I (Enter)

Center of circle: Choose the          Uses the INTersection option
Intersection button

INT of Pick on ⑤

<Circle radius>/Diameter:
.25 (Enter)
```

*Clean up the drawing, as illustrated
in figure 9.12*

Save the drawing

Figure 9.12

The table bracket.

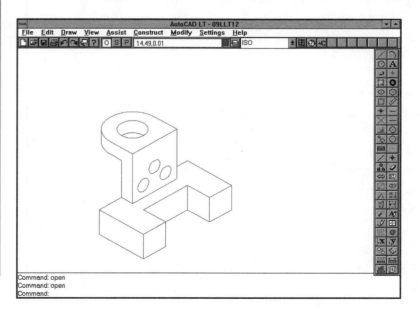

Creating Isometric Text

Another useful tool available in AutoCAD LT is the ability to give text an obliquing angle. This means you can tilt the letters uniformly—this is best described through an exercise. You will add some text to the table bracket in the next exercise.

Creating Isometric Text

Exercise

Continue with the BRACKET.DWG.

Command: *Choose* **S**ettings, *then* **T**ext Style	Opens the Select Text Font dialog box
Double click on the Romans Simplex box	
Style name (or ?)<STANDARD>: romans New style Font file <txt>:romans Height <0.00>: (Enter)	These settings establish the font appearance
Width factor<1.00>: (Enter)	
Obliquing angle <0>: **30** (Enter)	
Backwards? <N>: (Enter)	
Upside-down? <N>: (Enter)	
Vertical? <N>: (Enter)	
Romans is now the current style	
Command: *Choose* **S**ettings, *then* **D**rawing Aids	Opens the Drawing Aids dialog box
Set the Snap Angle to 30, *and turn Isometric off*	
Command: *Choose the Text button*	Issues the DTEXT command
_DTEXT Justify/Style/<Start point>: *Pick near* ①	
Height <0.20>: .3 (Enter)	Sets the height
Rotation angle <30>: (Enter)	

continues

```
Text: TABLE BASE Enter

Text: Enter
```

Rotate this text string 60° in order to
create vertical text on the left plane

```
Command: Choose Settings, then              Opens the Select Text Font
Text Style                                   dialog box
```

Double click on the Romans Simplex box

```
Style name (or ?)<STANDARD>:romans          Establishes the font settings
New style
Font file <txt>:romans
Height <0.00>: Enter

Width factor<1.00>: Enter

Obliquing angle <0>: -30 Enter

Backwards? <N>: Enter

Upside-down? <N>: Enter

Vertical? <N>: Enter
```

Romans is now the current style

```
Command: Choose Settings, then Drawing      Opens the Drawings Aids
Aids                                          dialog box
```

Set the Snap Angle to -30

```
Command: Choose the Text button             Issues the DTEXT command

_DTEXT Justify/Style/<Start
point>: Pick near ②

Height <0.30>: Enter

Rotation angle <330>: Enter

Text: ISOMETRIC Enter

Text: Enter
```

Rotate this text string 60° in order to
create text on the top plane

Save the drawing (see fig. 9.13)

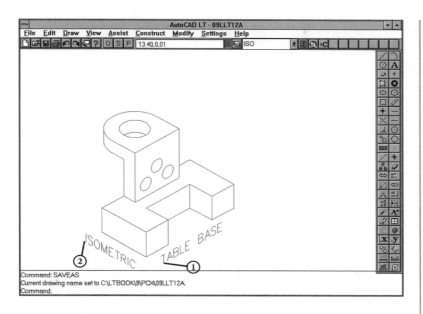

Figure 9.13

Adding isometric text to the drawing.

Use this style of text to label and dimension isometric drawings. It creates an interest to the drawing and follows the manual drafting convention.

Understanding Dimetric Projection

An isometric drawing is a version of a dimetric drawing. In a *dimetric drawing*, two of the axes angles are equal to each other, while the third is a greater or smaller angle (see fig. 9.14).

Isometric drawings are certainly the most popular form of pictorial drawing, but it may be necessary to construct a drawing dimetrically due to an object's geometry. Typically, a dimetric drawing will emphasize one of the faces more than the others. Therefore, this projection method could prove useful in your drawing arsenal.

Figure 9.14

A dimetric diagram.

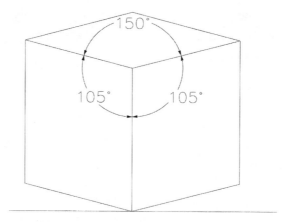

Working with the Scale

The scale of an dimetric drawing is based on the fact that the new angles present a different degree of foreshortening. The angles in an isometric drawing are equal so each axis is scaled the same. Similarly, the two angles that are the same are scaled equally, while the third axis is scaled differently. This third axis is typically scaled at 3/4 or 75 percent of full scale. Ultimately, however, the foreshortened distance will be dictated by the severity of the angle. Figure 9.15 displays some examples of the effect that the angle has on the scaling factor.

Understanding Symmetrical and Unsymmetrical Dimetric Drawings

As mentioned earlier, most dimetric drawings are drawn with equal angles to the picture plane. These drawings are known as *symmetrical dimetric drawings.* As long as these angles are equal, the right and front side of the view will be treated equally. The larger this angle, the more emphasis is placed on the top view. The smaller the angle, the more visible the side views (see 1 and 2 in fig. 9.15).

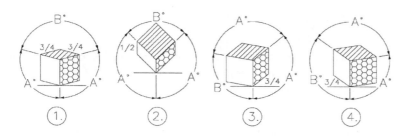

Figure 9.15

Examples of symmetrical and unsymmetrical dimetric drawings.

If the side view in a dimetric drawing needs the emphasis, then an unsymmetrical dimetric drawing will be constructed. This means that the angles from the picture plane are unequal, usually 45° and 15°. The angles created by the axes must continue to be equal in two cases and greater or smaller in the third (see 3 and 4 in fig. 9.15)

Try using this method to construct the table base. Figure 9.16 illustrates the table base as a symmetrical and unsymmetrical dimetric drawing.

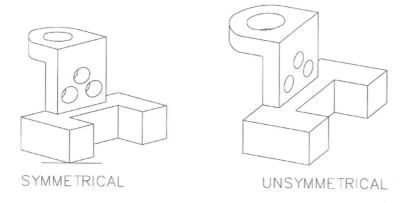

SYMMETRICAL   UNSYMMETRICAL

Figure 9.16

The table base as symmetrical and unsymmetrical dimetric drawings.

Understanding Trimetric Projection

The last type of axonometric projection is the trimetric drawing. None of the axis angles in a trimetric projection are equal, thus creating an infinite number of positions. None of the angles will be less than 90°; therefore, the drawing is described as three angles or trimetric. The typical method uses a 30/60/90 triangle (see fig. 9.17).

Working with the Scale

Each side of the trimetric drawing has a different degree of foreshortening. In order to scale a trimetric drawing, you must first establish how the scale will affect the drawing. Figure 9.17 illustrates the simplest method in which to construct a trimetric drawing.

Figure 9.17

Developing a trimetric drawing.

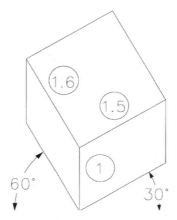

As you can see, the only length that will remain true scale will be the vertical axis. The other two axes will be scaled 1.5 and 1.6 times their original size.

Creating a Trimetric Drawing

With the scale in mind, the table base will be constructed in the same fashion as the dimetric drawing. Figure 9.18 illustrates the results of the table base drawn as a trimetric projection.

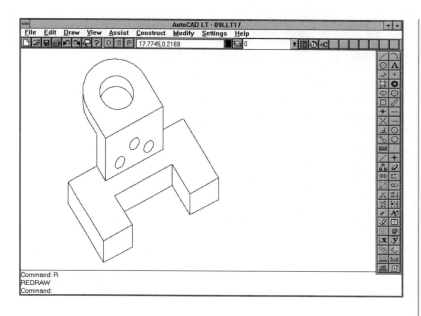

Figure 9.18

The trimetric table base.

Chapter

9

Understanding Assembly Drawings

Up to this point, you have been constructing drawings of individual parts. It is certainly easier to understand orthographic and axonometric concepts using a single object, but real-world circumstances are rarely so simple. Even though parts can be constructed on individual sheets, at some point the parts must be drawn as an assembled unit—this is called an *assembly drawing*. This drawing is constructed of several parts that represent the larger assembled object. These drawings can also be sub-assembly drawings of an even larger object. You can imagine the number of sub-assembly drawings it takes to fully detail an automobile, for example.

Assembly drawings can be quite complex, but they function as another method of communicating the idea of the object to the machinists or carpenters that are working with the drawings.

The assembly drawing is generated to show the relationship of the parts involved. These drawings can also include the techniques of sectioning with which you experimented earlier in the book, as well as all of the orthographic drawing techniques. Everything that is included in the drawing has the purpose of explaining the assembly. One of the greatest benefits of the CAD environment is the ability to take several ortho drawings, possibly created by different drafters, and combine them together in one assembly.

Another technique that is incorporated into an assembly drawing is *keynoting.* Keynotes are numbers or letters that are conventionally circled on the drawing. They are used to cross-reference other orthographic drawings within the set to keep the drawing from becoming too crowded with information.

Creating an Assembly Drawing

In this section, you combine your newfound axonometric skills with your orthographic drawing knowledge to create the exploded isometric drawing of figure 9.19.

Figure 9.19

A linking bar.

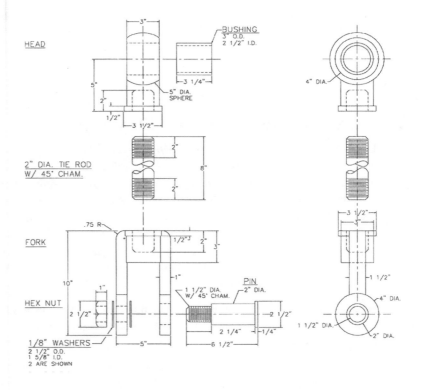

An exploded assembly is different in that the isometric is pulled apart to reveal how the individual objects are actually assembled. This is another drawing that will require great visualization skills, but the results are really quite striking.

For the purposes of this next exercise, the individual parts will be constructed as separate isometric drawings and finally combined together as one. You will be guided through the development of the head and fork, and are expected to clean up the drawings and create isometric drawings of the other parts shown in figure 9.19.

The drawings should be titled as follows:

◆ HEAD.DWG

◆ TIE-ROD.DWG

◆ FORK.DWG

◆ BUSHING.DWG

◆ PIN.DWG

◆ NUT-WASH.DWG

NOTE

Spheres that are drawn in isometric must have a diameter 1.22 times larger than its orthographic dimension in order to accommodate the foreshortening that occurs to the rest of the drawing.

The balance of the chapter will be used to create the isometric assembly of the linking bar. Each exercise will deal with a specific part. Pay attention to the notes given at the beginning of each exercise for direction. In the next exercise, you draw the head in isometric.

Creating the Isometric Head

Exercise

Begin the exercise by creating a new drawing.

Set the Snap style to Isometric and begin

Set the Isoplane to Top and choose the Line button to draw an isometric X on the screen

continues

Set the running osnap to INTersection and ENDpoint

Command: *Choose the Ellipse button*	Issues the ELLIPSE command
`<Axis endpoint 1>/Center /Isocircle:` **i** `Enter`	
`Center of circle:` *Pick on* ①	Sets the center of the ellipse
`<Circle radius>/Diameter:` **D** `Enter`	Selects diameter option
`Diameter:` **3.5** `Enter`	Creates 3 1/2" isocircle
Command: `Enter`	Reissues the ELLIPSE command
`<Axis endpoint 1>/Center /Isocircle:` **i** `Enter`	
`Center of circle:` *Pick on* ① *(see fig. 9.20)*	Sets the center of the ellipse
`<Circle radius>/Diameter:` **D** `Enter`	Selects the Diameter option
`Diameter:` **3** `Enter`	Creates a 3" isocircle
Command: *Choose the Copy button*	Issues the COPY command
`Select objects:` **ALL** `Enter`	Highlights the objects
`Select objects:` `Enter`	Accepts the selection set
`<Base point or displacement >/Multiple:` *Pick anywhere not on the object*	Sets the base point
`Second point of displacement:` **@.5<90** `Enter`	Copies the objects
`Second point of displacement:` *Choose the Copy button*	Issues the COPY command
`Select objects:` *Pick the X lines*	Highlights the objects
`Select objects:` `Enter`	Accepts the selection set

`<Base point or displacement>` `/Multiple:` *Pick anywhere not on* *the object*	Sets the base point
`Second point of displacement:` `@5<90` (Enter)	Sets the sphere center point
`Second point of displacement:` *Choose the Circle button*	Issues the CIRCLE command
`CIRCLE 3P/TTR/<Center point>:` *Pick on* ②	Sets the center point
`Diameter/<Radius>:` `D` (Enter)	Selects Diameter option
`Diameter:` `6.12` (Enter)	Creates an isosphere
`Command:` *Choose the Save button on* *the toolbar*	Opens the Save Drawing As dialog box

Save the drawing as HEAD.DWG

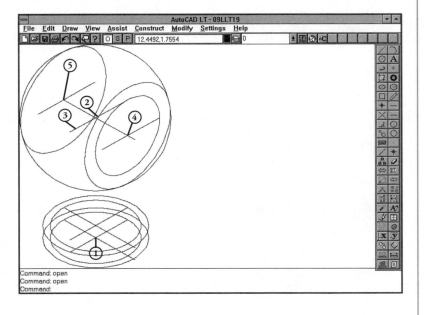

Figure 9.20

Developing the head.

Notice that the sphere created had a diameter 1.22 larger than the dimensioned piece. This is necessary because of the foreshortening that exists in an isometric drawing. AutoCAD LT will not automatically make this adjustment for you.

Exercise

Continuing to Create the Isometric Head

Continue working in the HEAD.DWG.

Command: *Choose the Copy button*	Issues the COPY command
Select objects: *Pick on* ③	Highlights the line
Select objects: `Enter`	Accepts selection set
<Base point or displacement> /Multiple: **M** `Enter`	Uses the Multiple option
Base point: *Pick anywhere not on the object*	
Second point of displacement: **@1.5<330** `Enter`	
Second point of displacement: **@1.5<150** `Enter`	
Second point of displacement: *Choose the Ellipse button* *Press Ctrl+E to change the Isoplane Right*	Issues the ELLIPSE command
<Axis endpoint 1>/Center /Isocircle: **i** `Enter`	
Center of circle: *Pick on* ④	Sets the center of the ellipse
<Circle radius>/Diameter: **D** `Enter`	Selects the Diameter option
Diameter: **4** `Enter`	Creates a 4" isocircle
Command: *Choose the Ellipse button*	Issues the ELLIPSE command
<Axis endpoint 1>/Center /Isocircle: **i** `Enter`	

Center of circle: *Pick on* ⑤	Sets the center of the ellipse
<Circle radius>/Diameter: **D** (Enter)	Selects the Diameter option
Diameter: **4** (Enter)	Creates a 4" isocircle
Command: *Choose the Ellipse button*	Issues the ELLIPSE command
<Axis endpoint 1>/Center /Isocircle: **i** (Enter)	
Center of circle: *Pick on* ④	Sets the center of the ellipse
<Circle radius>/Diameter: **D** (Enter)	Selects the Diameter option
Diameter: **3** (Enter)	Creates a 3" isocircle

Zoom Extents in the drawing area and save the drawing

Clean up the drawing to create a HEAD.DWG that looks like the part illustrated in figure 9.21

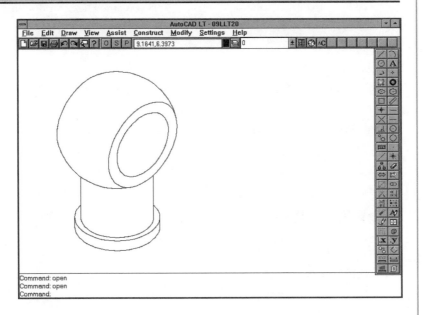

Figure 9.21

The completed head isometric.

In the next exercise, you lay out the FORK.DWG. The top of the fork will contain threads that link the fork to the head. The fork arms will drop down the side of the part and attach to isocircles at the bottom.

Exercise

Creating the Isometric Fork

Create a new drawing with the same parameters as those used for the HEAD.DWG.

Set the Isoplane to Top and use the Line button to draw an isometric X on the screen

Set the running osnap to INTersection and ENDpoint

Command: *Choose the Ellipse button*	Issues the ELLIPSE command
<Axis endpoint 1>/Center/Isocircle : **i** (Enter)	
Center of circle: *Pick on* ① *(see fig. 9.22)*	Sets the center of the ellipse
<Circle radius>/Diameter: **D** (Enter)	Selects the Diameter option
Diameter: **3.5** (Enter)	Creates a 3 1/2" isocircle
Command: (Enter)	Issues the ELLIPSE command
<Axis endpoint 1>/Center /Isocircle: **i** (Enter)	
Center of circle: *Pick on* ①	Sets the center of the ellipse
<Circle radius>/Diameter: **D** (Enter)	Selects the Diameter option
Diameter: **3** (Enter)	Creates a 3" isocircle
Command: (Enter)	Issues the ELLIPSE command
<Axis endpoint 1>/Center /Isocircle: **i** (Enter)	
Center of circle: *Pick on* ①	Sets the center of the ellipse

`<Circle radius>/Diameter:` `D` (Enter)	Selects the Diameter option
`Diameter:` **2** (Enter)	Creates a 2" isocircle
`Command:` *Choose the Copy button*	Issues the COPY command
`Select objects:` *Pick on* ②	Highlights the isocircle
`Select objects:` (Enter)	Accepts selection set
`<Base point or displacement>` `/Multiple:` *Pick anywhere not on the object*	Sets the base point
`Second point of displacement:` `@.5<270` (Enter)	
`Second point of displacement:` *Right click the mouse twice*	Reissues the COPY command
`COPY Select objects:` *Pick on* ③ *and the center lines*	Highlights the isocircle and lines
`Select objects:` (Enter)	Accepts selection set
`<Base point or displacement>` `/Multiple:` *Pick anywhere not on the object*	Sets the base point
`Second point of displacement:` `@3<270` (Enter)	
`Second point of displacement:` *Right click on the mouse twice*	Reissues the COPY command
`COPY Select objects:` *Pick on* ④	Highlights the line
`Select objects:` (Enter)	Accepts selection set
`<Base point or displacement>` `/Multiple:` **M** (Enter)	Selects the Multiple option
`Base point:` *Pick anywhere not on the object*	
`Second point of displacement:` `@2.5<150` (Enter)	

continues

```
Second point of displacement:
@2.5<330 (Enter)

Second point of displacement:              Reissues the COPY command
Right click on the mouse twice

COPY Select objects: Pick on ⑤           Highlights the line

Select objects: (Enter)                    Accepts selection set

<Base point or displacement>               Selects the Multiple option
/Multiple: M (Enter)

Base point: Pick anywhere not on
the object

Second point of displacement:
@.75<30 (Enter)

Second point of displacement:
@.75<210 (Enter)
```

Extend these lines and the original line
out to the first lines copied
(see fig. 9.22)

```
Command: Choose the Copy button           Issues the COPY command

COPY Select objects: Pick on ⑥           Highlights the line

Select objects: (Enter)                    Accepts selection set

<Base point or displacement>               Selects the Multiple option
/Multiple: M (Enter)

Base point: Pick anywhere not on the
object

Second point of displacement:
@.75<30 (Enter)

Second point of displacement:
@.75<210 (Enter)

Second point of displacement:             Issues the LINE command
Choose the Line button

LINE From point: Pick on ⑦               Sets the first point
```

| To point: **@8<270** (Enter) | Sets center point for lower isocircles |
| To point: (Enter) | Ends the LINE command |

Zoom Extents in the drawing area and save the drawing as FORK.DWG

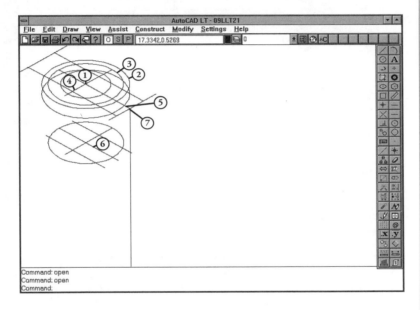

Figure 9.22

Developing the fork.

In the next section of the exercise, you work on the arms of the fork.

Creating the Isometric Fork Arms

Continue with the FORK.DWG.

Set the Isoplane to Right and set the running osnap to INTersection and ENDpoint

Command: *Choose the Ellipse button*   Issues the ELLIPSE command

continues

Exercise

Figure 9.23

Developing the fork arms.

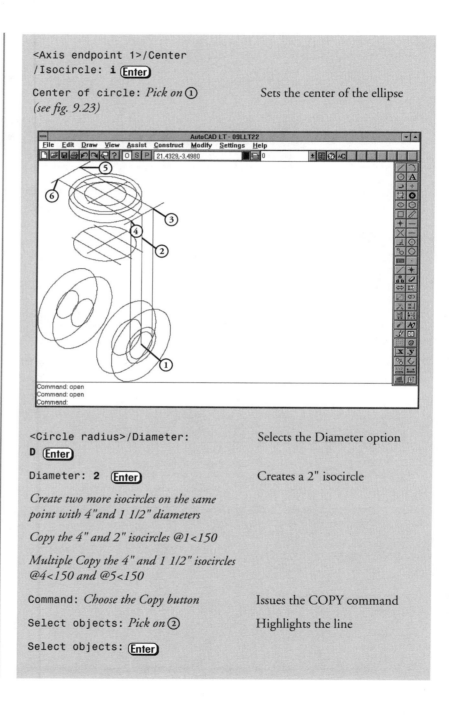

```
<Axis endpoint 1>/Center
/Isocircle: i Enter

Center of circle: Pick on ①      Sets the center of the ellipse
(see fig. 9.23)
```

```
<Circle radius>/Diameter:        Selects the Diameter option
D Enter
```

```
Diameter: 2 Enter                Creates a 2" isocircle
```

*Create two more isocircles on the same
point with 4" and 1 1/2" diameters*

Copy the 4" and 2" isocircles @1<150

*Multiple Copy the 4" and 1 1/2" isocircles
@4<150 and @5<150*

```
Command: Choose the Copy button   Issues the COPY command

Select objects: Pick on ②         Highlights the line

Select objects: Enter
```

```
Base point or displacement>
/Multiple: M (Enter)

Base point: Pick on ②

Second point of displacement:           Copies the lines
Pick on ③④⑤⑥

Second point of displacement:
(Enter)
```

*Clean up the FORK.DWG to the level
that is illustrated in figure 9.24*

*Zoom Extents in the drawing area and
save the drawing as FORK.DWG*

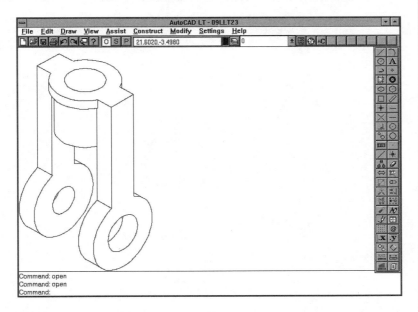

Figure 9.24

*Cleaning up the
fork arms.*

The last item to finish the drawing is the threads in the top portion of the fork. This will be constructed with the ARRAY command.

The ARRAY command enables you to copy objects in an XY pattern or around a center point. This is quite useful in a variety of applications. The two major options of the ARRAY command are described as Rectangular and

Polar. A rectangular ARRAY will prompt you for the number of rows (horizontal) and the number of columns (vertical), along with their distance from the original. If you specify four rows and four columns with equal distance, for example, the result will be 16 objects in a square pattern. If you choose the polar option, the objects will be copied around a specified center point. Within the polar option, objects either maintain their orientation as they are rotated or they can rotate around the centerpoint.

The threads in the FORK.DWG will be copied straight down, which is easy to do because the snap recognizes the y axis. If threads need to be arrayed along a projector line, the SNAP must be rotated within the Drawing Aids dialog box. Rotate the snap to the appropriate direction (either 30° or -30°) and perform the array.

Continuing with the exercise, you will create the threads.

Exercise

Creating the Isometric Fork Arms

Continue with the FORK.DWG.

Set the Isoplane to Top and set the running osnap to NONE

Command: *Choose* **C**onstruct, *then* **A**rray	Issues the ARRAY command
Select objects: *Pick on* ① *(see fig. 9.25)*	Highlights the isocircle
Select objects: (Enter)	Accepts the selection set
Rectangular or Polar array (R/P)<R>: (Enter)	Uses the Rectangular option
Number of rows (--) <1>: 8 (Enter)	Sets eight rows
Number of columns (III) <1>: (Enter)	Accepts one column
Unit cell or distance between rows (---): -.2 (Enter)	

Trim the threads to create the drawing in figure 9.25

Chapter 9

Zoom Extents in the drawing area and save the drawing as FORK.DWG

Complete the other isometric parts before continuing with the exploded assembly

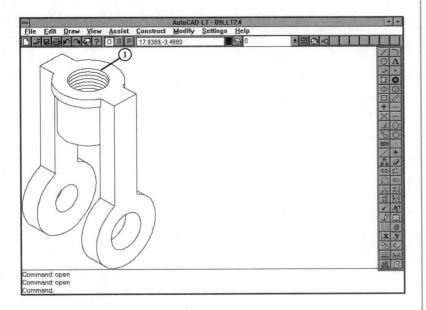

Figure 9.25

Adding threads to the fork arms.

The drawing names listed in the next exercise are derived from the list given earlier. This exercise assumes that all of the drawings are complete.

Setting Up the Isometric Assembly

Exercise

Begin the drawing using the HEAD.DWG.

Open the HEAD.DWG

Command: *Choose the Line button*	Issues the LINE command
LINE From point: *Choose the Nearest button*	Issues the NEArest osnap
NEA to *Pick on* ① *(see fig. 9.26)*	

continues

`To point:` *Choose the Perpendicular button*	Issues the PERpendicular osnap
`PER to` *Pick on* ②	
`To point:` *Right click on the mouse twice*	Reissues the LINE command
`LINE From point:` *Choose the Midpoint button*	Issues the MIDpoint osnap
`MID of` *Pick on* ③	
`To point:` `@4<-90` (Enter)	Creates a 4" center line
`To point:` *Choose* **D***raw, then Insert Bloc***k**	Issues the DDINSERT command
Choose the File button and path to the FORK.DWG	
Choose OK *and insert the drawing anywhere*	
Zoom Extents in the drawing area	
Save the drawing as ASSEMBLY.DWG	

Figure 9.26

Setting up the isometric assembly.

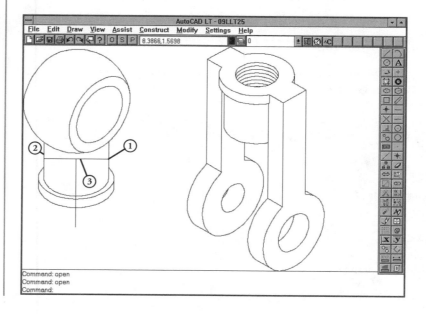

The next step in this exercise is to put these two objects in line vertically. The center line you created in the last exercise will function as the anchor point. Ellipses have endpoints all along their circumference because they are made up of many arcs, and they can be quite useful when adding tangent lines to isometric ellipses or as "center" points of the ellipse. If you guess where the tangent line will be, there will be an endpoint there. This is not the best way to find the tangent point, but it is certainly the quickest. Continue with the exercise by moving the FORK block.

Setting Up the Isometric Assembly

Exercise

Continue the ASSEMBLY.DWG.

Set the running osnap to ENDpoint

Command: *Choose the Move button*	Issues the MOVE command
Select objects: **L** (Enter)	Selects the FORK block as the last object
1 found	
Select objects: *Right click on the mouse*	
Base point or displacement: *Pick on* ① *(see fig. 9.27)*	
Second point of displacement: *Pick on* ②	
Command: *Choose the Line button*	Issues the LINE command
LINE From point: *Pick on* ③	Sets the first point
To point: *Pick on* ④	Sets the second point
To point: *Right click on the mouse twice*	Reissues the LINE command
LINE From point: *Choose the Midpoint button*	Issues the MIDpoint osnap
MID of *Pick on* ⑤	Selects the MIDpoint of the last line
To point: **@4<330** (Enter)	

continues

To point: *Right click on the mouse twice*   Reissues the LINE command

*Perform the same operation on the
FORK hole, as illustrated in 9.27*

Zoom Extents and save the drawing

Figure 9.27

Adding center lines.

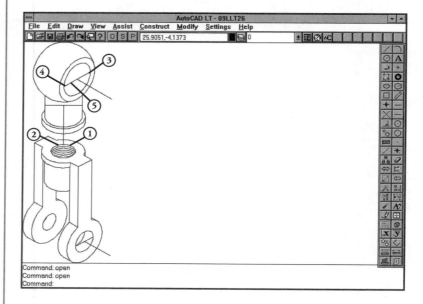

Now that the setup is done, it is time to add the other drawings. The FORK.DWG may need to be moved down with ortho mode on in order to accommodate the TIE-ROD.DWG.

Each one of the drawings that will be inserted must have construction lines that determine their center lines. This enables you to move the objects anywhere along the projector lines to achieve the assembly.

Inserting the **BUSHING.DWG** into the Isometric Assembly

Exercise

Continue with the ASSEMBLY.DWG and add the other objects.

Set the running osnap to ENDpoint

Command: *Choose* **D**raw, *then* Insert Bloc**k**   Issues the DDINSERT
command

*Choose the File button and path to the
BUSHING.DWG*

Choose OK *and insert the drawing anywhere*

Zoom Extents in the drawing area

*Move the BUSHING from the center
endpoint along its circumference to
the centerline constructed in the last
exercise (see fig. 9.28)*

Save the ASSEMBLY.DWG

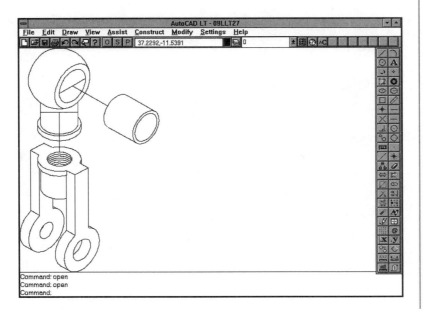

Figure 9.28

*Adding the
BUSHING.DWG.*

Now you will add the PIN into the assembly drawing using the DDINSERT command.

Exercise

Inserting the PIN.DWG into the Isometric Assembly

Continue with the ASSEMBLY.DWG.

Command: *Choose* **D**raw, *then* Insert Bloc**k**   Issues the DDINSERT command

Choose the File button and path to the PIN.DWG

Choose OK *and insert the drawing anywhere*

Zoom Extents in the drawing area

Move the PIN from the center endpoint along its circumference to the centerline constructed in the last exercise (see fig. 9.29)

Save the ASSEMBLY.DWG

Figure 9.29

Adding the PIN.DWG.

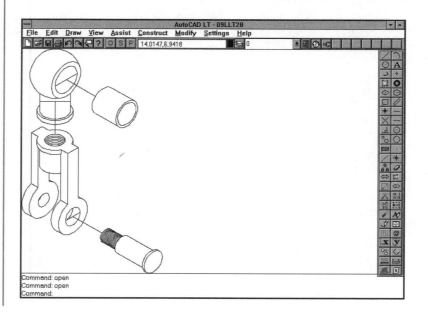

Using the previous method, insert the NUT-WASH drawing.

Inserting the NUT-WASH.DWG into the Isometric Assembly

Continue with the ASSEMBLY.DWG.

Command: *Choose* **D**raw, *then* Insert Bloc**k**   Issues the DDINSERT command

Choose the File button and path to the NUT-WASH.DWG

Choose the Explode box and then OK; *insert the drawing anywhere*

Zoom Extents in the drawing area

Move the NUT-WASH parts from the center endpoints along their circumferences to the extended fork centerline (see fig. 9.30)

Save the ASSEMBLY.DWG

Finish the drawing by inserting the TIE-ROD and cleaning up the odds and ends.

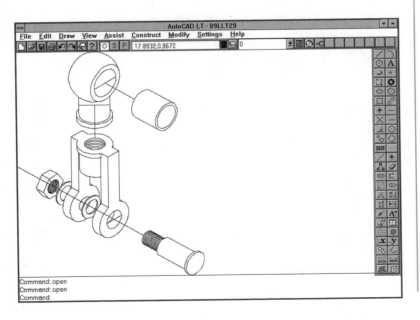

Figure 9.30

Adding the NUT-WASH.DWG.

Exercise

Inserting the TIE-ROD.DWG into the Isometric Assembly

Continue with the ASSEMBLY.DWG.

Command: *Choose* **D**raw, *then* Insert Bloc**k**   Issues the DDINSERT command

Choose the File button and path to the TIE-ROD.DWG

Choose OK *and insert the drawing anywhere*

Zoom Extents in the drawing area

Move the FORK and assorted parts down to accommodate the TIE-ROD

Move the TIE-ROD from the center endpoint along its circumference to the centerline of the assembly (see fig. 9.31)

Clean up the drawing and add the appropriate centerlines, as illustrated in 9.32

Save the drawing

Figure 9.31

Adding the TIE-ROD.DWG.

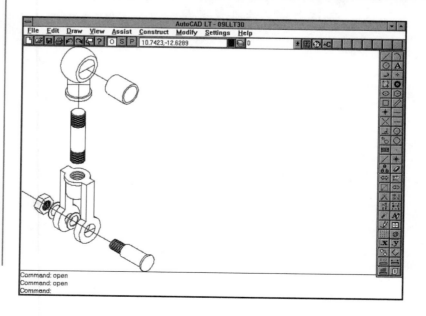

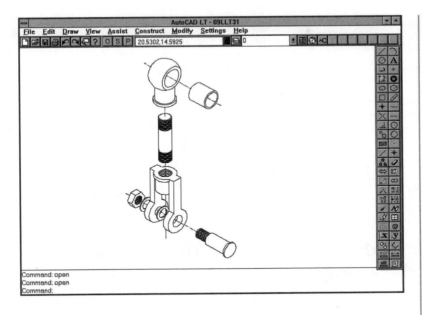

Figure 9.32

The completed assembly.

Developing Oblique Projections

An oblique projection is the simplest form of pictorial drawing. These drawings are created by extruding an orthographic view along a 30° or 45° projector line. This is a very simple construction in the CAD environment as the view can be copied as one selection set and projector lines are added. Some trimming and erasing may be required as the drawing is completed.

Although projector lines can be drawn at any angle, the conventional 30° or 45° angles work best.

TIP

There are a couple of ways to handle the scale along the projector lines. If the distance is full scale, it is commonly called a *cavalier projection*. In figure 9.33, you can see that this type of projection is rather unnatural. It has been attributed to a method of drawing used by medieval draftsman to illustrate castles; it gave the illusion of size and projected the appearance of dominance. If the distance is half scale, this is referred to as a *cabinet projection*. The correction of size has been attributed to early furniture makers looking for a quick method of illustrating a design or detail.

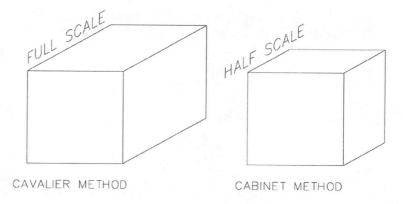

CAVALIER METHOD   CABINET METHOD

Understanding Elevation Oblique Drawings

If the front view is selected to create the oblique, it is described as an *elevation oblique.* Each face of the front view that is parallel with the view will be drawn at full scale and can simply be copied along the projector and connected with projector lines. To create the elevation oblique, use the orthographic drawing shown in figure 9.34.

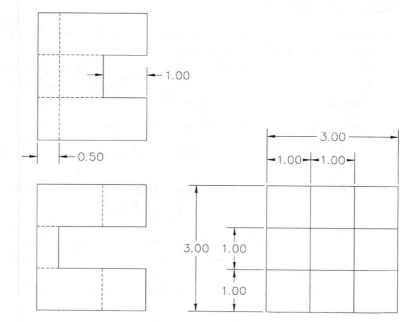

Creating an Elevation Oblique

Start a new drawing; accept all the defaults.

Recreate the orthographic front view shown in figure 9.34 and begin the drawing

Command: *Choose the Copy button*

COPY Select objects: **ALL** (Enter)

Select objects: (Enter)

<Base point or displacement>/Multiple: **M** (Enter)

Base point: *Pick anywhere not on the object*

Second point of displacement: **@.5<30** (Enter)

Second point of displacement: **@1<30** (Enter)

Second point of displacement: **@1.5<30** (Enter)

Set the osnap to ENDpoint and clean up the drawing, as shown in figure 9.35

The drawing is rather complicated at this stage, so remove the hidden lines and clean it up to resemble figure 9.36.

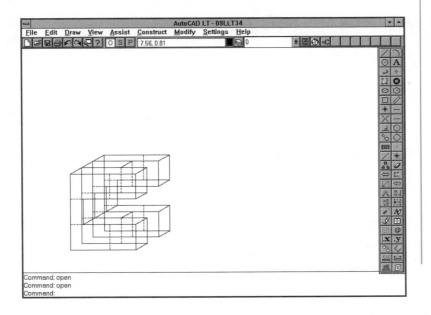

Figure 9.35

Developing the elevation oblique.

Figure 9.36

Cleaning up the elevation oblique.

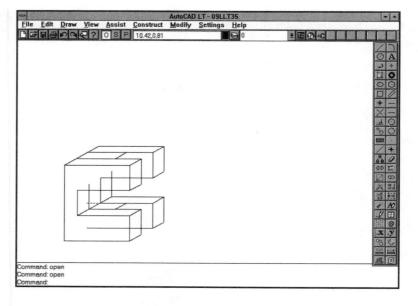

Pay close attention to the orthographic drawing and use the TRIM command to clean up the drawing, as illustrated in figure 9.37.

Figure 9.37

Continuing to clean up the oblique with the trim tool.

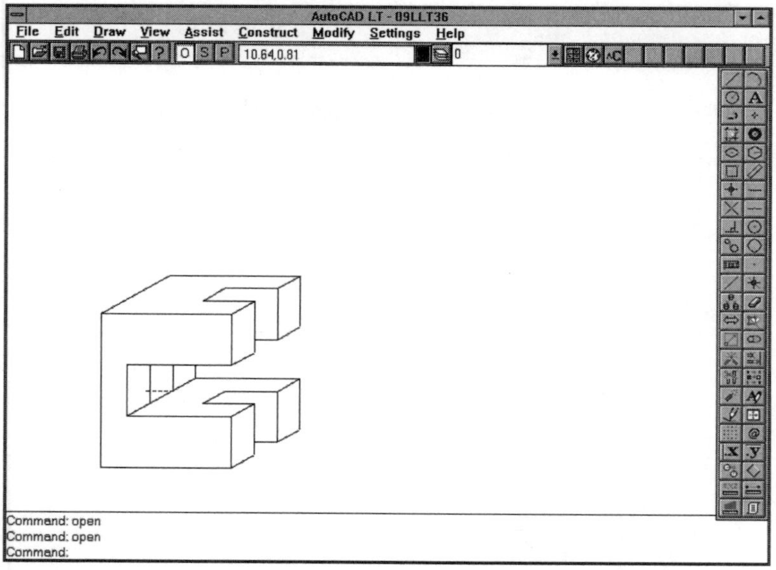

Finally, the square hole in the center needs to be trimmed and cleaned up to resemble the final drawing displayed in figure 9.38.

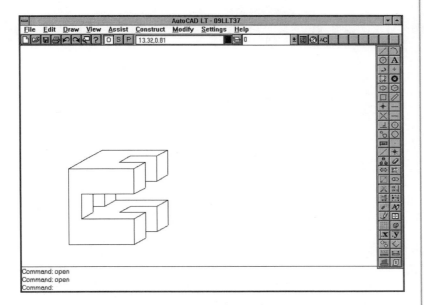

Figure 9.38

The elevation oblique.

This is a very simple technique, enabling you to create pictorial drawings quite quickly. As with all pictorial drawings, mastering this technique involves a confidence with orthographic drawing.

Understanding Plan Oblique Drawings

Plan oblique drawings are constructed in the same fashion—they are drawn using the top view. Architects employ this presentation technique to create interest and to add the illusion of depth to the floor plan. Figure 9.39 is drawn as a section with the walls cut four feet above the floor. This is a plan oblique of the sample floor plan that is included in the drawings when loading AutoCAD LT.

Figure 9.39

*The AutoCAD LT
sample floor plan in
plan oblique.*

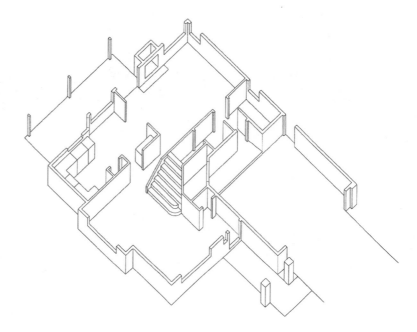

The plan was rotated 45° and copied up 4' at an angle of 90°. The corners of the walls were connected, and after trimming and erasing appropriate objects, the results illustrate the floor plan with depth. Plan obliques are perfect drawings that cover two aspects of a part or building. They display the sectional qualities and dimensional aspects in a drawing.

You are encouraged to pursue these pictorial drawing techniques. Not only will they help non-technical people understand complicated objects, but these drawings will also help you strengthen necessary visualization skills.

Summary

In this chapter, you have explored the aspects of pictorial drawing. A firm grasp and proven skill with this form of drawing will undoubtedly make your visualization skills strong. It has been suggested that some people are born with the ability to visualize spatial concepts. If you find this type of drawing difficult, you will need to work that much harder to increase your skills. As always, the more exposure you have to these pictorial concepts, the better off you will be. Many firms do not dabble with pictorial drawings until they are asked. The advent of CAD and its capabilities has these firms struggling to keep up with the demand for drawings that laypeople can understand.

You have been presented information on developing axonometric projections. These drawings include isometric, dimetric (two equal angles), and trimetric (three unequal axes angles). Isometric is the most common form and the easiest to assemble within AutoCAD LT.

You have also been introduced to assembly drawings and isometric assembly drawings, which prove to be very useful in visualizing the assembled parts. Those firms that do utilize pictorial drawing almost always use them in isometric assemblies. These drawings can cut out a lot of verbal instruction by displaying a logical assembly sequence of the parts.

You have also learned how to complete oblique drawings using two methods. The elevation oblique using the front view is most often incorporated into quick pictorials when time is short. Plan obliques are utilized frequently by architects when a more visual drawing of the floor plan is necessary.

In the next chapter, you will change gears to study the various means of plotting and printing your drawings to paper.

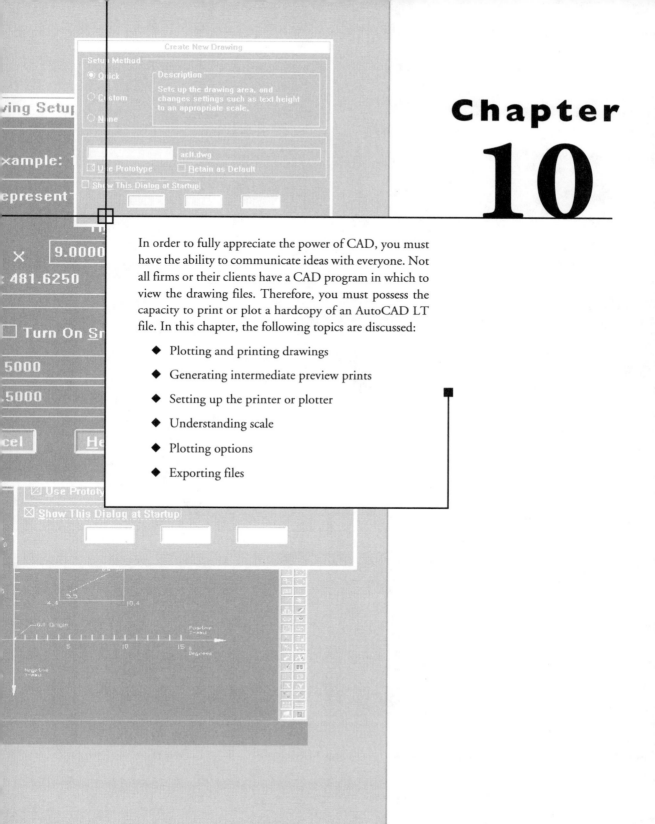

In order to fully appreciate the power of CAD, you must have the ability to communicate ideas with everyone. Not all firms or their clients have a CAD program in which to view the drawing files. Therefore, you must possess the capacity to print or plot a hardcopy of an AutoCAD LT file. In this chapter, the following topics are discussed:

◆ Plotting and printing drawings

◆ Generating intermediate preview prints

◆ Setting up the printer or plotter

◆ Understanding scale

◆ Plotting options

◆ Exporting files

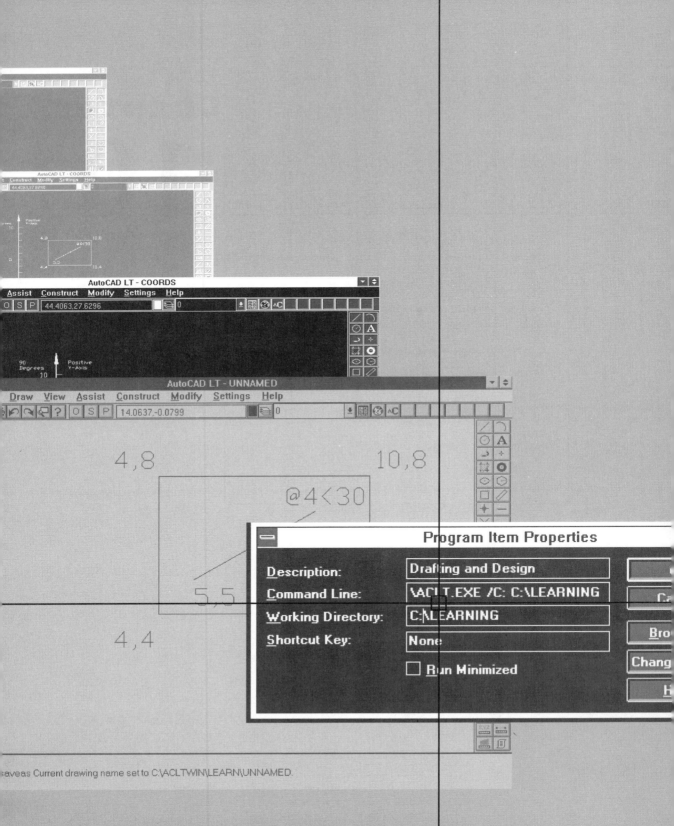

Reproducing Drawings

This chapter concentrates on the techniques used to bring your electronic ideas into the traditional realm by printing them on paper. With that in mind, let's first look at plotting and printing drawings.

Plotting and Printing Drawings

The ability to create a physical product from your AutoCAD LT file is referred to as producing hardcopy. *Hardcopy* is the piece of paper that contains the drawing that will be reproduced by traditional means. Although technology has made the process more efficient, surprisingly, not much has changed within the reproduction process.

The drawings that you create will need to be reproduced for a variety of reasons, as follows:

◆ **Documentation.** In order to keep a physical record of the project, it will be necessary to produce hardcopy documenting the steps along the way. Many firms do this for billing purposes or to preserve design ideas that inevitably are compromised during the design phase of the project.

◆ **Internal needs.** Many people or departments within a firm will need to have copies of the drawings for review or in order to progress with their portion of the project. Copies are sent to the machinists on the shop floor in order to produce parts, or to the interior designers in order to determine how much paint will be needed to finish a specific room.

◆ **Bidding.** All projects require varying amounts of outside need. Whether it is the steel supplier anticipating a large manufacturing order or a plumbing contractor offering a subcontracted bid on a 30-story apartment building, all projects require reproduction to enable others enough time to keep the project on schedule and under budget.

◆ **Renderings.** At times, depending on the discipline, a reproduction may be required for an artist to create a marketing piece. Illustrations help clients or investors understand the project.

◆ **Final Sets.** After all the drawings within a set are complete, the drawings are reproduced for the shop floor or the construction crew. They assess the drawings and determine how much material is needed, how much time a project will take, and how many people will be required to perform the job.

Although technology is evolving to eliminate the need to generate hardcopy drawings, the majority of machine shops and builders still require a set of drawings. For example, many machine shops utilize CNC, or computer controlled milling machines, which can create a part based on coordinates developed directly from a CAD file.

Architects who develop drawings in the CAD environment often need a hardcopy in order to survey the drawing. Trying to find discrepancies in a drawing on a 14" monitor can be time-consuming and difficult to do correctly.

Hardcopy will always play a role in the development of technical drawings.

Although a paperless office would be difficult to accomplish, it is easier to maintain electronic documentation. Work toward proper electronic file maintenance as a way to record projects, instead of keeping multiple hardcopies.

Understanding the Methods Used to Reproduce Drawings

The processes used to reproduce drawings have changed as technology has improved. As archaic as it seems, many firms still employ most of these methods to some extent. Depending on the size of the firm, it might have a reproduction department on the premises. As needs and reproduction volume change, most firms contract a service bureau to create copies for them. These bureaus make it their business to keep up with the latest technology, ensuring their clients efficient reproduction. Some are sophisticated enough to receive electronic files over the phone, print them, and deliver them all within a few hours. The following section is a brief history of the methods used in the past and the present to duplicate hardcopy drawings.

Examining Reproduction Methods

Original drawings are treated with care and are rarely distributed. The time that each drawing, sketch, or specification represents is worth more than the paper on which it was created. The material is conceived, designed, produced, revised, and finalized before any actual physical items are produced. Therefore, reproductions are made to save the quality of the original. Reproduction technology has changed over the years, and computers provide unimaginable storage potential. The following is a short journey through the world of reproduction:

◆ **Tracing.** Before the advent of reproduction machines, drawings were duplicated by the tedious process of tracing. Although a certain amount of tracing still occurs within firms that still draft on the boards, tracing was the only way to duplicate a set of drawings. A transparent paper was placed over the original drawing, and every line was painstakingly redrawn. Typically, only one set of drawings existed, and those drawings were guarded closely. Certainly, most of the work created was *design built*, meaning the designer was on hand to make particular decisions on the spot.

◆ **Blueprinting.** The process of creating a blueprint is the oldest of the reproduction processes. Creating a *blueprint* of a drawing is similar to developing a negative to produce a photograph. The original drawing blocks the light under the linework and leaves an impression on chemically treated paper that is sensitive to light.

The two sheets are fed into the machine together and are exposed to intense light. The blueprint paper is fed through a developing bath, washed with water, passed through a fix bath, rinsed with water, and dried. The remaining chemical turns the paper a grayish blue with white lines, thus reproducing the drawing.

Brownprints are often created in the same fashion, utilizing a different chemical treatment on a translucent paper. This process is often used to create a master from which other prints are generated.

◆ **Diazo.** The same process of exposing the paper to light is used in a Diazo printer, but the development process is much faster. Instead of a fix bath, the paper is exposed to a developer that dampens the treated side of the paper. By changing the paper type, you can create blue, brown (sepia), black, or even red reproductions on a white background. This process is called *Diazo-Moist*.

Diazo-Dry is the most popular method of reproduction. If you have a "blueprint" machine at your firm, chances are it is actually a Diazo printer. Again, the same process is used, but the print is exposed to ammonium vapors in order to fix it to the sheet. The print emerges from the machine virtually dry. The final duplicate will be a reverse of a "true" blueprint—dark lines on a white background.

◆ **Microfilm/Microfiche.** Although these two methods have different final states (microfilm is a roll of film, whereas microfiche is a separate film card), they utilize the same technology to store and reproduce drawings at a fraction of their true size. This is a rather outdated process of storing drawings now that computer technology is so advanced. You can still find firms that used these methods for older drawings, but most firms with this capacity were on the forefront of technology and were first to accept computers.

In order to view this film, a special reader is required. Depending on the system, a print can be made directly from the reader or the film is sent out to a service bureau for prints.

◆ **Xerography.** With the advent of Xerox machines, another reproduction process emerged. Just like a typical Xerox machine, the lines on the reproduction paper are electrostatically charged, and negatively charged ink is then spread over the sheet. The ink adheres to the lines and is baked on to produce the copy. This type of reproduction has blossomed into a thriving business for service bureaus. The reproduction itself is cheap and fast, but is expensive to maintain. Most firms will not have a large Xerox copier.

In manual drawing, corrections and updates are often made to the duplicated drawing, which now serves as the new original. Strategically, this may save time in the short run, but nothing beats a single original. You can avoid costly mistakes by maintaining one master drawing.

The CAD environment is the best solution to this issue. The file is constantly updated and given a new date each time it is revised, the hardcopy prints are easy to reproduce and very clean, and the electronic file is relatively easy to store.

Plotting or Printing the Drawings

Understanding the aspects of reproduction is important, but the drawing has to be generated on the paper before any duplication can occur.

In the past, whenever a hardcopy was made, it was referred to as a *plot*. Most hardcopy was generated on a plotter that had from one to eight ink pens of varying width (see fig. 10.1). The paper is fed in and out while the pen moves right and left to create all of the entities in the drawing.

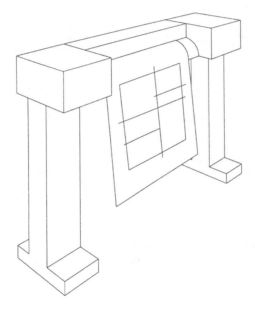

A pen plotter.

Although many firms still employ these plotters, many have switched to more expensive and larger laser plotters or printers. These machines commonly contain a large roll of paper that is cut after the plot is created. The features depend on the sophistication of the machine. Without the pen mess and with the ability to plot one drawing after another, these printers are improving yearly and will eventually come down in price.

With the popularity and affordability of desktop laser printers, hardcopy also is referred to as the *print*. To avoid the confusion between plotting and printing, the term "print" is used in this chapter to describe the hardcopy. All of the actions involved in printing the drawing are controlled through the Plot Configuration dialog box.

In the next exercise, you load a drawing and prepare to print it.

Exercise

Opening the Plot Configuration Dialog Box

Open the CAM.DWG that comes with AutoCAD LT.

Command: *Choose the Print/Plot button from the toolbar (see fig. 10.2)*   Opens the Plot Configuration dialog box

Figure 10.2

The Plot Configuration dialog box.

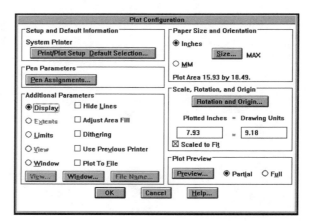

The CAM.DWG is in paper space, which was discussed in Chapter 2, "An Overview of Drafting Tools and Techniques," but will be reviewed later in this chapter.

The next several sections will utilize the Plot Configuration dialog box in various ways. Like most of AutoCAD LT, its complexity will seem minimal after you have an understanding of the various functions and settings. These quick plots are designed for a laser printer in order to reduce materials waste on a plotter. Before you begin these exercises, make sure your printer is on and has been recognized while the computer was booting up.

Printing to a laser printer is much different than plotting. A laser printer typically will convert the vectors in your drawings to raster output. Vectors are defined by their endpoints and a line is generated in between. Raster output is essentially a series of dots (pixels) that simulates the line that would have been plotted. Look closely at an angled line on a laser print and you will see the stair-stepping generated by pixels. This stair step, or alias, can be greatly improved by working with a laser printer with a higher resolution. This effect is not found on a plotted drawing.

Understanding Review Plots

As you are working on a design, or as the production drawings are being completed, you will often do quick check plots. These plots may or may not be to scale. Their purpose is simply to record progress or to check the overall layout.

In this exercise, you create a check plot to determine that the system is working and the drawing is plotting correctly. It is best to create these types of plots on a laser printer. The AutoCAD LT default prints to the system printer, which is the same printer that your word processing documents are printed on. Continuing from the last exercise, create a check plot of the CAM.DWG.

Exercise

Previewing and Plotting the Drawing

Using the CAM.DWG, continue the exercise.

Within the Plot Configuration dialog box, choose E**x**tents	Sets the plot to include all entities

continues

In the Scale, Rotation, and Origin area, choose the Rotation and Origin *button*	Opens the Plot rotation and Origin dialog box
Under Plot Rotation, choose **9**0, *then* OK	Rotates the plot 90°
In the Scale, Rotation, and Origin area, choose Scale **t**o Fit	Ignores all scale settings
In the Plot Preview area, choose F**u**ll *and then the* P**r**eview *button*	Creates a quick preview of the drawing (see fig. 10.3)
Choose **E**nd Preview *in the Plot Preview*	Exits back to the Plot Configuration dialog box
Choose OK	Switches back to the drawing screen

`Effective plotting area:`
`7.94 wide by 10.28 high`
`Position paper is plotter.`	

`Press Enter to continue or S` `to Stop for hardware` (Enter)	Begins plotting and displays dialog box verifying that the plot is being sent to the system printer

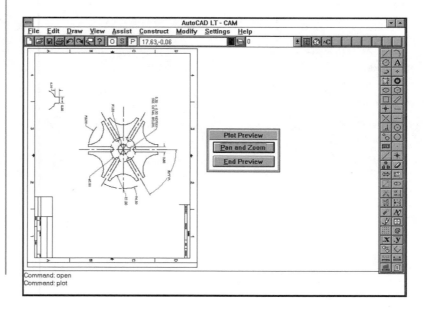

If the system is set up correctly, a print will be generated that matches the drawing preview displayed in figure 10.3.

As mentioned earlier, quick checks should be created on the laser printer, but drawings that are to be checked by a supervisor should be plotted. It is very difficult to review complex, D-sized mechanical or architectural drawing on A-sized paper.

Before the final plots are made for the set of drawings, many of these check plots will be generated.

Setting Up the Printer/Plotter

The Print/Plot Setup & **D**efault Selection button in the upper left corner of the Plot Configuration dialog box is used to specify the printing device. This setting can call a pen plotter or a laser printer. If the printer you use isn't listed, you can customize the list by adding your own device.

In order to add a plotter to your system, you must take into account that you are operating in Windows. All devices used with Windows applications have drivers that are maintained through the Control Panel, in the Main Group, found under Program Manager. The Control Panel maintains all the drivers that are needed to print from Windows. AutoCAD LT drivers must be set up here in order to specify them in the Plot Configuration dialog box.

All printer and plotter devices have drivers that are Windows-compliant. Most manufacturers will supply the procedures needed to load their particular driver. Many of the CAD programs that are traditionally DOS-based have moved to a Windows format, and device manufacturers are prepared for the transition.

Before devices can be specified in AutoCAD LT, their drivers must be loaded through the Control Panel in Windows. See your device manual for instructions for this procedure.

After choosing the Print/Plot Setup & **D**efault Selection button, the Print/Plot Setup & Default Selection dialog box appears (see fig. 10.4).

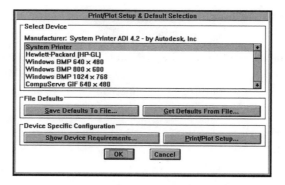

You are presented a scrolling list of alternatives for printing. The system printer is the first in the list. This is the printer currently being used by Windows. The next selection is the Hewlett-Packard (HP-GL) device. Most HP plotters are supported by this selection in AutoCAD LT Release 2. If you are using an earlier version of LT, you may experience some problems with the driver. Also listed are the raster formats that are available. BMP, GIF, PCX, and TIF are popular raster formats that can be created if the device is selected.

Highlight any of the devices and choose the Show Device Requirements button in the Device Specific Configuration area. If the Hewlett-Packard (HP-GL) device is selected, the requirements are displayed as shown in figure 10.5.

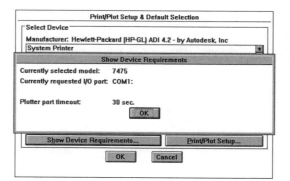

The device can be changed by selecting the Print/Plot Setup button. This displays the Print/Plot Setup dialog box, which changes for each device. The HP GL Driver Setup dialog box is shown in figure 10.6.

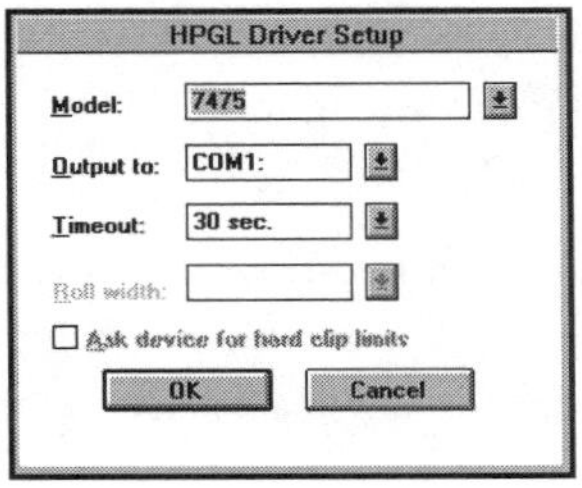

HPGL Driver Setup
Model: 7475
Output to: COM1:
Timeout: 30 sec.
Roll width:
☐ Ask device for hard clip limits
OK   Cancel

Figure 10.6

The HP GL setup options.

All of the current HP plotters are listed and their defaults can be changed at this juncture.

By highlighting any of the raster options, you have the ability to set the *bit depth* of the resulting image—either 2-, 4-, or 8-bit. Bit depth is defined by the number of colors or levels of gray that can be displayed by one pixel. Think of a bit as a switch that is either on or off—the more bits a pixel has, the more combinations of on and off. Therefore, a 1-bit pixel can be either black or white; a 1-bit image is a two-color image (2^1=2). A 2-bit image, 2 switches, allows each pixel to display four colors (2^2=4). The calculation of color depth is not imperative to understand color settings, but you can always calculate the color by multiplying two to the power of the bit depth.

You can also specify the background color from white (0) to black (255), which is 8-bits (256 colors). Any number in between will give you a shade of gray.

CAUTION

Entities that are created with the colors black or white will be adversely affected by the background color specified in the Print/Plot Setup. Your preferred background color in the drawing area will always display white on black or black on white.

After the printer or plotter is configured, you should be on your way to exploring the other options that are examined in the next section.

Configuring the Device

After the device is up and running, there are a number of settings that still need to be considered.

Pen weight or width is assigned in the Plot Configuration dialog box by color. If the color yellow is given a weight of .25mm, then all yellow lines will be drawn with a width of .25mm.

To set the Pen Parameters correctly, you or your firm should have a standard setup if you don't already. The technical drawing community at large has yet to agree on a standard, although at a local level, consultants and firms may be a little more organized.

If you are working with a multi-pen plotter, the pen assignments should be irrelevant. The pen tip size should dictate the weight of the line. The terms "pen weight" and "width" are used interchangeably within this chapter. Both terms refer to the thickness of the line. The only value to setting the pen widths, equal to the tips, is that AutoCAD LT will use the information when filling in solids or polylines with width. The setting will determine how close to place the lines during the fill. If you have an older model plotter with one pen, pen weights will save pen switching time, but the pen will tend to run out sooner due to overwork.

CAUTION

Some pen plotters will require you to configure the pen width and speed on the plotter itself. The settings within AutoCAD LT should match, or you will receive unexpected results, such as skipping, smudging, and so forth.

Line widths on a laser printer can only be verified on the hardcopy. Be consistent with the line weights that you choose. This is important while creating the drawing because it is time-consuming to go back into the drawing and make it conform to a new pen scheme. Service bureaus will ask for your pen assignments in order to produce a plot—you must be prepared; it might seem trivial, but a little research is worth the time, which, when drafting, is money.

To set the widths, choose the Pen Assignments button in the Pen Parameters area.

Figure 10.7 shows the default Pen Assignments dialog box. AutoCAD LT has assigned each of the colors a pen number; this idea may be familiar from Chapter 4, which discusses layers. Pen number one is assigned to color one, Red. In the Modify Values area, the pen number is displayed, and the Ltype (linetype) and Width can be adjusted.

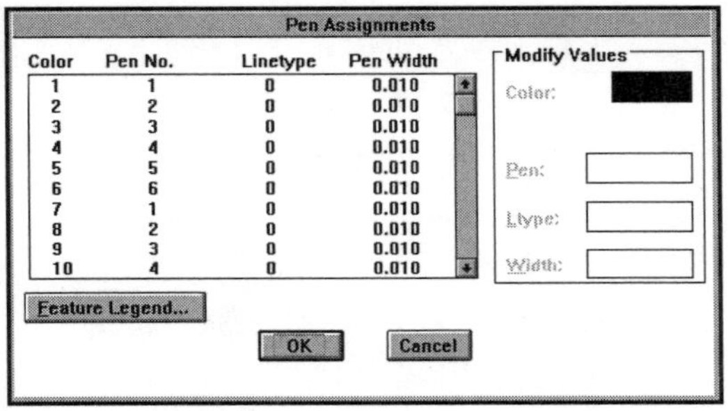

Figure 10.7

The Pen Assignments dialog box.

> Although it is necessary to have the ability to specify all of the pen functions, it is best the rely on the drawing settings to determine linetypes.

Experiment with the pen widths to get the proper adjustments. Returning to the main dialog box, there are numerous settings that need to be examined.

In the Paper Size and Orientation area, you have the ability to specify the unit of measurement for plotting. The default is inches; this is the unit with which the current paper size is displayed. Also on display in the Paper Size and Orientation area is the plot's area measure. The *plot area* defines the size that can be printed. This is generated from the paper size minus the margin that the printer needs to feed the paper.

In order to change the paper size, choose the **S**ize button. Figure 10.8 displays the Paper Size dialog box.

If you are working with a laser printer, the size is dictated by the Control Panel settings. In the next exercise, you create a new paper size that assumes your printer can handle 8 1/2" x 14" paper.

Figure 10.8

The Paper Size dialog box.

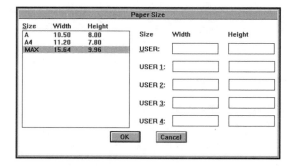

Paper Size		
Size	Width	Height
A	10.50	8.00
A4	11.20	7.80
MAX	15.64	9.96

Size   Width   Height

USER:

USER 1:

USER 2:

USER 3:

USER 4:

OK   Cancel

Exercise

Creating a New Paper Size

Continue with the CAM.DWG.

Choose the Print/Plot button from the toolbar	Opens the Plot Configuration dialog box
*Choose the **S**ize button within the Paper Size and Orientation area*	Opens the Paper Size dialog box
Choose the Width edit box on the User row and type **13.5**	
Press the Tab key and type **7.93** *in the Height edit box*	
Press (Enter)	The size will appear in the list
Press OK	The new size is the current plot area
*Choose the **S**ize button within the Paper Size and Orientation area (See fig. 10.9)*	Opens the Paper Size dialog box
Highlight the original setting and choose OK	Returns the plotting area to the default

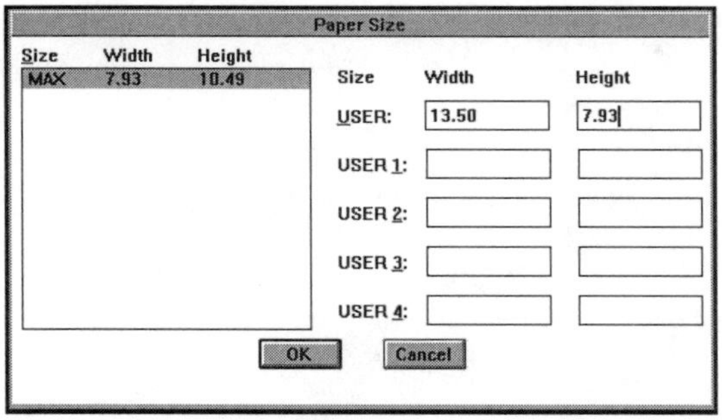

Paper Size					
Size	Width	Height	Size	Width	Height
MAX	7.93	10.49	USER:	13.50	7.93
			USER 1:		
			USER 2:		
			USER 3:		
			USER 4:		

OK   Cancel

Figure 10.9

Adding a new paper size.

Remember, the size of the paper is controlled by the Control Panel within the Windows environment. If you change the paper size, you must adjust the settings in the Control Panel to avoid getting unexpected results.

CAUTION

In order to rotate the plot, you must adjust the settings in the Scale, Rotation, and Origin area of the Plot Configuration dialog box. This area contains a button that will adjust the rotation and origin of the plot. The current paper orientation of the system printer will affect the rotation settings; it is assumed to be Portrait. In the next two exercises, you adjust the CAM.DWG with these settings and quickly view the results.

Exercise

Rotating the Plot

Continue with the CAM.DWG.

Command: *Choose the Print/Plot button in the toolbar*	Opens the Plot Configuration dialog box
Choose the Rotation and Origin *button in the Scale, Rotation, and Origin area*	Opens the Plot Rotation and Origin dialog box
Choose **0** *under Plot Rotation*	Sets rotation to 0°
Choose OK	
Choose Scale **to** Fit	Checks the Scale to Fit check box

continues

Choose F**u**ll *under Plot Preview, then choose the* P**r**eview *button (see fig. 10.10)*	Displays a preview of the CAM
Choose **E**nd Preview	Exits the preview mode
Choose the Rotation and Ori**g**in *button in the Scale, Rotation, and Origin area*	Opens the Plot Rotation and Origin dialog box
Choose **9**0 *under Plot Rotation*	Sets rotation to 90°
Choose OK	
Choose the P**r**eview button *(see fig. 10.11)*	Displays a preview of the CAM
Choose **E**nd Preview	Exits the preview mode
Repeat these steps for 180° and 270° rotations (see figures 10.12 and 10.13)	

Figure 10.10

Rotated 0 degrees.

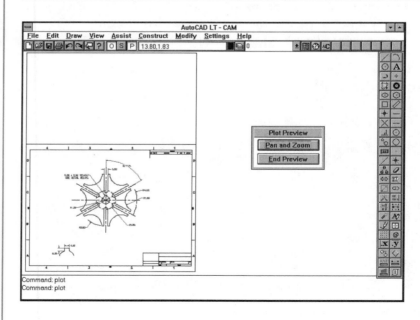

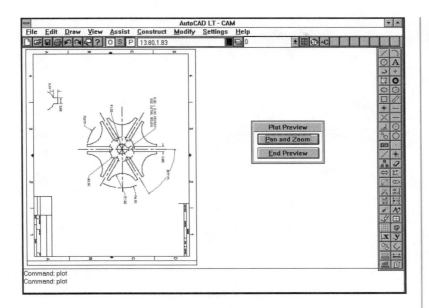

Figure 10.11

Rotated 90 degrees.

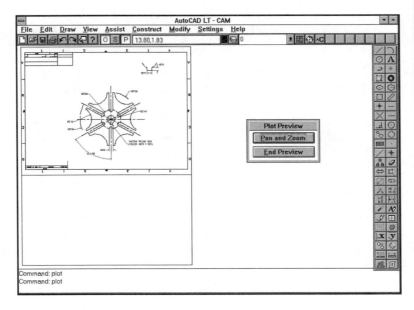

Figure 10.12

Rotated 180 degrees.

Figure 10.13

Rotated 270 degrees.

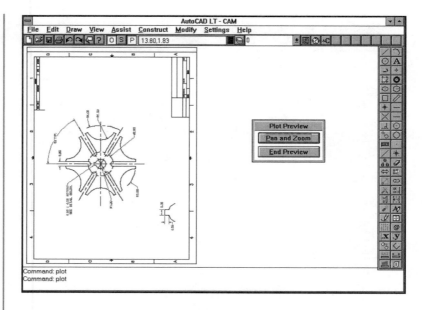

Again, depending on the orientation and size of the paper, these results will vary. This method of checking the plot enables you to avoid the frustration of waiting for a plot, only to find a rotation error.

Never rotate your drawing to achieve the desired orientation. The Plot Rotation tools are included to produce the desired results while printing.

Some plotters are designed to create plots with a portrait orientation. Typically, the rotation can be adjusted using the menu pad on the plotter itself.

In the next exercise, you adjust the origin of the plot in order to observe the effects. It is important to realize that these settings are adjusting the plot only—the drawing does not change. The plot origin is always located in the bottom left corner of the page.

Adjusting the Plot Origin

Continue with the CAM.DWG.

Command: *Choose the Print/Plot button in the toolbar*	Opens the Plot Configuration dialog box
Choose the Rotation and Origin *button in the* Scale, Rotation, and Origin *area*	Opens the Plot Rotation and Origin dialog box
Choose **9**0 *under Plot Rotation*	Sets rotation to 90°
Type **1** *in both the* **X** Origin *and* **Y** Origin *edit boxes*	Sets the plot origin to 1,1
Choose OK	
Choose the P**r**eview button *(see fig. 10.14)*	Displays a preview of the CAM

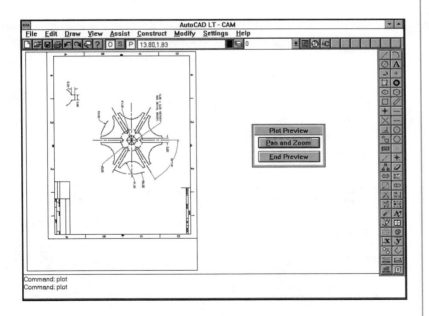

Figure 10.14

Changing the plot origin.

When the Scaled to Fit option is selected on the Plot Configuration dialog box, AutoCAD LT will force the whole drawing on the page. In the next section, you examine the features used to scale the plot.

Understanding Scale During Plotting

As a rule, drawings are created in full scale. AutoCAD LT has the capability to view a building that is 100 feet long on a 14" monitor with the ZOOM options. Plotting the building is another aspect of the CAD environment that must be investigated.

When setting up drawings manually, draftspeople had to take into consideration what was to be included on the drawing and the scale of the views. By blocking out the drawings with various scales, the drafter was able to determine the best scale suitable for the drawing and adjusted the measurements accordingly. Because objects are drawn at full scale in CAD environments, the flexibility of plot scale is infinite.

The CAM.DWG that is used as an example does not fit correctly on an A-size sheet when scaled properly. Part of the border is cut off in order to maintain a half scale. The mview viewport is scaled 50 percent in order to plot correctly and to scale. The MVIEW command is discussed at length in Chapter 11, "Creating a Mechanical Drawing."

The plot scale of the CAM.DWG will be adjusted in the next exercise. To this point, you have plotted this drawing by scaling it to fit the paper size. Even though this is common practice while creating check plots, final plots must be done to a specific scale.

Exercise

Plotting to Scale

Continue with the CAM.DWG.

Command: *Choose the Print/Plot button in the toolbar*	Opens the Plot Configuration dialog box
Choose Scaled to Fi**t** *under Scale, Rotation, and Origin*	Removes the check from the Scaled to Fit check box
Type **1** *in the Plotted Inches edit box*	

Type **2** *in the Drawing Units edit box*

This means for every 2 units drawn on the CAM, the printer will draw one inch. All entities are given a unit measurement, until they are plotted. The real units aren't given to an entity until it is plotted to paper. This is essentially plotting the drawing in half scale. If you measure the 1/2" slot in the CAM, it will be 1/4".

Choose Ex*tents under Additional Parameters*	Highlights the Extents radio button
In the Plot Preview area, choose Fu*ll and then the* P*review button (see fig. 10.15)*	Creates a quick preview of the drawing
Choose E*nd Preview in the Plot Preview dialog box*	Exits back to the Plot Configuration
Choose OK	
Plot the drawing	

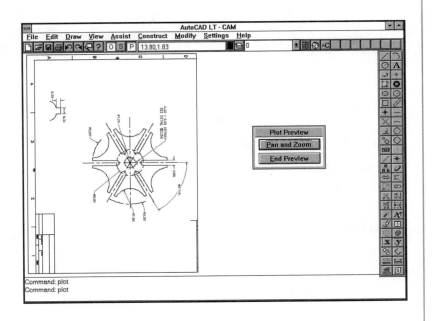

Figure 10.15

The CAM plotted to scale.

Plotting in Paper Space

When the TILEMODE system variable is set to 0, you find yourself in paper space. Paper space is used to lay out the plotted page by creating viewports of the model (drawing). When the model changes, the paper space views are automatically updated to reveal the change. All entities created in paper space should be drawn full scale or 1:1, according the sheet size of the final plot. All of the viewports are scaled using the ZOOM command with the XP option to reflect the final scale of the drawing. The CAM.DWG is currently in paper space and contains two mview viewports. The viewports are zoomed to the appropriate size based on the final plot, which will be half scale.

In the next exercise, you re-create the paper space viewpoints of the CAM.DWG.

Exercise

Creating and Plotting Paper Space Viewports

The current CAM.DWG contains two mview viewports.

Command: *Choose* **F**ile, *then* Save **A**s	Opens the Save Drawing As dialog box
Call the drawing MYCAM *in your user directory*	
Command: *Choose the Layers button in the toolbar*	Opens the Layer Control dialog box
Choose the layer PS, then the On button	Turns the viewport borders on
Choose Current, *then* OK	Sets the current layer to PS
Command: *Choose the Erase button*	Issues the ERASE command
Select objects: *Choose the two viewport borders*	Highlights the borders
Select objects: **Enter**	Removes the viewports
Command: *Choose* **V**iew, Viewp**o**rts, *then* **M**ake Viewports	Issues the MVIEW command

`ON/OFF/Hideplot/Fit/2/3/4` `/Restore/ <First point>:` **1.5,3** (Enter)	Sets first point of viewport
`Other corner:` **14,14** (Enter)	Sets other corner of the first viewport
`Command:` (Enter)	
`MVIEW ON/OFF/Hideplot/Fit` `/2/3/4/Restore/ <First` `point>:` **15,9** (Enter)	Sets first point of viewport
`Other corner:` **20,13** (Enter)	Sets other corner of the second viewport
Choose the Paper Space button in the toolbar	Issues the MSPACE command
Choose inside the large viewpoint	Highlights the viewport border
`Command:` *Choose the Zoom button in the toolbar*	Issues the ZOOM command
`All/Center/Extents/Previous` `/Window <Scale(X/XP)>:` **E** (Enter)	Zooms the view to include all visible entities
`Command:` (Enter)	Reissues the ZOOM command
`All/Center/Extents/Previous` `/Window <Scale(X/XP)>:` **1XP** (Enter)	Magnifies the viewport to the proper size of 1:1
Choose the other viewport	Highlights the small viewport border
`Command:` *Choose* **V**iew, Viewport **L**ayer Visibility, *then* **F**reeze	Issues the VPLAYER command
`?/Freeze/Thaw/Reset/Newfrz` `/Vpvisdflt:_freeze`	
`Layer(s) to Freeze:` **ART,DIM,** **0** (Enter)	These are the layers to freeze

continues

`All/Select/<Current>:` (Enter)	Freeze in current viewport
`Command:` *Choose the Zoom button in the toolbar*	Issues the ZOOM command
`All/Center/Extents/Previous /Window <Scale(X/XP)>:` **E** (Enter)	Zooms the view to include all visible entities
`Command:` (Enter)	Reissues the ZOOM command
`All/Center/Extents/Previous /Window <Scale(X/XP)>:` **2XP** (Enter)	Magnifies the viewport to the proper size of 1:2
Choose the Pan button to move the view, as illustrated in figure 10.16	
`Command:` *Choose the Layers button in the toolbar*	Opens the Layer Control dialog box
Choose the layer 0, then the Current button	Sets 0 as the current layer
Choose the Clear All *button, choose layer PS, then* Off	Turns the PS layer off
Choose OK	
Choose the Paper_Space button in the toolbar	Highlights the button
`Command:` *Choose the Change _Properties button*	Issues the DDCHPROP command
`Select objects:` *Pick the two viewport borders*	Highlights the borders
`Select objects:` (Enter)	Opens the Change Properties dialog box
Choose the Layer *button*	Opens the Select Layer dialog box
Choose the PS layer, then OK	Exits to the Change Properties dialog box
Choose OK	Exits to the drawing area
Follow the plotting procedures with the Print/Plot button to print the drawing	

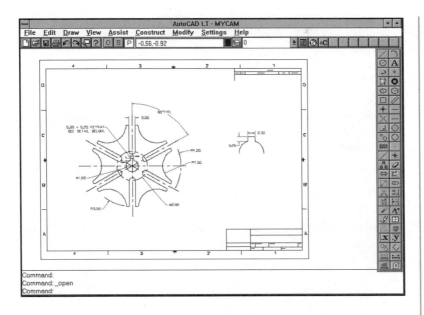

Figure 10.16

The MYCAM drawing.

10 *Chapter*

If you turn the PS layer back on, you can move the viewport border lines like blocks to rearrange the sheet. The view of the model inside the viewport will not change.

NOTE

Plotting in Model Space

By setting TILEMODE to 1, you will be returned to model space. This is the default view when beginning a drawing and where all the entities drawn in model space reside. Plotting out of model space requires adjustments under Scale, Rotation, and Origin. In the next exercise, you preview the drawing using various settings while in model space.

Plotting in Model Space

Exercise

Continue with the CAM.DWG.

Command: *Choose* **V**iew,   Sets TILEMODE to 1
then **T**ile Mode

continues

```
New value for TILEMODE <1>: 0
Regenerating drawing.
```

Command: *Choose the Print/Plot button in the toolbar*	Opens the Plot Configuration dialog box
Choose the Rotation and Ori**g**in *button in the Scale, Rotation, and Origin area*	Opens the Plot Rotation and Origin dialog box
Choose **0** *under Plot Rotation*	Sets rotation to 0°
Type **1** *in the* **X** Origin *and* **3** *in* **Y** Origin *edit boxes*	Sets the plot origin to 1,3
Choose E**x**tents *under Additional Parameters*	Highlights the Extents radio button
In the Plot Preview area, choose F**u**ll *and then the* P*review button (see fig. 10.17)*	Creates a quick preview of the drawing
Choose **E**nd Preview *in the Plot Preview dialog box*	Exits back to the Plot Configuration

Figure 10.17

The model space preview.

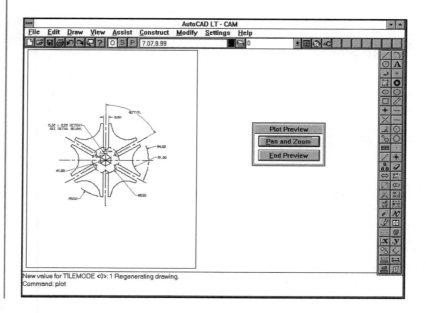

Plotting Text

In order to set the text height within paper space, insert the text at the correct size if the entities are drawn one to one. In the default CAM.DWG, the border was doubled in size so the inserted text would have to be doubled in height to reflect your required final plot text size. This is why it is important to draw in paper space at a 1:1 scale. In order for text to be 1/2" high in the final plot, it would need to be inserted as 1" high text. These conversions should be avoided to eliminate possible delays, mistakes, or reinsertion of the text.

Some areas of the drafting world have not accepted a standard on which to base text heights. Drawings that draftspeople create are quite personal. Many firms do have a set of standards that have been established and it is important to follow them, even when plotting from AutoCAD LT.

When creating text in paper space, insert the text at the final plotted height. In model space, you must make provisions, while drawing, to insert the text at a size that corresponds to the ZOOM XP (ZOOM times paper space) in the paper space drawing. For example, if you are drafting a floor plan of a building at 1/4" = 1 foot (quite common) and the drawing titles should be 1/2" on the plot, you should specify a text height of two feet.

These conversions will be tricky, and your check plots will likely identify potential problems within the final plot. Don't get too wrapped up in understanding these aspects because they will become clear to you through trial and error.

Exploring Plotting Options

To this point, you have been creating previews using the Extents plotting option. There are several more plot options to explore that are similar to the ZOOM command's options. The following options are listed under Additional Parameters in the Plot Configuration dialog box:

◆ **Display.** Choosing this option will plot only what appears in the current drawing area (on-screen).

◆ **Extents.** This option will plot all visible entities in the current drawing, even those not currently shown on-screen.

◆ **Limits.** Plots to the limits are specified in the drawing. Using this option cancels the effect of Scale to Fit.

◆ **View.** This option will print any saved views. There must be views established before this option will highlight. Choose the Views button and highlight a view in the list. If the list is blank, no views have been specified.

◆ **Window.** Choosing this option will enable you to form a window in the drawing area and plot it. After a window is established, this option will be available. The next exercise will step through this procedure.

Figures 10.18–10.22 display the results of each option. A view was saved of the CAM's front elevation and a window was formed around the keyway detail.

Figure 10.18

Plotting by Display.

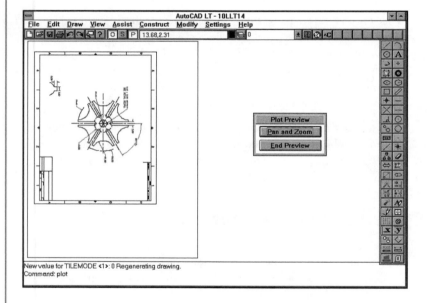

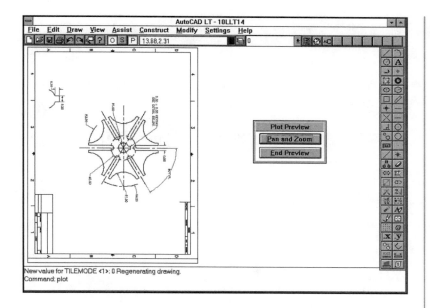

Figure 10.19

Plotting by Extents.

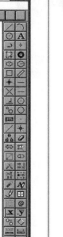

Figure 10.20

Plotting by Limits.

Figure 10.21

Plotting by View.

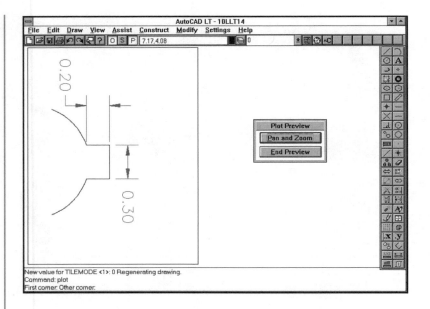

Figure 10.22

Plotting by Window.

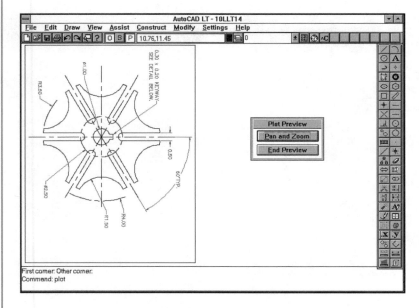

In the next exercise, you continue using the CAM.DWG to examine the different plotting options.

Exercise

Working with the Plotting Options

Continue with the CAM.DWG.

Command: *Choose the Print/Plot button in the toolbar*	Opens the Plot Configuration dialog box
Choose the Rotation and Origin *button in the Scale, Rotation, and Origin area*	Opens the Plot Rotation and Origin dialog box
Type **0** *in both the* **X** Origin *and* **Y** Origin *edit boxes*	Sets the plot origin back to 0,0
Choose OK	
Choose Display *under Additional Parameters*	Highlights the Display radio button
Choose Scale to Fit *under Scale, Rotation, and Origin*	Checks the Scale to Fit check box
In the Plot Preview area, choose Full *and then the* Preview *button (see fig. 10.23)*	Creates a quick preview of the drawing
Choose End Preview *in the Plot Preview dialog box*	Exits back to the Plot Configuration
Choose Extents *and* Limits *and view the preview*	
Choose the View *button under Additional Parameters*	Opens the View Name dialog box
If views are saved, highlight a name and choose OK	Highlights the View radio button
Preview the saved view	
Choose the Window *button under Additional Parameters*	Opens the Window Selection dialog box

continues

Choose the **P**ick < *button*	Exits to the drawing area
Form a window around the keyway detail	Opens the Window Selection dialog box with new First Corner and Other Corner settings
Choose the P**r**eview *button*	
Choose **E**nd Preview	

A few other settings are available that provide control during the printing and plotting process.

The Hide **L**ines option under Additional Parameters is used when working with 3D drawings. This toggle enables you to plot the file in the same state that is displayed, when performing the HIDE command.

Ad**j**ust Area Fill uses the Pen Assignments settings to move the pens over one half the width specified. When filling in solids and polylines with width, moving the pen over will ensure that the proper boundary is maintained.

Use Pre**v**ious Printer is self-explanatory. If you have switched printers while plotting, this toggle overrides the current printer in favor of the last device specified.

The next section describes the use of the other options listed under Additional Parameters.

Export Options for Alternative Printing

At times, a plot may be needed for incorporating into other Windows applications. Many of these applications will handle a .PLT (or plot) file. Instead of producing a print, AutoCAD LT produces a file with a series of instructions that are used to generate the entities within the drawing.

Many firms use a spooler to send drawings to a plotter or printer. A *spooler* is a utility or piece of hardware that enables connected users to continue with their tasks instead of waiting for the whole plotting process to complete. These spoolers queue up the plots, which are printed one at a time. Many firms will send all of their plots to the spooler at the end of the day and let the plotter churn out drawings all night. This works exceptionally well with a laser plotter, but can be hazardous on plotters with ink pens that might clog or dry up. Before you use a spooler, make sure you have some way of continuously feeding paper to your plotter.

In order to create this file, you must choose the Plot to **F**ile option in the Plot Configuration dialog box. In the next exercise, you create a .PLT file. Open the CAM.DWG that is located in the AutoCAD LT directory. The drawing is in paper space, so it *should* be plotted 1:1 or full scale.

Exercise

Creating a .PLT file

Continue with the CAM.DWG.

Set the running osnap to ENDpoint

Use the DISTANCE command to determine the height of the drawing. The result should be as follows:

Distance=17.00, Angle in XY plane: 90, Angle from XY plane: 0, Delta X= 0.00, Delta Y=17.00, Delta Z= 0.00	Double the proper size of 8.5"
Command: *Choose the* Print/Plot *button in the toolbar*	Opens the Plot Configuration dialog box
Choose Ex**t**ents *and* Plot to **F**ile	
Set Plotted Inches *to* **1** *and* Drawing Units *to* **2**	Corrects the doubled border size
Choose the P**r**eview *button*	See fig. 10.23
Revise rotation if incorrect	
Choose the File N**a**me *button*	Opens the Create Plot File dialog box
Type **CAM1**, *then choose* OK	
Choose OK	Regenerating message appears

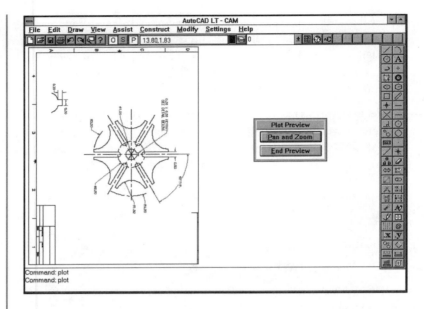

This .PLT file can now be imported into various Windows applications like Word or Excel for illustration purposes. As you can see, this is a great method of producing quite technical drawings to add interest to documents.

A .PLT file can also be copied to your printer if you are having difficulty printing files to a laser printer. If you are copying the file to a plotter on the LPT port, you may receive unexpected results. The plotter driver should be updated by the plotter manufacturer in order to run directly from AutoCAD LT.

TIP

.PLT files can be directly copied to your laser printer using a simple DOS function. This may help in an emergency when the printer is not responding properly and the drawing must be printed.

As you are a novice when printing files, it may be necessary at the beginning to satisfy your patience by circumventing a direct plot from AutoCAD LT. This little trick will help you during troubleshooting or if your systems manager is busy and a print must be done. In the next exercise, your laser printer is assumed to be connected to LPT1. Using the dreaded DOS, you copy the CAM1.PLT to the printer.

Printing the .PLT file from DOS

Continue with the CAM.DWG.

`Command:` *Hold down the Alt key, press Tab until the Program Manager icon appears, and then release*	Transfers you to Program Manager
From the Main window, double click on the MSDOS Prompt icon	
`C:\WINDOWS>CD\LEARNING` (Enter)	Changes to the LEARNING directory
`C:\WINDOWS>CD\ACLTWIN` (Enter)	Changes to the AutoCAD LT directory
`C:\ACLTWIN>COPY /B CAM1.PLT LPT1:` (Enter)	Copies the file to the laser printer
`C:\ACLTWIN>EXIT` (Enter)	Exits back to Windows
Hold down the Alt key, press Tab until the AutoCAD LT icon appears, and then release	

In the previous exercise, the /B in the DOS command automatically issues a form feed to the printer. If this is not included, the form feed must be done manually or you must wait until another document is printed.

Exporting Files

There are several ways to export a file from AutoCAD LT for use with many desktop publishing applications. As mentioned previously, these options can be utilized for marketing efforts or to spruce up a proposal document. Many firms are boosting their image by incorporating these files to display their technological capacity. These options are accessible through Import/Export under the File menu.

The first option, Make **S**lide, creates a screen capture of the active drawing area. Likewise, the **V**iew Slide option within the pull-down menu displays slides that have been created. Slides are particular to Autodesk Products.

D<u>X</u>F Out is the first exporting option to discuss. *Drawing interchange format* (DXF) is the de facto standard in the CAD world. Every decent CAD package has the ability to export and import a DXF file. This enables you to communicate with a majority of the people who utilize CAD. DXF files are ASCII text—lists of code that describe all of the entities and settings of a particular drawing. The commands **DXFOUT** and **DXFIN** are used to export and import DXF files.

The **B**lock Out option will create a WBLOCK. This was discussed in Chapter 3.

Blocks can be assigned attributes. *Attributes* are special entities that contain text information that is used in many applications. If you create a block of a bolt, attributes that are assigned can describe the bolt's material, size, or even cost. The **A**ttributes Out option enables you to extract the information from blocks that contain attributes. This information can be placed in a spread-sheet program to provide costs or other inventory tasks.

Pos<u>t</u>Script Out (PSOUT) enables you to create an EPS file from AutoCAD LT. If you are familiar with desktop publishing, Encapsulated PostScript files are commonly used as a method to transfer files between graphics programs.

PostScript is a vector-based language that has been used in illustration and presentation software for years. The ability to export this file increases the capability of AutoCAD LT to a multifunctioning drawing tool.

When using this command, you are asked questions similar to those asked while plotting, with the exception of a screen preview option. A screen preview can be included with the file in EPS or TIFF format. If you choose to include it, you are asked to determine the image size in pixels. The preview image is used by the importing software package to give you a preview of the EPS file.

NOTE

The larger the preview image size in pixels, the larger the EPS file.

The last exporting option is W<u>M</u>F Out or WMFOUT. This is the command to export a Windows Metafile Format. Although AutoCAD LT only exports WMF files as vectors, WMF files can contain vector- and raster-based graphics to be included in other Windows applications. When vector objects are placed in the Clipboard, they are temporarily stored as WMF files.

BM**P** Out is controlled by the device setting mentioned earlier in the chapter. By specifying any of the graphics files as a device, AutoCAD LT creates the graphics file like a plot. The Dith**e**ring option, listed under Additional Parameters in the Plot Configuration dialog box, creates *dithered* output. This means that the printer creates a grayscale to mimic the colors on the screen. Depending on the type of printer you have, the dithering may be controlled in the Print Manager. Experiment with these settings to determine what works best for your system.

Correcting Common Printing Errors

Unfortunately, there are a number of factors to consider when a plot does not work. If you are printing to a laser printer, the system printer should work if other applications are printing properly. The first step would be to check the obvious:

◆ Is all of the hardware (cables, computer, printer, plotter) connected properly and functioning? Hardware issues can be very tricky for new computer users. The most frustrating problem with plotting is that you assume you are the problem. Double-check your hardware—a bad cable can ruin your day.

◆ Is the device turned on?

◆ Does the device need to be initialized by rebooting the system? If so, properly exit all of the running applications and exit Windows before you reset your computer and printer. The printer may have a variety of reset levels accessible through its menu, in addition to simply turning the printer off and on. Also, your PC may require a full cold boot rather than a warm boot or reset to completely clear the printer interface.

◆ Is the paper loaded? Is it aligned properly?

◆ Is the plot area listing equal to the paper size?

◆ Are the pens clogged? Do they need more ink? Are they in the carousel correctly? Do the pens correspond with the color and width indicted by the pen assignments? Many plotters have settings for pen speed. Is the speed causing the pens to skip?

◆ Are the linetypes printing correctly? If not, check the LTSCALE setting.

In addition, you might need to review the previous discussions in the book regarding origin, paper space and model space, rotation, and scale. This may lead you to the source of the problem.

It is impossible to cover all of the problems that could be associated with printing. Consider it a blessing to have them happen when you have time to investigate. Over time, you will understand the problems that can occur and will develop the ability to resolve them quickly.

Summary

This chapter covered various aspects of plotting and printing AutoCAD LT files. You have explored the reproduction process and should feel comfortable with the Plot Configuration dialog box. Setting up the printing device and assigning pens should be understood. You have learned to scale the plots and to check the prints before they are executed to the plotter or printer. Finally, alternative methods of exporting files were evaluated.

Now that you can create plots of the .DWG files, you can proceed to the next chapter. The balance of the book is very specific to mechanical and architectural applications. Both disciplines rely heavily on hardcopy at various stages in the design and production drawing process.

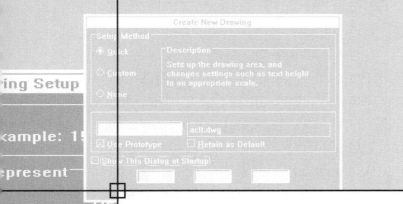

Part Four

Projects to Challenge Your Knowledge

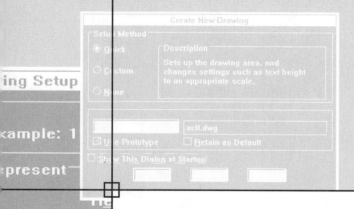

This chapter is devoted to creating an adjustable sliding support. You will explore the development of the part, from how it lays out on an A-size sheet to plotting the drawing at a proper scale. Although it is impossible to cover all the aspects you will encounter in mechanical drawing, this chapter should give an accurate description of a typical production drawing. The following topics are covered:

- ◆ Determining the layout of the sheet
- ◆ Developing the orthographic views
- ◆ Utilizing grips
- ◆ Annotating and dimensioning the drawing
- ◆ Printing the drawing

This chapter will utilize the knowledge you have gained about CAD to this point in the book. If you need to refresh your memory, return to the chapters specified in the examples. The exercises in this chapter are not the typical step-by-step instructions you are accustomed to reading. Each exercise assumes that verbal instructions and a few hints are all you need to complete each step.

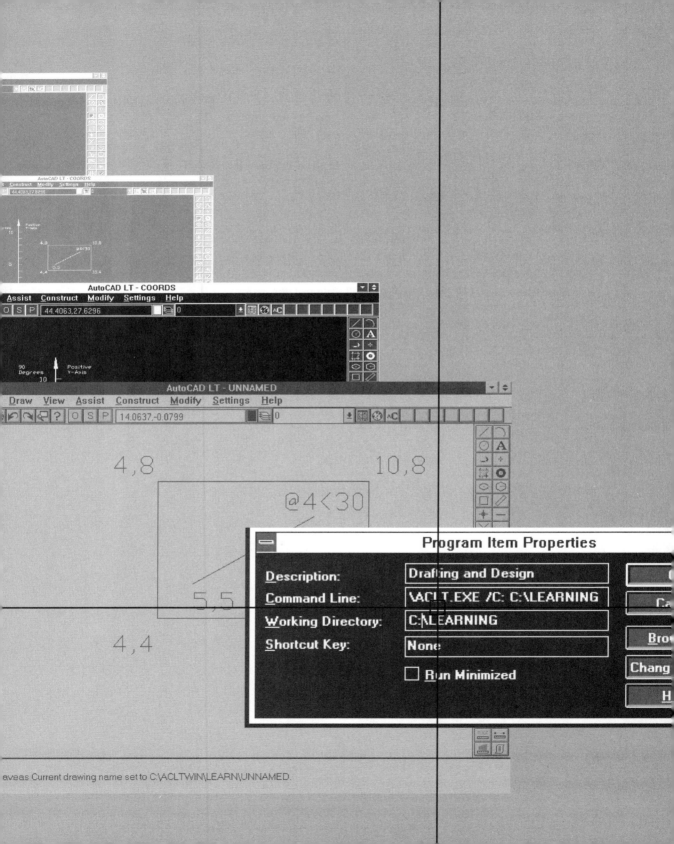

AutoCAD LT - COORDS

Assist Construct Modify Settings Help

44.4063,27.6296 0

90
Degrees   Positive
                Y-Axis
10

AutoCAD LT - UNNAMED

Draw View Assist Construct Modify Settings Help

14.0637,-0.0799 0

4,8   10,8

@4<30

5,5

4,4

Program Item Properties

Description:	Drafting and Design
Command Line:	\ACLT.EXE /C: C:\LEARNING
Working Directory:	C:\LEARNING
Shortcut Key:	None

☐ Run Minimized

Bro

Chang

H

Creating a Mechanical Drawing

This chapter will give you a glimpse into the real-world task of creating a part drawing. Designing the part is a task assigned to licensed engineers or seasoned draftspeople. The world of engineering encompasses so many disciplines that if someone is described as an "engineer," you might be talking about an individual who invents or designs machines, develops software, determines where the roads can be widened, designs cars, submarines, or bridges, or even drives a train. The world is full of specialized people, and engineering is no exception. Engineers are supported by people who bring the ideas to paper—the draftspeople. Depending on the size of the firm, there is typically a hierarchy followed that dictates who is responsible for each aspect of a production drawing.

The task at hand is to fully understand the part you are about to draw. Terminology that may or may not be familiar to you is discussed in this chapter, so an attempt will be made to define unique items.

Describing the Part

The part you will be working with is an adjustable sliding support, as shown in figure 11.1.

Figure 11.1

The adjustable sliding support.

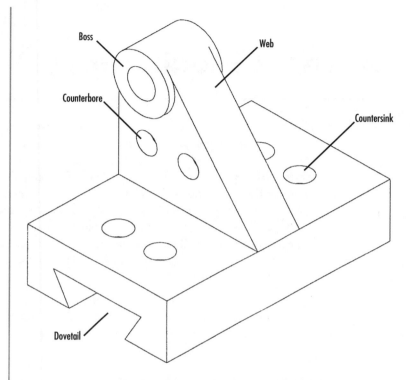

The part is molded from cast iron, and several machining techniques are used to create the features of the part, as follows:

◆ **Boss.** The boss that is noted is a cylindrical, knob-like projection on a casting. This projection must be noted while you are dimensioning. Typically, the boss will have a small fillet or radius against the part.

◆ **Dovetail.** The dovetail is formed in two steps. A rectangular channel is formed, followed by the angles, which are milled on separate passes. The angles and opening size will need to be dimensioned.

◆ **Counterbore.** A hole that contains a counterbore is machined with a larger diameter bit for a certain depth. The larger hole provides a recessed surface for a bolt or pin to rest. There are two methods in which to dimension the hole; both will be explained later in the chapter.

◆ **Countersink.** A countersunk hole has a cone-shaped recess drilled into the opening. This enables a cone-shaped flat head bolt or screw to remain flush with the surface of the part. There are two ways to dimension these holes, which will be explored later in the chapter.

Many terms are associated with the elements in machine parts. If you decide to pursue a career in drafting these parts, the terms will quickly become part of your vocabulary. One of the best references for element names are machinists—they know their work.

Before you begin to create the drawings, there are several factors that need to be considered. The next few sections will help you to develop an understanding of the decisions that should be made before proceeding with a project. Although one of the benefits of AutoCAD LT lies in the capability to manipulate a drawing after the fact, it is important to develop good strategic habits, at the beginning of a project, in order to avoid wasting time.

Laying Out the Page

There are many ways to organize the drawings on a page. Conventions will dictate some of the view's positions, but the overall purpose of the drawing is to translate an idea for a machinist. Many firms will have standards developed that make this task easier, but depending on the circumstances or the geometry of the part, you may need to divert from set conventions.

One of the first questions to ask is, what is the orientation of the paper? A landscape or horizontal orientation is conventional, but the drawing may be suited for a vertical or portrait arrangement. For this particular drawing, use a landscape orientation.

Review the page layout section in Chapter 2, "An Overview of Drafting Tools and Techniques," for additional information.

The next question concerns how many views or drawings will be included on the sheet. After that decision is made, the sheet size can be resolved.

Determining Sheet Size

There are many factors that need to be considered when determining sheet size, as follows:

- ◆ **You might be restricted due to a client preference.** It is not unheard of to provide interior design drawings on B-size sheets.

- ◆ **Convention may restrict your sheet size.** Many architectural firms create all of their drawings on D-size sheets.

- ◆ **Available resources may restrict the sheet size.** If your only plotting device is a printer, you may be restricted to A-size sheets.

This particular drawing will be constructed to fit on an A-size sheet. It is assumed that the computer on which you are working is connected to a system printer, which prints to A-size sheets. Also, this is just an exercise. When you start to do work that must portray a professional image, contact a service bureau for instructions or invest in a plotter.

Determining Layers

As mentioned earlier in the book, it is important to establish layers before developing the drawing. The drafting industry has a difficult time agreeing on a convention for layer names, but your firm or local market may have them developed.

In Chapter 4, "Multiview Drawings," you were introduced to the theory of layers. You will recall that the drawing will be broken down into layers. The support drawing will be assigned layers. They are listed with their color and the linetype, as follows:

- ◆ **Border:** Green/Continuous—Contains the border of the drawing.

- ◆ **PS-Border:** Green/Continuous—Contains the viewport borders of the drawing in paper space.

- ◆ **Object:** Green/Continuous—Contains the object lines of the drawings.

- ◆ **Hidden:** White/Hidden—Contains the hidden lines.

- ◆ **Dims:** Blue/Continuous—Contains the dimension lines.

- ◆ **Text:** Red/Continuous—Contains the text of the drawing.

- ◆ **Const:** Cyan/Continuous—Contains the construction lines of the drawing.

These are not conventional layer names; as mentioned earlier, prepare to follow a set guideline.

In the next exercise, you create a new drawing and set up the layers. The exercise takes you through the creation of the first layer.

Creating the Layers

Exercise

Create a new drawing.

`Command:` *Choose* **S**ettings, Li**n**etype Style, *then* **L**oad	Issues the LINETYPE command
`?/Create/Load/Set:` **L** `(Enter)`	Selects the Load option
`Linetype(s) to load:` ***** `(Enter)`	Selects all the linetypes available and opens the Select Linetype File dialog box
Choose OK	Accepts the file ACLT.LIN
`?/Create/Load/Set:` `(Enter)`	Ends the LINETYPE command
`Command:` *Choose the Layers button from the toolbar*	Opens the Layer Control dialog box
Type the word **BORDER** *and choose* Ne**w**	Places a new layer name in the Layer Name list
Click on the BORDER *listing, then choose the* Set Color *button*	Highlights the name in the list and opens the Select Color dialog box
Under Standard Colors, *choose green,* then OK	Sets the BORDER layer to green
Double-click the left mouse button in the text box and type the next layer name	Highlights the word BORDER
Repeat the process for the other layers listed previously (see fig. 11.2)	
Remember to set the Ltype on the hidden layer	

Figure 11.2

The completed listing of layers.

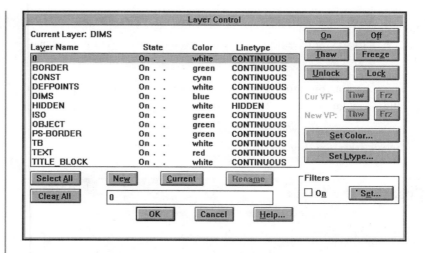

NOTE

The extra layers that are listed will be imported with the title block that is inserted in the next exercise. When you insert a wblock into your drawing, all the layers associated with that wblock will be added to your list. The 0 layer is a default layer in all AutoCAD LT drawings.

The next step is to insert and manipulate a title block.

Creating a Title Block

Chapter 2 contained a brief overview of title blocks and how they are used in the industry. Title blocks in some form are required on all drawings. These blocks of text display a lot of information about the drawing on the page. You will insert a pre-drawn title block and fill in the necessary information.

Chapter 4 covered the idea of limits and how limits are used in AutoCAD LT. In the next exercise, you also determine the limits of the support drawing.

Continuing from the last exercise, insert a title block and manipulate it into position within paper space.

Inserting the Title Block

Continue with the same drawing from the last exercise.

Command: *Choose* **V**iew, *then* **T**ilemode	
New value for Tilemode<1>: 0 Regenerating drawing	Sets the TILEMODE to paper space
Command: *Choose* **D**raw, *then* Insert Bloc**k**	Issues the DDINSERT command
Choose the **F**ile *button*	Opens the Select File dialog box
Select the file ISO_A3.DWG *from the ACLTWIN directory, then* OK	
Choose **E**xplode *then* OK *in the Insert dialog box*	
Insertion point: 0,0 (Enter)	Specifies the insertion point
Scale factor<1>: (Enter)	
Rotation angle<0>: (Enter)	
Zoom Extents	
Command: **CHPROP** (Enter)	
Select objects: **ALL** (Enter)	Selects all visible entities
Select objects: (Enter)	
Change what property(Color /LAyer/LType/Thickness)? **LA** (Enter)	Prepares to change the layer of the selected entities
New layer <varies>: **BORDER** (Enter)	
Change what property(Color /LAyer/LType/Thickness)? (Enter)	

continues

Figure 11.3

*Stretching the title
block into proportion.*

*Use the DISTANCE command to
determine height of the current title
block page marks*

*SCALE all of the entities using the
Reference option. 273" (current size
as shown by the previous Distance
command) should equal 8.5" (new
size after Scale)*

Zoom Extents

Zoom .8x

*Use the DISTANCE command to
determine width of the current title
block page marks*

Command: *Choose the Stretch button*   Issues the STRETCH command

First corner: *Choose* ①   Sets first point of crossing window
(see fig 11.3)

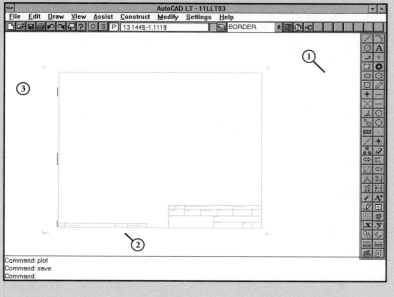

`Other corner:` *Choose* ②	Sets diagonal corner of crossing window
`Select objects:` (Enter)	Accepts the selection set
`Base point or displacement:` *Pick anywhere*	Sets the base point
`Second point of displacement:` `@1.3297<180` (Enter)	
`Command:` (Enter)	Reissues the STRETCH command
`First corner:` *Choose* ①	Sets first point of crossing window
`Other corner:` *Choose* ③	Sets diagonal corner of crossing window
`Select objects:` (Enter)	Accepts the selection set
`Base point or displacement:` *Pick anywhere*	Sets the base point
`Second point of displacement:` `@.1<270` (Enter)	
`Command:` *Choose* **S**ettings, Dra**w**ing, *then* **L**imits	Issues the LIMITS command
`ON/OFF/<Lower left corner>:` `.5,.25` (Enter)	Sets the lower left corner
`Upper right corner:` `10.7,8.2` (Enter)	Sets the upper right corner
`Command:` *Press* F7 *to check the limits*	Grid displays to limits only
`Command:` *Choose the Print/Plot button*	Issues the PLOT command
Set the plot to Limits, 270° rotation, Y Offset = .2, then Preview the full drawing	View the border as it will appear
Save the drawing as SUPP-1 *in the* ACLTWIN *directory*	

TIP

There are a variety of title blocks that ship with AutoCAD LT. They are located in the ACLTWIN directory. Use these templates as a guide toward what to expect in a title block in the "real world."

In Chapter 5, "Text and Annotation," you examined the methods of inserting text. Those procedures will be used to insert text into the title block. The following is an explanation of each part of the title block and border:

- **Itemref.** This is an internally generated number. It might be used to identify the client for billing purposes. Use **1-956**.

- **Quantity.** This indicates how many of these pieces are needed. Use **4**.

- **Title/Name,designation,etc.** Typically, the part title is located at the top of the block. In this case, list the material used for the support. Use **CAST IRON**.

- **Article No./Reference.** Another internally generated number for organizing and billing. This is the fourth project for this particular client. Use **1-956-4**.

- **Designed by.** This text is either inserted or manually initialized after the plot. At times, the designer is also the drafter. Insert the initials, **PAM**.

- **Checked by.** All drawings need to be checked by a supervisor or competent associate. This text is added after the plot unless the approval process is completed on-screen. Use your initials.

- **Approved by-date.** Another set of eyes will be examining the drawing and adding the date. This may occur after all the pieces within a drawing set are completed. The person who signs this off might be considered the project manager. Use your initials again.

- **Filename.** Providing unique names will help keep track of the specific drawings. Use **SUPP-1**.

- **Date.** Insert the date that the initial drawing was completed, before revisions.

- **Scale.** Although there can be many drawings with different scales on a sheet, there is generally one scale that can describe the main drawing. On sheets with several different scales, use VARIES. On this particular drawing, use **1=.75** or **3/4**.

◆ **Owner.** The firm or project for which this drawing has been created. It is also common to list your firm on the drawing. In this case, use **New Riders Publishing, 201 West 103rd St, Indianapolis, IN, 46290**.

◆ **Title/Name.** List the name of the part. In this case, insert **ADJ. SLIDING SUPPORT**.

◆ **Drawing number.** Another internally generated number. Usually this number is generated for cross-referencing purposes. If this part is depicted in an assembly drawing, there will be a cross-reference to this drawing sheet. Use **4437-55**.

◆ **Edition.** This is the first edition. Use **1**.

◆ **Sheet.** There can be many sheets in a drawing set, particularly a large assembly drawing. This is considered the first set in this set. Use **1**.

The next portion of the title block belongs to revisions. After the drawing has been given an issue date, chances are there will be adjustments to the drawing. Anyone receiving the drawing will be able to see which sheet has been adjusted.

Each revision is given a number. It is not uncommon to draw a balloon around the change in order to identify exactly where the drawing has been adjusted and to add a cross-reference to the revision number. The revision box provides a space for a short description. These descriptions can be cryptic due to limited space.

The revision should be checked by another associate and a date should be indicated for reference. The revisions listed are created from the bottom up. There are no revisions for this particular drawing.

Along the bound edge of the title block are three strings of text. This text should be edited to describe the drafter, the latest revision date, and the file name.

NOTE Remember, most firms have established title blocks that have been utilized for years. Title blocks tend to be as individual as the companies who use them. Often, it is a chance to prominently place the company's logo for quick recognition.

Continuing with the exercise, you will fill in the title block with text that is full scale. If your final text height is 1/4" on the plotted page, insert the text at 1/4" high. The following list describes the areas of the title block and the information that should be filled in.

Exercise

Inserting Text into the Title Block

Continue with the SUPP-1.DWG.

Command: *Choose* **S**ettings, *then* **T**ext Style	Opens the Select Text Font dialog box
Choose Romans Simplex, then OK	Sets the text style to Romans
Accept all the defaults with **Enter**	
Set the current layer to TEXT	
Command: *Choose the Edit_Text button*	Issues the DDEDIT command
<Select a TEXT or ATTDEF object>/Undo: *Choose the vertical USER text in the lower left corner*	
Change the text to your name, then OK	
<Select a TEXT or ATTDEF object>/Undo: **Enter**	
Command: *Choose the Zoom button on the toolbar*	Issues the ZOOM command
Zoom into the title block with a Window	
Command: *Choose the Text button*	Issues the DTEXT command
Justify/Style/<Start point>: *Pick a point in the lower left corner of the Itemref box*	
Height<0.2000>: **.05** **Enter**	Sets the height

```
Rotation angle<0>: (Enter)          Accepts the rotation as 0

Text: 1-956 (Enter)

Text: (Enter)

Command: (Enter)                    Reissues the DTEXT command

Justify/Style/<Start point>:
```
Pick a point in the lower left corner of the Quantity box

```
Height<0.0500>: .1 (Enter)          Sets the height

Rotation angle<0>: (Enter)          Accepts the rotation as 0

Text: 4 (Enter)

Text: (Enter)
```

Use DDEDIT to change the next title to read Material *(see fig. 11.4)*

Continue inserting the text using the information listed previously

The height may need to be adjusted based on the information needed in each box

You may need to MOVE the text in order to place it properly in the title boxes

Save the drawing

Now that the title block is done, move back to model space and begin the task of creating the views for the support.

Figure 11.4

The completed title block text.

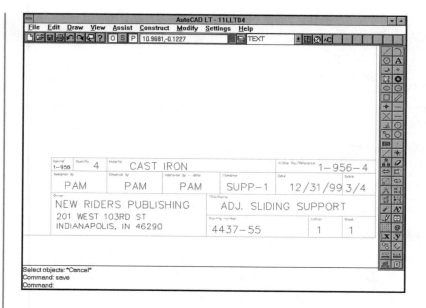

Blocking Out the Drawing

Creating a rough shape for the views of the drawing will help you establish the layout of the sheet. Figure 11.5 illustrates the basic size of the part.

Use these dimensions for the next exercise, in which you generate boundary boxes for each view. The current drawing is in paper space. When the part is laid out, it will be developed in model space.

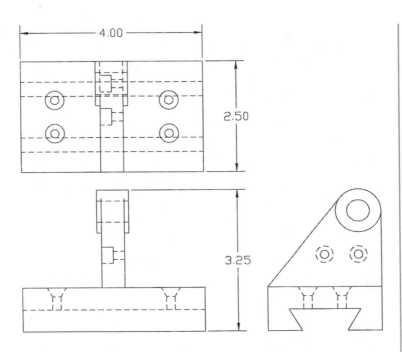

Figure 11.5

The rough dimensions of the support.

Exercise

Blocking Out the Three Views

Continue with the SUPP-1.DWG.

Command: *Choose* **V**iew, *then* **T**ilemode

New value for Tilemode<0>: 1   Sets the TILEMODE to model
Regenerating drawing   space

Change the current layer to CONST

Draw a horizontal line from 0,0, 10" in the 0° direction

Draw a vertical line from 0,0, 7" in the 90° direction

Command: *Choose* **C**onstruct, *then*   Issues the OFFSET command
Offset

continues

`Offset distance or Through` `<Through>:` **4** (Enter)	Sets the distance to 4
`Select object to offset:` *Pick* ① *(see fig. 11.6)*	Highlights the line
`Side to offset?` *Pick to the right*	Duplicates the line
`Select object to offset:` *Right* *click the mouse twice*	Ends and reissues the OFFSET command
`Offset distance or Through` `<4.0000>:` **1** (Enter)	Sets the distance to 1
`Select object to offset:` *Pick* ②	Highlights the line
`Side to offset?` *Pick to the right*	Duplicates the line
`Select object to offset:` *Right* *click the mouse twice*	Ends and reissues the OFFSET command
`Offset distance or Through` `<1.0000>:` **2.5** (Enter)	Sets the distance to 2.5
`Select object to offset:` *Pick* ③	Highlights the line
`Side to offset?` *Pick to the right*	Duplicates the line
`Select object to offset:` *Right* *click the mouse twice*	Ends and reissues the OFFSET command

Use this technique to establish the horizontal construction lines. Place a 1" space between the views

Save the drawing

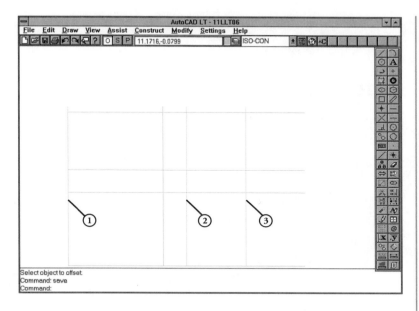

Figure 11.6

The blocked-out views.

Now the construction lines for the three views are created. The next step is to develop the views.

Developing the Views

In Chapter 3, "CAD Drawing Techniques," you were introduced to the concept of orthographic drawing techniques. These techniques will be used to generate the views of the support. You will generate the front, side, and top views. Begin by constructing the front view of the support.

Creating the Front View

In this section, you create a front view of the support. The front view is currently blocked out on the drawing in the lower left corner. Figure 11.7 illustrates the dimensioned front view.

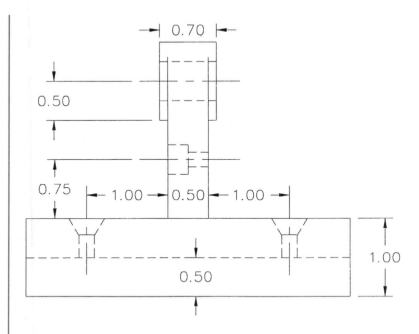

In the next exercise, you create the major elements for the front view.

Exercise

Creating the Front View

Continue with the SUPP-1.DWG.

Command: *Choose the Trim button*	Issues the TRIM command
Select cutting edge(s):	
Select objects: **ALL** (Enter)	Selects all entities
Select objects: (Enter)	Accepts the selection set
<Select object to trim>/Undo: *Choose* ①②③④ *(see fig. 11.8)*	Trims the lines
<Select object to trim>/Undo: (Enter)	Ends the TRIM command
Command: *Choose* **C**onstruct, *then* **O**ffset	Issues the OFFSET command

`Offset distance or Through` `<2.5000>:` **1** (Enter)	Sets the distance to 1
`Select object to offset:` *Pick ⑤*	Highlights the line
`Side to offset?` *Pick above the line*	Duplicates the line
`Select object to offset:` *Pick ⑥*	Highlights the line
`Side to offset?` *Pick below the line*	Duplicates the line
`Select object to offset:` *Right click the mouse twice*	Ends and reissues the OFFSET command
`Offset distance or Through` `<1.0000>:` **1.75** (Enter)	Sets the distance to 1.75
`Select object to offset:` *Pick ⑦*	Highlights the line
`Side to offset?` *Pick to the right*	Duplicates the line
`Select object to offset:` *Right click the mouse twice*	Ends and reissues the OFFSET command
`Offset distance or Through` `<1.7500>:` **.5** (Enter)	Sets the distance to .5
`Select object to offset:` *Pick ⑧*	Highlights the line
`Side to offset?` *Pick to the right*	Duplicates the line
`Select object to offset:` *Pick ⑤*	Highlights the line
`Side to offset?` *Pick above the line*	Duplicates the line
`Select object to offset:` *Pick ⑥*	Highlights the line

continues

`Side to offset?` *Pick below the line*	Duplicates the line
`Select object to offset:` *Right click the mouse twice*	Ends and reissues the OFFSET command
Use this technique to establish the width of the bossed piece. Use .1 offsets for a total width of .7	
Zoom Window on the front view	
Save the drawing	

Figure 11.8

Constructing the front view.

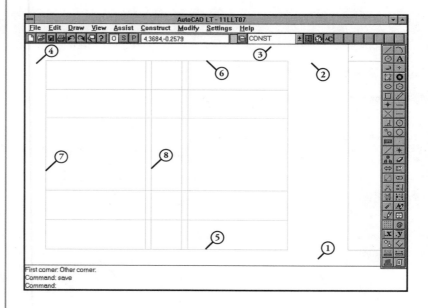

TIP

Remember, each view has information that another view requires. Keep the parts lined up for easy reference.

The next step in the exercise is to draw some of the object lines for the part.

Creating the Object Lines

Exercise

Continue with the SUPP-1.DWG.

Set the object snap to Intersection and the current layer to OBJECT

Command: *Choose the Line button* — Issues the LINE command

From point: *Pick ① (see fig. 11.9)* — Sets first point of line

To point: *Pick ②*

To point: *Pick ③*

To point: *Pick ④*

To point: **c** (Enter) — Closes the line by choosing the first point specified

Command: (Enter) — Issues the LINE command

From point: *Pick ⑤* — Sets first point of line

To point: *Pick ⑥* — Sets second point

To point: *Right click the mouse twice* — Ends and reissues the LINE command

From point: *Pick ⑦* — Sets first point of line

To point: *Pick ⑧* — Sets second point

To point: *Right click the mouse twice* — Ends and reissues the LINE command

From point: *Pick ⑨* — Sets first point of line

To point: *Pick ⑩*

To point: *Pick ⑪*

To point: *Pick ⑫*

To point: **c** (Enter) — Closes the line by choosing the first point specified

continues

Command: *Choose the Layers button*	Opens the Layer Control dialog box
Highlight the CONST layer, then choose the Free**z**e *button, then* OK	Hides the CONST layer
Save the drawing	

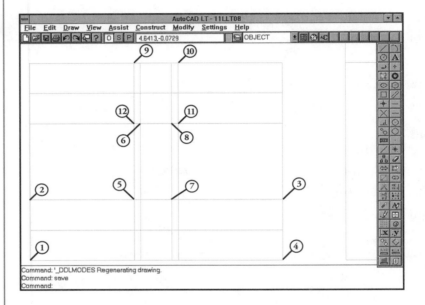

The front view needs to be refined with additional information. Continuing the exercise, you add the hidden lines and locate the countersunk set screws.

Exercise

Refining the Front View

Continue working on the SUPP-1.DWG.

Thaw all the layers and make the HIDDEN layer current

Command: *Choose the Line button*   Issues the LINE command

```
From point: Pick ① (see fig. 11.10)

To point: Pick ②

To point: (Enter)
```

Command: *Choose* **C***onstruct, then* **O***ffset*	Issues the OFFSET command
`Offset distance or Through <Through>:` **.25** (Enter)	Sets the distance to .25 for half the bore width
`Select object to offset:` *Pick ③*	Highlights the line
`Side to offset?` *Pick below the line*	Duplicates the line
`Select object to offset:` *Pick ③*	Highlights the line
`Side to offset?` *Pick above the line*	Duplicates the line
`Select object to offset:` *Choose the Line button*	Issues the LINE command

```
From point: Pick ④

To point: Pick ⑤
```

Repeat for the top edge of the bore

Offset the vertical boundaries of the front view toward its center .75"

Save the drawing

NOTE Countersunk holes are used to keep the set screws flush.

Next, you zoom into one of the countersunk set screws and create it as a hidden object. The set screw holes are 3/16" diameter with 60° countersink angle. The widest circle of the countersunk hole will be .44" in diameter. You also adjust the linetype scale to display the hidden lines more accurately.

Figure 11.10

*Refining the
front view.*

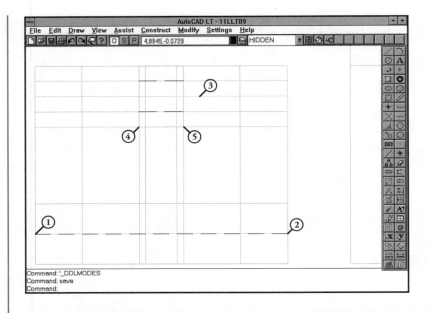

Exercise

Generating the Countersunk Holes

Continue with the SUPP-1.DWG.

Command: *Choose the Zoom button on the toolbar*	Issues the ZOOM command
Use the Window option	
First corner: *Pick* ① *(see fig. 11.11)*	
Other corner: *Pick* ②	
Command: *Choose* **C**onstruct, *then* **O**ffset	Issues the OFFSET command
Offset distance or Through <Through>: **.22** (Enter)	Sets the distance to .22 for half the countersunk hole width
Select object to offset: *Pick* ③	Highlights the line
Side to offset? *Pick to the right*	Duplicates the line

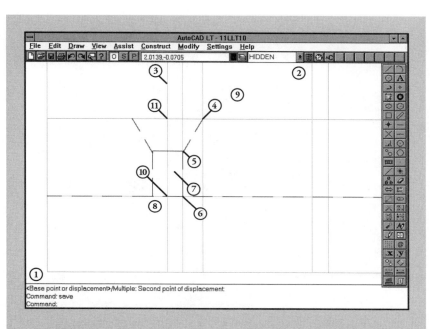

Figure 11.11

Creating the countersunk holes.

Select object to offset: *Right click the mouse twice*

Offset distance or Through <Through>: **3/32** (Enter)	Sets the distance to 3/32 for half the hole width
Select object to offset: *Pick* ③	Highlights the line
Side to offset? *Pick to the right*	Duplicates the line
Select object to offset: *Choose the Line button*	Issues the LINE command
From point: *Pick* ④	
To point: **@.5<240**	Creates the countersink angle
To point: *Right click the mouse twice*	Ends and reissues the LINE command
From point: *Pick* ⑤	

continues

To point: *Choose the Perpendicular button*	Sets the object snap to PERpendicular
PER to ③	
To point: *Right click the mouse twice*	Ends and reissues the LINE command
From point: *Pick* ⑤	
To point: *Pick* ⑥	
To point: *Choose the Trim button*	Issues the TRIM command
Select cutting edge(s):	
Select objects: **L** (Enter)	Highlights last entity created
Select objects: (Enter)	Accepts selection set
<Select objects to trim> /Undo: *Pick* ⑦	
<Select objects to trim> /Undo: *Choose* **C**onstruct, *then* **M**irror	Issues the MIRROR command
Select objects: *Pick* ⑧	
Other corner: *Pick* ⑨	
Select objects: (Enter)	
First point of mirror line: *Pick* ⑩	
Second point: *Pick* ⑪	
Delete old objects? <N>: (Enter)	
Command: *Choose* **S**ettings, *then* Li**n**etype Scale, *then* Li**n**etype Scale	Issues the LINETYPE command
New scale factor <1.0000>: **.5** (Enter)	Sets the scale to .5

Zoom Previous

Copy the entities of the set screw to the other side, as illustrated in figure 11.12

Save the drawing

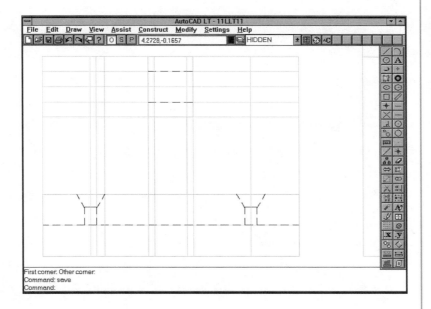

Figure 11.12

Copying the countersunk holes.

In the next exercise, you create the counterbore hole. This is quite similar to the countersunk holes. Begin the exercise by setting a center construction line, then zoom into the area in order to make the hole. The drill hole is 3/16" in diameter, while the counterbore is 3/8" in diameter and 1/4" deep. In addition, grips will be used to select a line at one point in the exercise.

Developing the Counterbore Holes

Continue with the SUPP-1.DWG.

Command: *Choose* **C**onstruct, *then* **O**ffset	Issues the OFFSET command
Offset distance or Through <Through>: **.75** (Enter)	Sets the distance to .75 for half the countersunk hole width
Select object to offset: *Pick* ① *(see fig. 11.13)*	Highlights the line
Side to offset? *Pick above the line*	Duplicates the line
Select object to offset: (Enter)	
Zoom Window	
Command: *Choose* **C**onstruct, *then* **O**ffset	Issues the OFFSET command
Offset distance or Through <Through>: **.25** (Enter)	Sets the distance to .25 for half the bore width
Select object to offset: *Pick* ②	Highlights the line
Side to offset? *Pick to the right*	Duplicates the line
Select object to offset: *Right click the mouse twice*	Ends and reissues the OFFSET command
Offset distance or Through <Through>: **3/32** (Enter)	Sets the distance to 3/32 for half the hole width
Select object to offset: *Pick* ③	Highlights the line
Side to offset? *Pick above the line*	Duplicates the line
Select object to offset: *Right click the mouse twice*	Ends and reissues the OFFSET command

`Offset distance or Through <Through>:` **3/16** `(Enter)`	Sets the distance to 3/16 for half the counterbore width
`Select object to offset:` *Pick* ③	Highlights the line
`Side to offset?` *Pick above the line*	Duplicates the line
`Select object to offset:` `(Enter)`	Ends the command
`Command:` *Choose the Line button*	Issues the LINE command
`From point:` *Pick* ④	
`To point:` *Pick* ⑤	
`To point:` *Pick* ⑥	
`To point:` *Pick* ⑦	
`To point:` `(Enter)`	
Mirror the top half of the counterbore down	
`Command:` *Form a small window around the line created between* ⑤ *and* ⑥	Grips are displayed
Pick the bottommost grip point	Highlights the grip point
Drag the grip to ⑧	
Zoom Previous	
Save the drawing	

NOTE

Counterbore holes are used to recess the heads of bolts and screws, particularly when there is a chance that the bolt or screw could cause an obstruction within a part or assembly.

The front view is basically complete. The only outstanding issue is the way the web meets the boss. A *web* is a thin extrusion, sheet, or plate that functions as a structural element. In order to establish the intersection, you need to construct the side view.

Figure11.13

Creating the counterbore holes.

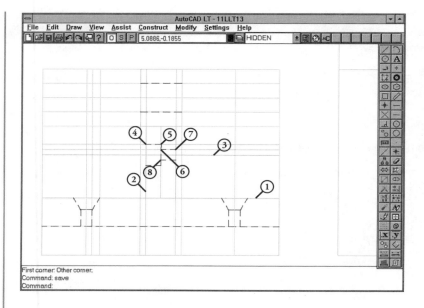

Creating the Side View

The front view will help determine the major elements of the side view. Figure 11.14 displays the dimensioned side view, which you create in the next series of exercises.

Begin the next exercise by offsetting some of the construction lines. These lines are used to locate center points in the side view.

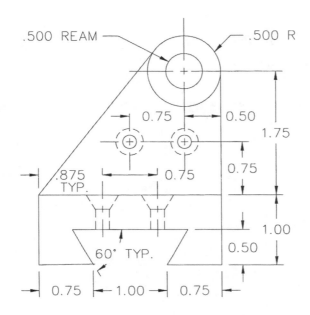

Figure 11.14

The side view.

Laying Out the Side View

Continue working on the SUPP-1. DWG.

Zoom Extents, then Zoom Window into the Side view

Command: *Choose* **C**onstruct, *then* **O**ffset	Issues the OFFSET command
Offset distance or Through <Through>: **.5** Enter	Sets the distance to .5
Select object to offset: *Pick* ① *(see fig. 11.15)*	Highlights the line

continues

`Side to offset?` *Pick to the left*	Creates the bore center line
`Select object to offset:` *Pick ②*	Highlights the line
`Side to offset?` *Pick below the line*	Creates the bore center line
`Select object to offset:` *Right click the mouse twice*	Highlights the line
`Offset distance or Through <Through>:` **7/8** `[Enter]`	Sets the distance to .875
`Select object to offset:` *Pick ③*	Highlights the line
`Side to offset?` *Pick to the left*	Creates the set screw center line
`Select object to offset:` *Pick ①*	Highlights the line
`Side to offset?` *Pick to the right*	Creates the set screw center line
`Select object to offset:` *Right click the mouse twice*	Highlights the line
`Offset distance or Through <Through>:` **1** `[Enter]`	Sets the distance to 1
`Select object to offset:` *Pick ④*	Highlights the line
`Side to offset?` *Pick above the line*	Duplicates the line
`Select object to offset:` `[Enter]`	Ends the command
Save the drawing	

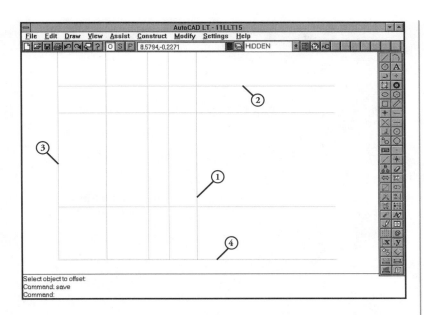

Figure 11.15

Beginning the side view.

The next step involves creating the boss and the bore through it. In the last exercise, you developed the center mark for the bore. Proceed with this exercise by creating the bore and some of the other object lines on the side view.

Refining the Side View

Exercise

Continue with the SUPP-1.DWG.

Change the current layer to OBJECT

Command: *Choose the Circle button*	Issues the CIRCLE command
3P/TTR/<Center point>: *Pick* ① (see fig. 11.16)	Sets the center point
Diameter/<Radius>: *Pick* ②	Creates the boss
Command: **Enter**	Reissues the CIRCLE command
3P/TTR/<Center point>: *Pick* ①	Sets the center point
Diameter/<Radius>: **.25 Enter**	Creates the bore

continues

Command: *Choose the Line button*	Issues the LINE command
From point: *Pick ③*	
To point: *Pick ④*	
To point: *Right click the mouse twice*	Ends and reissues the LINE command
LINE From point: *Pick ⑤*	
To point: *Pick ⑥*	
To point: *Choose the Tangent button*	Selects the TANgent object snap
TAN to *Pick ⑦*	
To point: *Right click the mouse twice*	Ends and reissues the LINE command
LINE From point: *Pick ⑥*	
To point: *Pick ⑧*	Creates the base
To point: (Enter)	Ends the command
Save the drawing	

Figure 11.16

Adding object lines to the side view.

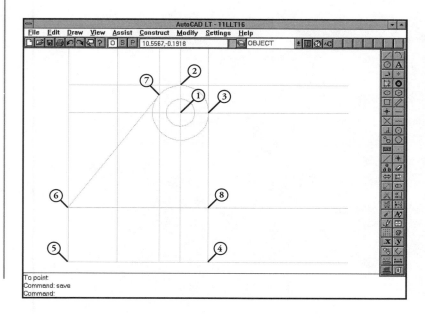

Most of the drawing has been created with existing entities because the dimensions originate from these object lines. The next step in the exercise will create the dovetail, and locate and draw the counterbore holes.

Drawing the Dovetail and Counterbore Holes

Continue with the SUPP-1.DWG.

Leave OBJECT as the current layer

Command: *Choose* **C***onstruct, then* **O***ffset*	Issues the OFFSET command
Offset distance or Through <Through>: **.75** (Enter)	Sets the distance to .75
Select object to offset: *Pick* ① *(see fig. 11.17)*	Highlights the line
Side to offset? *Pick to the right*	Creates the dovetail opening
Select object to offset: *Pick* ②	Highlights the line
Side to offset? *Pick to the left*	Creates the second counterbore center line
Select object to offset: *Pick* ③	Highlights the line
Side to offset? *Pick above the line*	Creates the counterbore center lines
Select object to offset: (Enter)	Ends the command
Select object to offset: *Right click the mouse twice*	Highlights the line
Offset distance or Through <Through>: **.5** (Enter)	Sets the distance to .5
Select object to offset: *Pick* ④	Highlights the line

continues

`Side to offset?` *Pick above the line*	Creates the top of the dovetail
`Select object to offset:` (Enter)	Ends the command
`Command:` *Choose the Circle button*	Issues the CIRCLE command
`3P/TTR/<Center point>:` *Pick* ⑤	Sets the center point
`Diameter/<Radius>:` **D** (Enter)	Selects the diameter option
`Diameter:` **3/8** (Enter)	
`Command:` (Enter)	Reissues the CIRCLE command
`3P/TTR/<Center point>:` *Pick* ⑤	Sets the center point
`Diameter/<Radius>:` **D** (Enter)	Selects the diameter option
`Diameter:` **3/16** (Enter)	

Change the larger diameter to the HIDDEN layer and copy both circles .75 to the left to form the other counterbore hole

`Command:` *Choose the Line button*	Issues the LINE command
`From point:` *Pick* ⑥	
`To point:` **@.7<60** (Enter)	Creates the angle on the dovetail
`To point:` *Choose the Trim button*	Issues the TRIM command
`Select cutting edge(s):`	
`Select objects:` *Pick* ⑦	
`Select objects:` (Enter)	
`<Select object to trim>/Undo:` *Pick* ⑧	

Mirror the dovetail to the other side using the MIDpoint of the base and draw a line to for the top of the dovetail

Copy the set screws from the front view

Temporarily freeze the CONST layer to examine the side view. Make any necessary changes to complete the view

Save the drawing

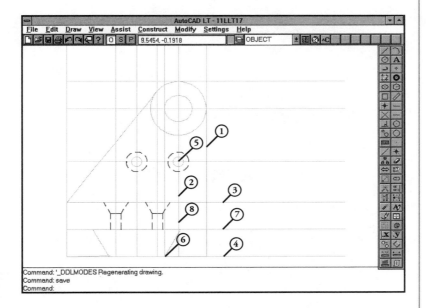

Figure 11.17

Creating the dovetail and counterbore holes.

The side view is now complete. The next step of the exercise is to determine the point at which the web strikes the boss in the front view. The side view has determined this point, so it is just a matter of extending the line to the front view.

Finding the Web and Boss Intersection in the Front View

Continue working on the SUPP-1.DWG.

Zoom Extents

Change the layer property on any stray construction lines to the CONST layer

Command: *Choose the Line button*	Issues the LINE command
From point: *Choose the* Endpoint *button*	Selects the ENDpoint object snap
END of *Pick* ① *(see fig. 11.18)*	
To point: *Choose the Perpendicular button*	Selects the PERpendicular object snap
PER to *Pick* ②	
To point: *Choose the Extend button*	Issues the EXTEND command
Select boundary edge(s)	
Select objects: L **Enter**	Selects the last entity drawn
Select objects: **Enter**	
<Select object to extend> /Undo: *Pick* ③④	Extends the lines
<Select object to extend> /Undo: **Enter**	
Command: *Choose the Erase button*	Issues the ERASE command
Select objects: L **Enter**	Erases the last entity drawn
Select objects: *Choose the Trim button*	Issues the TRIM command

```
TRIM Select cutting edge(s):

Select objects: Pick ③④          Highlights the lines

Select objects: Enter

<Select object to trim>
/Undo: Pick ⑤

<Select object to trim>
/Undo: Enter
```

Save the drawing

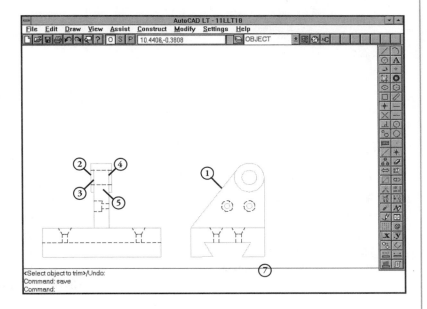

*The completed front
and side views.*

Now that these two major views are drawn, proceed to the next section and
create the top view.

There is no tried-and-true method to determine which view should be
constructed first. All the views are needed to explain the part.

NOTE

Creating the Top View

Like the side view, many of the construction lines used to generate the front view can be used to set up the layout of the top view. Figure 11.19 illustrates the dimensioned top view.

Figure 11.19

The top view.

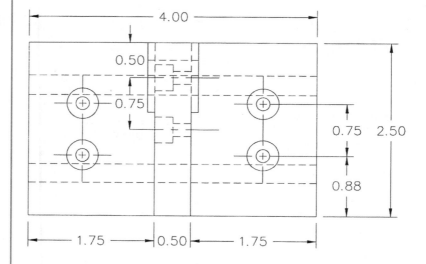

In Chapter 1, "Welcome to the World of Computer-Aided Drafting," you were exposed briefly to grips. You have probably seen these blue control points when selecting entities without issuing a command. Grips can be great tools for doing many tasks without commands. After the grips are highlighted, there are five commands that can be utilized, each with a series of subcommands. Each command is listed in the order they appear when pressing the spacebar. After the grips are selected and a grip has focus (turned red), the following commands are available:

◆ **STRETCH.** `<Stretch to point>/Base point/Copy/Undo/ eXit`: This is the default grip command. This option works like the STRETCH command.

◆ **MOVE.** `<Move to point>/Base point/Copy/Undo/eXit`: This grip option works in the same fashion as the MOVE command.

◆ **ROTATE.** `<Rotation angle>/Base point/Copy/Undo/Refer- ence point/eXit`: This works like the ROTATE command.

◆ **SCALE.** `<Scale factor>/Base point/Copy/Undo/Reference point/eXit:` This works like the SCALE command.

◆ **MIRROR.** `<Second point>/Base point/Copy/Undo/eXit:` This works just like the MIRROR command.

Experience will prove when they are useful. Like most commands, it will take some real-world experience to determine how and when to use grips. The UNDO command will work wonders during this trial-and-error period. Don't be afraid to experiment. Grips will tend to be a quick fix when you are in a bind. In the next exercise, you will see the unique power of grips.

After the grips are highlighted, they turn red (this is called a *hot grip*). By pressing the spacebar, you cycle through the variety of commands available when using grips.

NOTE

Start the top view exercises by using grips to create the construction lines.

Using Grips to Create Construction Lines

Exercise

Continue with the SUPP-1.DWG.

Thaw the CONST layer and make it current

Freeze the OBJECT layer

Command: *Form a crossing window by choosing ① then ② (see fig. 11.20)*	Forms a crossing window
All crossed entities display their blue controls points	
Command: *Hold the Shift key down and select ③④⑤⑥⑦⑧*	Turns the control points red
Command: *Pick any of the red points*	Issues the GRIP command
`** STRETCH **`	

continues

`<Stretch to point>/Base point /Copy/Undo eXit:` *Choose the Perpendicular button*	Selects the PERpendicular object snap
`PER to` *Pick* ⑨	Stretches the points through the top view
Save the drawing	

Figure 11.20

Generating the construction lines.

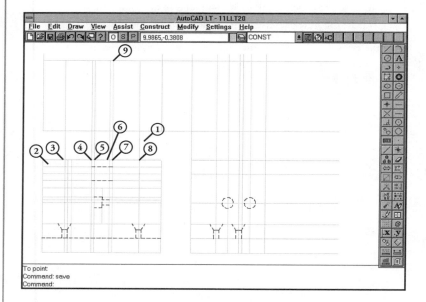

NOTE

If you issue a non-editing command like LINE, the grips will be cleared. If you issue an editing command like ERASE, the highlighted entities and their grips will be deleted. Remember, invoking grips is like preselecting entities.

Continuing with the top view exercises, use the OFFSET command to develop the hidden dovetail lines and the set screw center lines.

Creating Construction Lines for the Top View

Continue with the SUPP-1.DWG.

Command: *Choose* **C**onstruct, *then* **O**ffset	Issues the OFFSET command
Offset distance or Through <Through>: **.75** Enter	Sets the distance to .75
Select object to offset: *Pick* ① *(see fig. 11.21)*	Highlights the line
Side to offset? *Pick above the line*	Creates the dovetail opening
Select object to offset: *Pick* ②	Highlights the line
Side to offset? *Pick below the line*	Creates the dovetail opening
Select object to offset: *Right click the mouse twice*	Ends and reissues the OFFSET command
Offset distance or Through <Through>: **7/8** Enter	Sets the distance to .875
Select object to offset: *Pick* ①	Highlights the line
Side to offset? *Pick above the line* *Pick* ①	Creates the set screw center line
Select object to offset: *Pick* ②	Highlights the line
Side to offset? *Pick below the line*	Creates the set screw center line
Select object to offset: Enter	Ends the command

Save the drawing

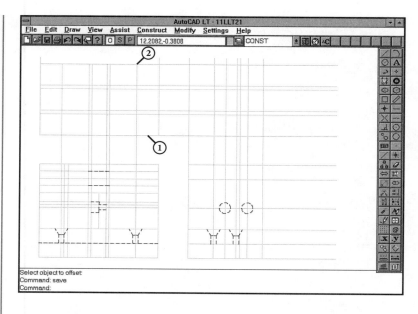

Now you will begin to add object lines to the top view. In the next exercise, you create the set screws, base outline, and a portion of the hidden dovetail.

Exercise

Creating Construction Lines for the Top View

Continue with the SUPP-1.DWG.

Thaw the OBJECT layer and make it current

Command: *Choose the Circle button*	Issues the CIRCLE command
3P/TTR/<Center of circle>: Pick ① *(see fig. 11.22)*	Sets the center of the set screw hole
Diameter/<Radius>: **D** (Enter)	Selects the diameter option
Diameter: **3/16** (Enter)	Sets the diameter
Command: (Enter)	Reissues the CIRCLE command
3P/TTR/<Center of circle>: *Pick* ①	Sets the center of the set screw hole

```
Diameter/<Radius>: D (Enter)          Selects the diameter option
Diameter: .44 (Enter)                 Sets the diameter
```

*Copy multiple set screw holes to
the four locations*

```
Command: Choose the Line button       Issues the LINE command
From point: Pick ②
To point: Pick ③
To point: Pick ④
To point: Pick ⑤
To point: C (Enter)                   Forms the base
```

*Change the current layer to
HIDDEN*

```
Command: Choose the Line button       Issues the LINE command
From point: Pick ⑥
To point: Pick ⑦
To point: Right click the mouse       Ends and reissues the LINE
twice                                 command
LINE From point: Pick ⑧
To point: Pick ⑨
To point: (Enter)                     Ends the LINE command
```

Save the drawing

Figure 11.22

Refining the top view.

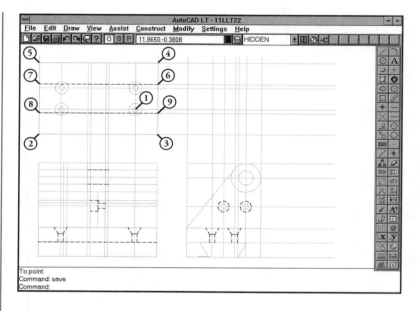

The next exercise locates the boss and the hidden top edge of the dovetail.

Creating the Boss and Dovetail Top

Continue with the SUPP-1.DWG.

Freeze the CONST layer

Command: *Choose the Line button*	Issues the LINE command
From point: *Choose the Midpoint button*	Selects the MIDpoint object snap
MID of *Pick ① (see fig. 11.23)*	
To point: *Choose the Midpoint button*	Selects the MIDpoint object snap
MID of *Pick ②*	
To point: *Choose the Copy button*	Issues the COPY command
Select objects: *Pick ③ then ④*	

```
Select objects: (Enter)
```

```
<Base point or displacement>
/Multiple: Choose the Endpoint button
```
Selects the ENDpoint object snap

```
END of Pick (5)
```

```
Second point of displacement:
Choose the Perpendicular button
```
Selects the PERpendicular object snap

```
PER to Pick (6)
```

```
Command: Choose the Extend button
```
Issues the EXTEND command

```
Select boundary edge(s):
```

```
Select objects: Pick (7)
```

```
<Select objects to extend>
/Undo: Pick (8)(9)
```

```
<Select objects to extend>
/Undo: (Enter)
```
Ends the command

```
Command: Choose Assist, then Distance
```
Issues the DISTANCE command

```
First point: Pick (10)
```

```
Second point: Pick (11)
```

Delta X = .2818

Offset the hidden dovetail lines in the top to reflect the top surface of the dovetail

Save the drawing

11
Chapter

Figure 11.23

Locating the boss and dovetail.

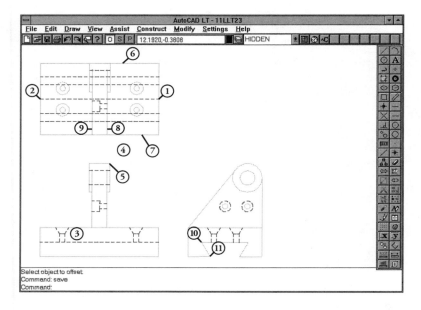

Finally, you will locate the counterbore holes, the intersection of the web, and the boss in the top view. Using a 45° angle from the front view, you will easily locate the points.

Cleaning Up the Top View

Continue with the SUPP-1.DWG.

Thaw the CONST layer and make it current

Command: *Choose the Line button*

Issues the LINE command

From point: *Choose the Endpoint button*

Selects the ENDpoint object snap

END of *Pick* ① *(see fig 11.24)*

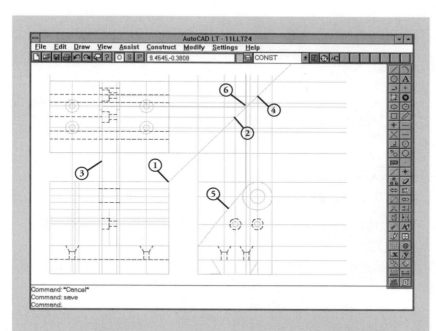

Figure 11.24

Adding the finishing touches to the top view.

To point: **@6<45** (Enter)	Creates a 45° construction line
To point: *Right click the mouse twice*	Ends and reissues the LINE command
LINE From point: *Pick* ②	
To point: *Choose the Perpendicular button*	Selects the PERpendicular object snap
PER to *Pick* ③	Creates the counterbore hole center line
To point: *Right click the mouse twice*	Ends and reissues the LINE command
LINE From point: *Pick* ④	
To point: *Choose the Perpendicular button*	Selects the PERpendicular object snap
PER to *Pick* ③	Creates the counterbore hole center line

continues

`To point:` *Right click the mouse twice*	Ends and reissues the LINE command
`LINE From point:` *Choose the Endpoint button*	Selects the ENDpoint object snap
`END of` *Pick* ⑤	
`To point:` *Choose* **A**ssist, *then* **O**bject Snap	Opens the Running Object Snap dialog box
Uncheck the Intersection box, then OK	
`To point:` `@4<90` (Enter)	
`To point:` *Right click the mouse twice*	Ends and reissues the LINE command
`LINE From point:` *Choose the Intersection button*	Selects the INTersection object snap
`INT of` Pick ⑥	
`To point:` *Choose the Perpendicular button*	Selects the PERpendicular object snap
`PER to` *Pick* ③	Creates the intersection of the web and the boss

Trim the web

Copy the counterbore hole to the proper center line

Zoom .9x

Turn off the construction layer and clean up the drawing, as illustrated in figure 11.25. Don't forget the hidden lines on the web through the boss

Save the drawing

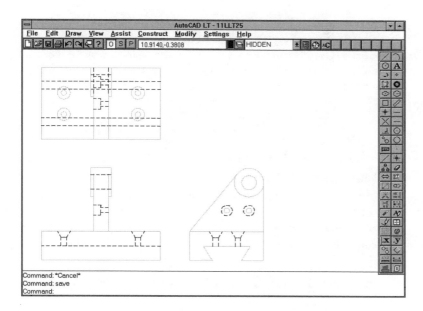

Figure 11.25

The completed views.

The three views are complete, but the drawing is not. The next section discusses dimensioning the support drawing

Dimensioning and Annotating the Drawing

The most critical information on a production drawing is the dimensions and notes. Even though AutoCAD LT is used to draw and plot accurately, machinists must never scale the plot to determine a dimension, nor should they assume anything about an object's material or its position in an assembly. If this information is not provided, the drawing should go back to the designers and the discrepancy resolved.

In the next few sections, the concepts you examined in Chapter 7, "How to Modify Dimensions," are reviewed. As a beginning drafter, be prepared to accept a lot of corrections when dimensioning a drawing. No matter how much study is involved, nothing beats experience. The nuances of dimensioning will come over time.

NOTE

Proper dimensioning will be accomplished through experience. Remember to observe the drawing like a machinist to identify which dimensions are necessary. Overdimensioning a drawing can also be frustrating for those in the shop who are trying to decipher the drawing. Stay conservative with the dimensions and make sure everything adds up.

Setting up DimVars

Like layers, most firms have established guidelines for the dimension variables or *dimvars*. In Chapter 6, you were introduced to the concept of dimension styles. The Dimension Style interface is easy to understand, but locating the proper variable can be frustrating. You can quickly set variables by typing them in. Here are the settings that should be used before dimensioning the support:

◆ **DIMASZ.** Set the arrowhead size to 1/8".

◆ **DIMDLI.** Set this to 0 in order to keep continuous dimensions running on the same line.

◆ **DIMEXE.** Set the extension line to extend past the dimension line 3/32".

◆ **DIMEXO.** The extension lines should be 3/32" off the origin point.

◆ **DIMTIX.** Always place the dimension inside the extension lines. Turn this ON or enter 1.

◆ **DIMTEXT.** Set the dimension height to 1/8".

TIP

Power users access the dimension variables by typing **DIM** at the Command prompt. This will bring up a Dim: prompt. Type **STAT** and AutoCAD LT will return a list of the dimension variables and a short description.

In the following exercise, you set the variables listed previously.

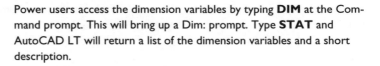

Setting the Dimension Variables

Continue with the SUPP-1.DWG.

Command: **DIM** (Enter)	Opens the dimension prompt
Dim: **DIMASZ** (Enter)	Displays the current settings for the arrow size
Current value <0.1800> New value: **.125** (Enter)	Sets the arrow size to 1/8
Use this same process to set the other variables listed previously	
Dim: *Press Ctrl+C*	Returns to the Command: prompt
Save the drawing	

All the other dimension variables will work according to their defaults. If the dimension does not match what appears in the figures, other variables may need adjustment.

Dimensioning the Views

As mentioned earlier, dimensioning is the most crucial aspect of the drawing. You must be accurate and logical in determining where the dimensions should be located. There are a variety of options to use when locating dimensions on a drawing. Space and clarity are usually the controlling factors. Space restrictions tend to decide many obstacles when dimensioning; if a dimension can be illustrated on another view more clearly, always dimension the other view. Clear logical dimensions are the goal.

In the next few exercises, you are guided toward dimensioning the side view of the support. You are expected to complete the dimensioning for the other two views.

Exercise

Adding Linear Dimensions to the Side View

Continue with the SUPP-1.DWG.

Set the layer to DIMS

Zoom Window on the side view
(see fig 11.26)

Set the Units to round off to two places

Command: *Choose* **D**raw, *then* **L**inear Dimensions, *then* **V**ertical	Issues the DIM command
First extension line origin or RETURN to select: *Pick* ① *(see fig. 11.26)*	Sets the first point
Second extension line origin: *Pick* ②	Sets the second point
Dimension line location (Text/Angle): @.75<0 (Enter)	Sets the distance
Dimension text <0.50>: (Enter)	Accepts the distance
Dim: (Enter)	
VER First extension line origin or RETURN to select: *Pick* ③	Sets the first point
Second extension line origin: *Choose the Center button*	
CEN of *Pick* ④	Sets the second point
Dimension line location (Text/Angle): *Choose the Endpoint button*	
END of *Pick* ⑤	Sets the distance
Dimension text <0.75>: (Enter)	Accepts the distance
Dim: (Enter)	
VER First extension line origin or RETURN to select: *Pick* ①	Sets the first point

`Second extension line origin:` *Pick* ③	Sets the second point
`Dimension line location(Text/Angle):` `@.75<0` (Enter)	Sets the distance
`Dimension text <1.00>:` (Enter)	Accepts the distance
`Dim:` *Choose* **D**raw, *then* **L**inear Dimensions, *then* **C**ontinue	Continues the dimension string
`Second extension line origin:` *Choose the Center button*	
`CEN of` *Pick* ⑥	Sets the second point
`Dimension text <1.75>:` (Enter)	Accepts the distance
`Dim:` *Choose* **D**raw, *then* **L**inear Dimensions, *then* **H**orizontal	Continues the dimension string
`HOR First extension line origin` `or RETURN to select:` *Pick* ⑦	Sets the first point
`Second extension line origin:` *Pick* ⑧	Sets the second point
`Dimension line location(Text/Angle):` `@3/8<270` (Enter)	Sets the distance
`Dimension text <0.75>:` (Enter)	Accepts the distance
`Dim:` *Choose* **D**raw, *then* **L**inear Dimensions, *then* **C**ontinue	Continues the dimension string
`Second extension line origin:` *Pick* ⑨	Sets the second point
`Dimension text <1.00>:` (Enter)	Accepts the distance
`Dim:` *Choose* **D**raw, *then* **L**inear Dimensions, *then* **C**ontinue	Continues the dimension string
`Second extension line origin:` *Pick* ①	Sets the second point
`Dimension text <0.75>:` (Enter)	Accepts the distance

Save the drawing

Figure 11.26

*Dimensioning the
side view.*

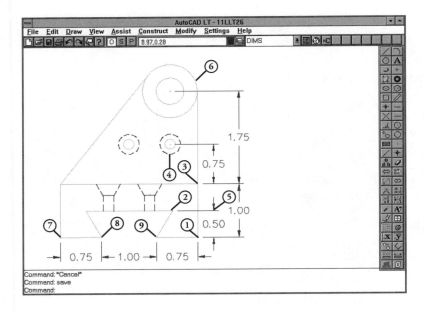

The next step in the exercise adds the angular dimension to the dovetail and
the center marks on the holes.

Exercise

Adding the Angular Dimension and Center Marks

Continue with the SUPP-1.DWG.

Set the running object snap to ENDpoint

Command: *Choose* **D**raw, *then* Ang**u**lar
Dimensions

Select arc, circle, line, or
RETURN: (Enter)

Angle vertex: *Pick* ① *(see fig 11.27)*

First angle endpoint: *Pick* ②

Second angle endpoint: *Pick* ③

Dimension arc line location   Highlights the arc and drags
(Text/Angle):   it to a new location

Drag the arc to a location within the dovetail and left click

```
Dimension text <60>: 60%%d
TYP. Enter
```

```
Command: DIMCEN Enter          Issues DIMCEN variable
                               setting
```

```
New value for DIMCEN <0.20>:   Sets DIMCEN to .4
.4 Enter
```

Command: *Choose* **D**raw, *then* Radial Di**m**ensions, *then* **C**enter Mark

```
Select arc or circle: Pick ④
```

Command: *Choose* **D**raw, *then* Radial Di**m**ensions, *then* **C**enter Mark

```
Select arc or circle: Pick ⑤
```

```
Command: DIMCEN Enter          Issues DIMCEN variable
                               setting
```

```
New value for DIMCEN <0.20>:   Sets DIMCEN to .75
.75 Enter
```

Command: *Choose* **D**raw, *then* Radial Di**m**ensions, *then* **C**enter Mark

```
Select arc or circle: Pick ⑥
```

Change all the center mark lines to center lines. Although it is preferable to change the attributes of an entity BYLAYER, use the CHANGE command to change the properties of these entities individually

Add center lines to the set screw holes

Figure 11.28 displays all the appropriate dimensions. Take a few moments and complete the dimensions on each view

Save the drawing

11
Chapter

Figure 11.27

The dimensioned side view.

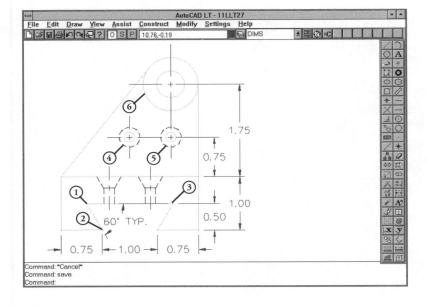

Figure 11.28

The dimensioned drawing.

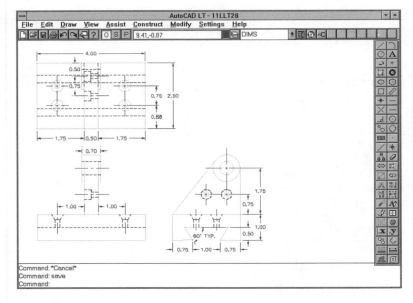

All of the dimensions to create the part are on the drawing. The next step is to add any notation that will be needed to direct the machinists. Missing in particular are the callouts for the holes. The next section explains each of the callouts.

Determining Appropriate Notes

Each of the callouts on the drawing has a specific function. Aside from the major elements, the boss and dovetail, the method of creating the holes must be annotated and dimensioned for the machinist. Typically, the counterbored holes are noted as follows: the diameter of the drill, the diameter of the counterbore, the depth of the counterbore, and the number of holes.

The countersunk holes are dimensioned as follows: the diameter of the drill, the angle of the countersink, the diameter of the large hole, and the number of holes.

Figure 11.29 illustrates the alternative ways to dimension a counterbore and countersunk hole with graphic depictions.

```
3/16" DRILL, 60° CSK to 0.44          3/16" DRILL, 3/8" CBORE
4 HOLES                               3/8" DEEP, 4 HOLES
             (OR)                                  (OR)
Ø .1875" THRU ⌵60 Ø 0.44              Ø .1875" , ⌴ Ø .375"
4 HOLES                               ⤓ .375", 4 HOLES
```

Figure 11.29

Alternative methods of dimensioning holes.

In the next exercise, you dimension the holes and call out the dovetail and boss.

Adding Notes to the Drawing

Exercise

Continue with the SUPP-1.DWG.

Zoom Window into the top view (see fig. 11.30)

continues

Figure 11.30

Adding the notes.

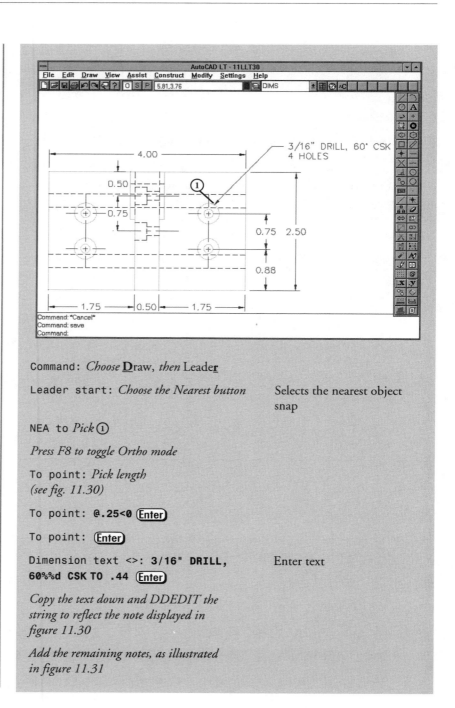

Command: *Choose* **D**raw, *then* Leade**r**

Leader start: *Choose the Nearest button* — Selects the nearest object snap

NEA to *Pick* ①

Press F8 to toggle Ortho mode

To point: *Pick length*
(see fig. 11.30)

To point: **@.25<0** (Enter)

To point: (Enter)

Dimension text <>: **3/16" DRILL,** — Enter text
60%%d CSK TO .44 (Enter)

Copy the text down and DDEDIT the string to reflect the note displayed in figure 11.30

Add the remaining notes, as illustrated in figure 11.31

Many of the notes that are needed on this
particular drawing have been covered by
the title block text. You should call out
the three separate views. These labels should
also indicate a scale if it varies from the
scale given in the title block. Figure 11.32
displays the drawing labels that should be
inserted on the TEXT layer. The top view may
need to be moved up in order to accommodate
the text.

Save the drawing

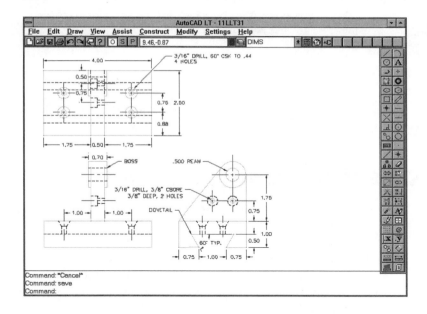

Figure 11.31

*Completing the
callouts.*

Figure 11.32

Labeling the views.

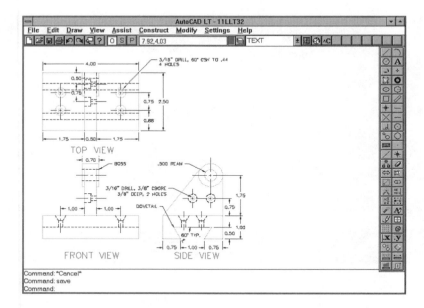

The model space drawing is essentially complete. In the next section, you organize the paper space drawing to ready the drawing for plotting.

Checking the Drawing

In a typical firm, you would now plot your drawing for review. This process is time-honored as the most efficient way to avoid costly mistakes. If the part has a dimension missing and the machinists are on a tight schedule or they are working into the night to finish a project, the last thing they need is a discrepancy. It could delay a project from being shipped on time. To avoid such situations, drawings must always be reviewed by a trained engineer or senior draftsperson. The project manager may also play a role in the review process in order to keep the project on schedule.

Often, drawings are plotted at various stages in order to get a fresh perspective on its state of completion. This also gives you an opportunity to do a preliminary review of the drawing.

Assuming that the drawing you have created is free of errors, proceed to the last section of this chapter, which is devoted to plotting the support on paper.

Plotting the Drawing

The final exercise is a review of the ideas presented in Chapter 10, "Reproducing Drawings." After the drawing has been approved, the next step is to get the final plot on paper. As mentioned earlier, this drawing is intended to be printed on an A-size sheet in a portrait orientation. Convert the drawing to paper space and create some mviews for the final plot.

Exercise

Chapter **11**

Plotting the Drawing

Finally, the last step in the SUPP-1.DWG will be to plot.

Set the current layer to PS-BORDER

Command: *Choose* **V**iew, *then* **T**ilemode	Sets the drawing to paper space
New value for TILEMODE <1>: 0	
Command: *Choose* **V**iew, *then* Viewp**o**rts, *then* **M**ake Viewports	Issues the MVIEW command
ON/OFF/Hideplot/Fit/2/3/4/Restore/ <First point>: *Pick* ① *then* ② *(see fig 11.33)*	Forms a viewport
Command: *Choose the Paper_Space button*	Enter the viewport
Command: *Choose the Zoom button*	
All/Center/Extents/Previous /Window/<Scale(X/XP)>:**.75xp** (Enter)	Scales the viewport to 3/4 of the full scale

Use the PAN command to center the drawing

Choose the Paper_Space button on the toolbar

Set the current layer to 0 and freeze the PS-BORDER layer

Save the drawing

All the plot settings were established at the beginning of the chapter so you can plot the drawing

Figure 11.33

The final plot.

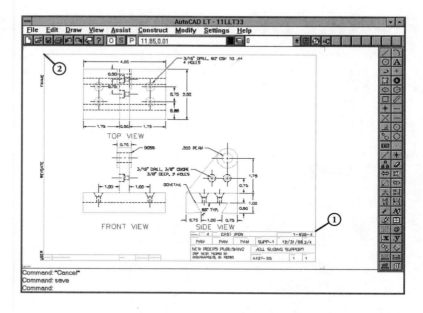

Summary

This chapter has been devoted to drawing an adjustable sliding support. You have completed a series of exercises that accomplished the following tasks:

- ◆ Set up the sheet

- ◆ Generated and inserted text into the title block

- ◆ Laid out the drawing

- ◆ Created the orthographic views in model space

- ◆ Dimensioned and annotated the views

- ◆ Checked the drawing

- ◆ Plotted the drawing in paper space

Although all projects do not follow this same procedure, the intent of the chapter was to examine a real-world example and utilize the drawing techniques you have developed while reading this book.

The next chapter will follow a similar style to this one, using an architectural drawing as an example.

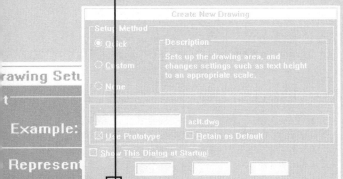

The process of producing a complete set of architectural drawings can be complicated. In this chapter, this process is explored, and is broken down into the fundamental steps needed for drafting two types of primary architectural drawings—the *floor plan* and the *elevation*. While it should be clear that a full set of architectural drawings includes far more than just the two mentioned, it is the basis of these drawings on which many other drawings in the set are created. Architectural projects evolve from a conceptual design stage into "construction documents" or "working drawings." Working drawings, which include notes and dimensions, are the final set of documents that enable your design to become reality. They communicate your ideas on paper to the contractor. The approach taken in this chapter is to learn the steps necessary to create a *basic* floor plan and elevation using AutoCAD LT. Through this procedure, this chapter investigates the following subjects:

◆ Setting up a prototype drawing

◆ Designing the layout of the sheet

◆ Blocking out the drawing

◆ Drafting the floor plan

◆ Developing the elevation

◆ Printing or plotting the drawing

◆ Checking the drawing

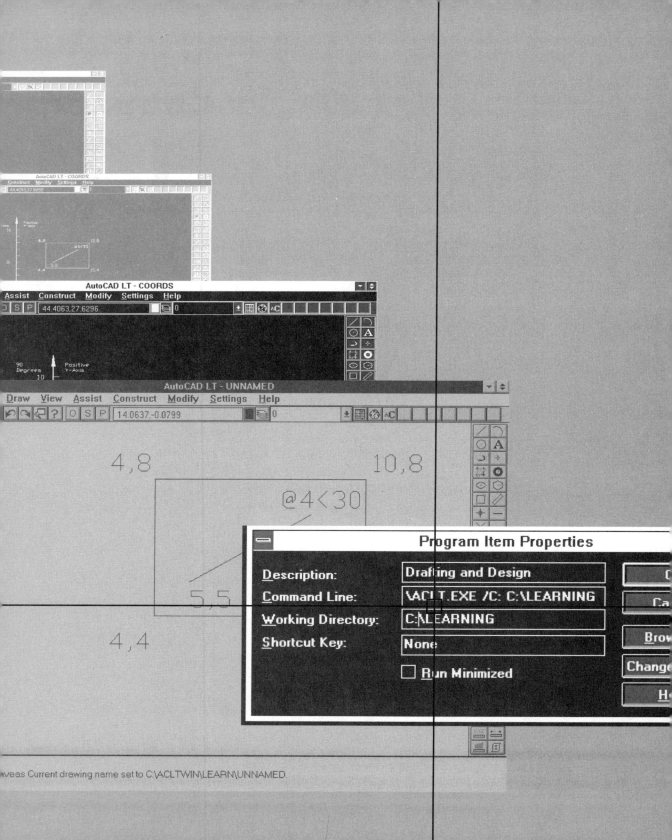

Architectural Drawing Project

As you proceed through these subjects, notice that you are also faced with examples, exercises, and further explanations to help you discover the interaction that exists between the draftsperson or architect and the client, contractor, and other professionals involved. The prototype drawing, layout, floor plan, and elevation are simply examples used to convey a series of concepts that one uses for any number of architectural projects. The ability for you to apply these concepts to other architectural projects is the ultimate goal.

The procedure of drafting any one architectural drawing in a set of drawings requires the draftsperson to have an overall idea as to how that drawing fits in to the "puzzle" of drawings. As previously stated, several working drawings are required for most architectural projects, even for a residential project. Many homes may require a site plan, electrical plan, plumbing plan, foundation plan, framing plan, roof plan, sections, and details, in addition to the floor plan and elevations that are discussed in this chapter. Although several different drawings are involved, each with its own purpose, there are also very strong similarities between many of these drawings. The most efficient way to draft an entire set, or partial set of drawings, is to first establish the common traits between the architectural drawings.

Your project, a small vacation home, provides the opportunity for you to learn the process of identifying and drafting the plan and at least one elevation. The project is intended to give you the basic skills needed in creating these two drawings, regardless of whether they are to be used for presentation drawings or working drawings. The common traits between different architectural drawings may not be evident at first. They are identified next so that you can begin setting up a prototype drawing.

Setting Up a Prototype Drawing

A *prototype drawing* is a drawing file that is used as a base, or starting point, for several other drawings. The purpose is to save time. While drafting any architectural project, you will discover that many aspects of each drawing are the same. This is particularly important on commercial projects that have several stories, such as an office building, where the basic floor plan is virtually identical from one floor to the next. The project in this chapter does not have multiple levels; there are, however, still areas in which a prototype drawing will be useful.

See "Creating Your Own Prototype Drawing" in Chapter 4, "Multiview Drawings," for more information on creating a prototype drawing.

Most architectural firms have their own style and graphics. Commonly, a firm uses the same title block and border for each drawing in the set of drawings, not unlike many other drafting professions. This is done for consistency and to save time. These title blocks can be used repeatedly throughout the office, and most definitely within the same project. The following sections discuss the settings used in this project that are common to many drawings in a given set of drawings.

Units

This system of measurement is common to all of your drawings. You will select "architectural" (see fig. 12.1). In this case, the fractional increment selected should be 1/4". This increment may vary, depending on the particular drawing you are drafting.

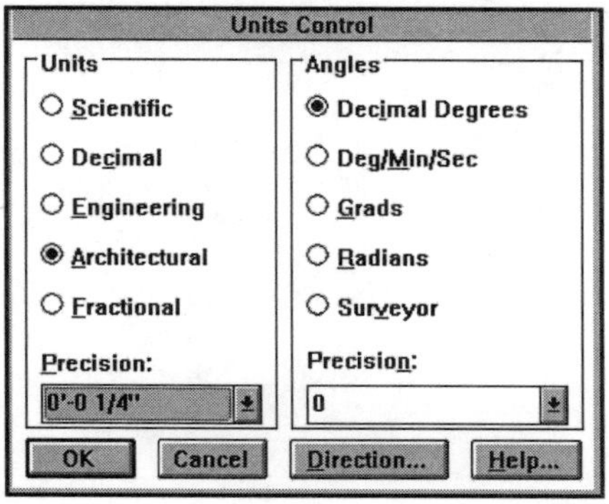

Figure 12.1

The Units Control settings.

Sheet Size

The most common sheet size for architectural firms is D size. While this size of paper is best for larger drawings such as floor plans and elevations, it is also used for smaller drawings at different scales, such as an entire sheet of details. As previously mentioned, consistency is a primary factor. A full set of working drawings is typically drafted on one sheet size, regardless of drawing size or scale. If the D-size sheet is more than is needed for the particular drawing, more than one drawing can be placed on that one sheet. The drawings covered in this chapter are on one sheet.

Limits

Having now selected the units, increments, and sheet size, the limits of your drawing must also be established. When determining the limits, it is important to consider both the scale of the drawing and the sheet size. The most common scale for residential floor plans and elevations is 1/4"=1'-0". The sheet, as mentioned, is D size. It is critical that you know which type of D-size paper you are working with. Many engineers also work with D-size sheets; however, an engineer's D-size sheet is typically 22" by 34". The D-size sheets most architects work with are actually 24" by 36". The latter should be selected for the following drawing (see fig. 12.2). The limits in this case are 0,0 for the lower left corner and 144',96' for the upper right corner. Note that the limits in the upper right corner are in feet, not the default unit of inches.

Figure 12.2

The limits for "D" size @ 1/4"=1'–0".

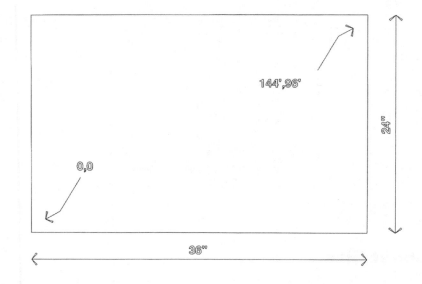

Layers

The selection of layers and their properties varies from one architectural firm to another. There are some architectural software packages available that automatically name the layers for you. This is convenient, but may not correlate with a particular firm's preference in naming layers. In some instances, it may also add several unneeded layers, possibly causing confusion. AutoCAD LT enables you to select the name and properties of layers.

The quantity of layers is also a factor to consider. In Chapter 4, you discovered how a drawing is broken down into layers. The exercise in this chapter has the following selection of layers and properties (see fig. 12.3):

- ◆ **ARDOORS:** Red/Continuous—Contains the doors of the house.

- ◆ **ARWALLS:** Green/Continuous—Contains the interior and exterior walls of the house.

- ◆ **ARWINDOWS:** Blue/Continuous—Contains the windows of the house.

- ◆ **BORDER:** Green/Continuous—Contains the border of the drawing.

◆ **HIDDEN:** White/Hidden—Contains hidden lines.

◆ **PLFIX:** Blue/Continuous—Contains the plumbing fixtures in the house.

◆ **TBLOCK:** Green/Continuous—Contains the title block lines and text.

◆ **TEXT:** Red/Continuous—Contains the general text of the drawing.

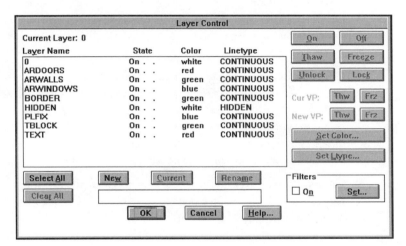

Figure 12.3

The completed listing of layers.

The previous layer names are only specific to the exercise covered in this chapter. You may notice that several of the layer names begin with "AR". This is a common method used in naming layers. It helps to distinguish between architectural data versus plumbing, electrical, or mechanical data. For future reference, you might find a need to have several more layers, depending on the complexity of the project and the purpose of the drawing file.

Border

The *border*, or outside boundary edge of the drawing and title block, varies from 3/8 to 1/2 inch from the edge of the paper. On the left side of the drawing sheet, the border ranges from 1 to 2 inches from the edge, depending on how many sheets there are in the complete set of drawings. This allows room for binding all of the sheets together, which typically is done on the left side of the drawing.

Title Block

There are as many designs for title blocks as there are architectural firms. The title blocks are often pre-printed, which saves additional time during drafting and plotting. While each firm may have their own particular design, here are some features common to most architectural title blocks:

◆ **Company name, address, phone number, fax number, etc.** This is for clearly identifying the company, or architect, responsible for creating the drawings. A company logo might be in the vicinity. On occasion, there may also be a space dedicated for a consultant's name.

◆ **Client name, project name, project address.** Typically, most residential projects have a client whose personal address is the same as the actual location of the project to be constructed. At other times, the client, perhaps a real estate developer, may have a different business address than the project address. At the very least, the project name and project address are displayed.

◆ **Drawing title.** This is the area dedicated to giving the drawing a name or title. In an architectural drawing, the drawing title may be *floor plan* or *front elevation*, or it may be a company code that identifies the project, sheet number, and drawing type. Most companies have their own methods of constructing a drawing title for identifying each sheet by name within a project.

TIP

The specific drawing name is not placed in this area in advance, enabling the draftsperson to title it once a specific drawing(s) has been selected for that particular sheet.

◆ **Sheet number.** Each sheet in a set of working drawings is to be numbered. The method of numbering the sheets also varies, depending on the firm or the size of the project. A common method of numbering sheets, particularly for smaller architectural projects, is to number them relative to the total number of drawings in the complete set. For example, if the entire collection of drawings for the project consists of six drawings, the first sheet should be labeled **1 of 6**. The next sheets in the series would then be numbered **2 of 6**, **3 of 6**, and so on.

◆ **Scale.** The *scale*, or ratio, of the drawing is typically determined prior to drafting. Architectural drawings vary in scale, depending on the actual size of the object being drafted. A large site plan, for example, might be drafted at a scale such as 1/16"=1'–0" in order to fit it on a standard 24" by 36" sheet of paper. Another sheet in a complete set might include a series of details of a stairway or handrail, each of which may be drafted as large as 3"=1'–0". Many standard drawings, such as a floor plan, are most typically drafted at 1/4"=1'–0".

◆ **Drafted by.** The name or initials of the draftsperson should always be listed. This is important—because several people may be working on one project, it is helpful to have an easy way of knowing which draftsperson was responsible for each drawing.

◆ **Date.** This should be the *completion* date of the sheet.

◆ **Checked by.** The name or initials of the person who reviewed the drawing should be included. The task of checking a drawing should be completed by someone other than the person who drafted the work. This person then identifies any areas of concern for revision. The process of checking a drawing is reviewed later in this chapter.

◆ **Revision date.** As you read in Chapter 11, "Creating a Mechanical Drawing," this date is a record of when the corrections were made.

The drawing completed in figure 12.4 is a typical prototype title block. You can access this particular drawing in AutoCAD LT by opening ARCHENG.DWG. You may choose to include additional drawing elements and settings to meet your specific needs, such as grid, ortho, snap, linetypes, and so forth—a prototype drawing should meet your specific needs. With this completed, you can now proceed with the next step—designing the layout of the sheet.

Chapter **12**

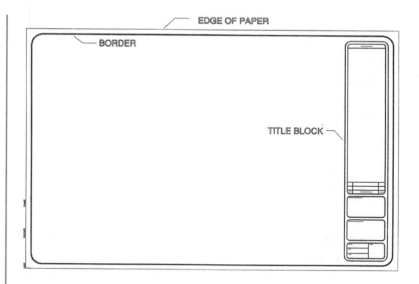

EDGE OF PAPER

BORDER

TITLE BLOCK

Designing the Layout of the Sheet

The purpose of designing the layout of the sheet is to make the process of drafting the project more efficient. The layout, sometimes referred to as the "cartoon set," is sketched by the architect, designer, project manager, or the draftsperson. As you read earlier, setting up a prototype drawing is a good way of saving time. Once the prototype with title block has been completed, it can be laser printed onto A-size paper. Several of these laser prints are then used to sketch the layout of each sheet projected in the complete set of working drawings. These are most likely freehand sketches that simply show which drawings will be located on each sheet. The location of some text or titles can also be included on these sheets. All of this gives the draftsperson a rough idea as to the general graphics of the final product. This process enables one person to organize an entire set of drawings.

TIP

Because all of the sheets are graphically organized ahead of time, there is a better chance that the graphics of the final product are consistent and correct. The time you spend in initially setting up the project is paid back as the project proceeds with a minimum of confusion and delays.

As illustrated in figure 12.5, the cartoon in this case shows a primitive sketch of the basic shapes of the floor plan and one elevation, to be placed together on one sheet. These sketches, along with some very basic text, are the extent to which you would go to create a cartoon set. It simple to do, easy to understand, and fast.

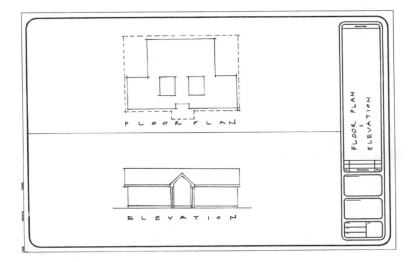

Figure 12.5

A sample cartoon sketch of the floor plan and front elevation.

Blocking Out the Drawing

Small residential projects often have several drawings on one sheet. This is one reason why blocking out, also called *laying*, the drawing is critical. At this point, you are working to scale, and a sense of balance needs to be evident in the final drawing. Most often, whether you are drafting floor plans, elevations, sections, or details, the draftsperson makes an attempt to center the drawing within the boundaries dedicated for that drawing. You must also be aware that titles, notes, and dimensions are added to the drawing, possibly causing the drawing to appear too crowded. The process of centering the drawings is far more critical when you are board drafting—once you start drafting lines, it becomes very time-consuming to erase and redraft just for the sake of centering the drawing. Of course, there is also the imminent danger of running out of space on the paper on which you are drafting.

Computer-aided drafting has eliminated this problem. If you are using AutoCAD LT and realize you are suddenly outside of your dedicated drawing boundaries due to the size of the object or the scale chosen, it is very easy to reorganize or rescale the project, leaving the software with most of the work. It is still wise, however, to determine the approximate size of the object you are drafting relative to the sheet of paper. The fewer surprises you have along the way, the more efficient you are with your time.

As you can see in figure 12.5, the cartoon sketch does show the general vicinity of the drawings on that sheet. The sketch is not to scale, however. The next exercise demonstrates where to begin blocking out the drawings. You must first have your prototype drawing with title block prepared. Use your prototype to start this exercise, and save the drawing as HOUSE.

Exercise

Blocking Out the Drawing

Use "ARCHENG.DWG" that comes with AutoCAD LT as a prototype to begin this exercise. Use figure 12.6 as a guide for the next exercise.

Set the units and layers as previously described and shown in figures 12.1 and 12.3.

Set the object snap to Midpoint

Set the current layer to HIDDEN

Set Ortho (F8) to on

Command: *Choose the Line button*	Issues the LINE command
From point: *Pick* MID *of* ① *(see fig. 12.6)*	Sets first point of line
To point: *Pick* PER *of* ② *and right click the mouse twice*	Draws line(s) and reissues the LINE command
To point: *Pick* MID *of* ③ *and right click on the mouse twice*	Draws line(s) and reissues the LINE command
From point: *Pick* ENDPOINT *of* ⑥	Sets first point of line
To point: *Pick* PER *of* ④	Sets the next point

`To point:` *Pick* END *of* ⑦ *and right click on the mouse once*	Draws line(s) and ends the LINE command
`Command:` *Choose* **C**onstruct, *then* **M**irror	Issues the MIRROR command
`Select objects:` *Pick diagonal lines*	Highlights the line(s)
`Select objects:` *Right click once*	Issues the next prompt
`First point of mirror line:` *Pick* MID *of* ⑤	Selects point and issues next prompt
`Second point:` *Pick below first point*	Duplicates line(s) and issues next prompt
`Delete old objects?<N>:` *Press Enter to accept the default No and right click on the mouse once*	Displays old and new objects and ends the MIRROR command
Save the drawing as HOUSE	

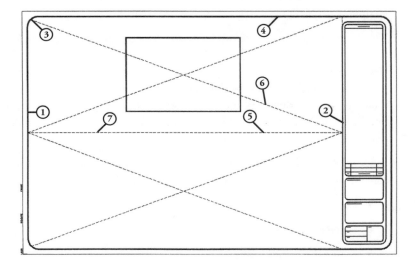

Figure 12.6

Blocking out the drawing.

As seen in figure 12.6, the drawing is divided into two areas. The upper area is for the floor plan and the lower area is for the elevation. In the next exercise, you continue with blocking out the floor plan. The approximate overall size of the floor plan is 42' by 28'.

Drafting the Floor Plan

Blocking out the floor plan is not a difficult task, but an understanding of the "principles" of floor plans is required first. The floor plan serves as a very important tool for all people involved in any architectural project. The floor plan, as a design and presentation tool for the designer and client, is often the first area of focus. It allows for a two-dimensional understanding of space planning, circulation, and traffic patterns, giving the client the opportunity to review the design for overall acceptability.

It is equally important for designers and clients to be focused on the exterior elevations and the volumes within the structure. This is a difficult task for some clients to thoroughly understand, however, mostly due to the fact that it involves three dimensions. The third dimension can be a difficult concept to convey, especially when presented in a two-dimensional format, such as elevations. Floor plans, on the other hand, are drawings that many people have seen before. By default, this attributes to some of the focus placed on the floor plan. This trend should slowly change as more and more draftspeople become better versed in three-dimensional computer-aided drafting and computer animation.

The floor plan is one of the most important drawings in a set of drawings—it is the basis for generating several other drawings in the set, such as the elevations, electrical and plumbing plans, roof plans, framing and foundation plans, and so on. The accuracy of the plan is thus essential due to the impact it has on the other drawings. It is also an important method of communication between the draftsperson and the contractor, as are all of the working drawings.

In addition to showing walls, doors, windows, and so forth, the floor plan also conveys important technical information, including the following:

- ◆ Dimensions
- ◆ Reference symbols
- ◆ Material specifications
- ◆ Construction methods
- ◆ Structural considerations

As a method of communication, therefore, it is critical that the floor plan stay well organized and easy to understand.

Making a Floor Plan Communicate

To some people, a floor plan is considered to be a drawing that represents the layout of a house or building, as though the roof had been removed and you could see inside it. This perception, however, is not quite correct. The floor plan is actually a *section* that is cut horizontally through all of the walls. More specifically, the floor plan is a horizontal section, cut through the entire structure, at four feet above the plane of the floor. This imaginary section is taken at this point so that doors, windows, and any other objects that are in the cutting plane are included.

As mentioned previously, a drawing is a method of communication. Another form of communication is the *language of line weights*, such as the weight, or thickness, of a line. The use of proper line weights is one method used to express specific information to the contractor. Because the cutting plane—set at four feet above the floor—is a standard used within the entire architectural profession, it provides the opportunity to make graphic decisions about the drawing, relative to the cutting plane. There are three major principles in establishing line weights or line types, as follows:

- ◆ **Objects at the cutting plane.** Objects (which are located directly through the imaginary cutting plane at four feet above the floor) such as the walls are normally drafted with a heavy line weight. This is not typically done by varying the thickness of the line, such as with the PLINE command. Most often, it is done by assigning a specific color to the line. That color will then be assigned a thicker pen when plotting.

- ◆ **Objects below the cutting plane.** When an object is located below the cutting plane, it is drafted in a line lighter than that of objects at the cutting plane. A *medium* or *light* pen thickness is used in this case. While it is not critical whether medium or light is chosen, it is important to pick only one, and remain *consistent*. Again, allow color to be the graphic difference on the computer screen, leaving the pen assignment and plotter to create the actual line thickness.

- ◆ **Objects above the cutting plane.** Some objects are located above the 4-foot cutting plane, such as a closet shelf, upper kitchen cabinets, and archways. There are also numerous objects that might be located higher than four feet. These objects are typically conveyed using a *hidden* line, with a light pen assignment.

CAUTION

While hidden lines are typically used to identify objects drawn above the cutting plane, you should remember that hidden lines are also used on other drawings for different reasons. Use different screen colors and line weights as needed to differentiate categories of objects.

Drawing the Exterior Walls of a Floor Plan

Two parallel lines are used to indicate walls on a floor plan. The width of a particular wall varies depending on the materials being used to construct the structure. In the following exercise, the house is constructed of 2×6 studs for the exterior walls, and 2×4 studs for the interior walls. It is important to realize that a 2×6 is actually not 2" by 6" wide. A 2×6 is usually 1-1/2" by 5-1/2", and a 2×4 is really 1-1/2" by 3-1/2". The terms 2×4 and 2×6 are used only in reference to the nominal size of the stud, not the actual size.

The following exercise uses a wall (stud) width of 5-1/2" for exterior walls and 3-1/2" for interior walls, representing the actual thicknesses of those materials. Therefore, your parallel lines will be 5-1/2 or 3-1/2 inches apart.

NOTE

If you wanted to include the width of your wall covering, such as brick on the outside or drywall on the inside, you would add that to the stud width.

Exercise

Blocking Out the Exterior Walls of the Floor Plan

Continue from the previous exercise; refer to figure 12.7 for pick points.

Set current layer to ARWALLS

Draft a rectangle, using the LINE command, and center it in the upper area, as seen in figure 12.6. The rectangle should be 42" by 28"

Command: *Choose* **C**onstruct, *then* **O**ffset	Issues the OFFSET command
Offset distance or Through	Sets the distance to 5.5"

```
<Through>: 5.5 (Enter)
```

`Select object to offset:` *Pick ① (see fig. 12.7)*	Highlights the line
`Side to offset:` *Pick below the* *line*	Duplicates the line

Repeat this process with the other
three outer lines of the rectangle,
then right click on the mouse
to discontinue the command

You are now ready to create the construction lines for the entryway. The entryway is 18' from each side wall and 18" deep.

`Command:` *Choose* **C**onstruct, *then* **O**ffset	Issues the OFFSET command
`Offset distance or Through` `<Through>:` **18'** (Enter)	Sets the distance to 18'
`Select object to offset:` *Pick ③*	Highlights the line
`Side to offset:` *Pick to the right*	Duplicates the line
`Select object to offset:` *Pick ④*	Highlights the line
`Side to offset:` *Pick to the right*	Duplicates the line
`Select object to offset:` *Pick ⑤*	Highlights the line
`Side to offset:` *Pick to the left*	Duplicates the line
`Select object to offset:` *Pick ⑥*	Highlights the line
`Side to offset:` *Pick to the left*	Duplicates the line
`Select object to offset:` *Right click on the mouse twice*	Ends and reissues the OFFSET command
`Offset distance or Through` `<Through>` **18"** (Enter)	Sets the distance to 18"
`Select object to offset:` *Pick ⑦*	Highlights the line

continues

Side to offset: *Pick above the line*	Duplicates the line
Select object to offset: *Pick* ⑧	Highlights the line
Side to offset: *Pick above the line*	Duplicates the line
Select object to offset: *Right click once*	Ends the OFFSET command
Save the drawing	

Figure 12.7

Blocking out the exterior walls.

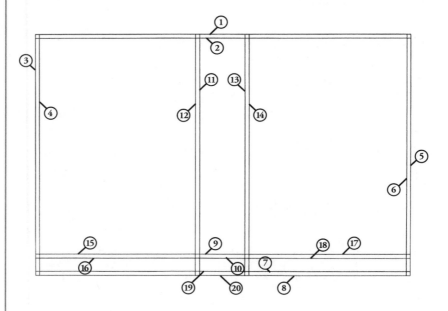

You now have all of the construction lines for the exterior walls. You need to trim and erase the extra lines to make the entryway. This is done in the next exercise.

Exercise

Trimming the Exterior Walls of the Floor Plan

Continue from the previous exercise; refer to figure 12.7 for pick points.

Command: *Choose the Trim button*	Issues the TRIM command

```
Select cutting edge(s):
Select objects: Pick ⑨          Highlights edge
(see fig. 12.7)
Select objects: Pick ⑩          Highlights edge
Select objects: Pick ⑪          Highlights edge
Select objects: Pick ⑫          Highlights edge
Select objects: Pick ⑬          Highlights edge
Select objects: Pick ⑭          Highlights edge
Select objects: (Enter)
<Select object to trim>/Undo:    Erases line(s)
Pick ⑪⑫⑬⑭⑮⑯⑰⑱⑲⑳
Save the drawing
```

Now that you have the outline of the exterior, you need to clean up the lines by filleting the overlapping lines in each corner.

Using the Fillet Command for Corners

Exercise

The basic layout of the exterior walls is now complete. Use figure 12.8 for the pick points with the following exercise. Figure 12.9 shows the final result after filleting the corners.

```
Command: F (Enter)              Issues the FILLET command

Polyline/Radius/<Select first   Selects the Radius option
object> R (Enter)

Enter fillet radius             Sets the fillet radius to zero
<0'-0 1/2">: 0(Enter)

Command: Choose the Fillet button   Issues the FILLET command

Polyline/Radius/<Select         Selects first entity
first object> Pick ①
(see fig. 12.8)
```

continues

`Select second object` *Pick* ② then *right click on the mouse again*	Fillets the two lines and reissues same command
`Polyline/Radius/<Select first object>` *Pick* ③	Selects first entity
`Select second object` *Pick* ④	Selects second entity

Repeat this process as needed throughout the exercise to avoid crossing lines at intersections of walls. Figure 12.9 shows the exterior walls after each corner has been filleted

Save the drawing

The layout of the exterior walls before filleting.

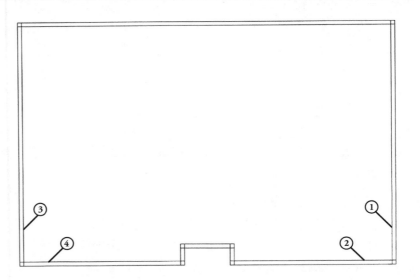

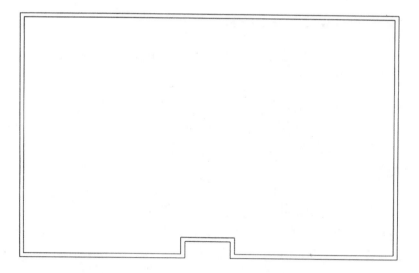

Figure 12.9

The exterior walls after using the FILLET command.

Drawing the Interior Walls on the Floor Plan

The process of completing the interior is slightly more complex than completing the exterior—there are more construction lines to create and more corners to fillet, trim, extend, and erase. The following exercise leads you through one half of the house.

TIP

The trick to keeping confusion to a minimum and your speed to a maximum is to use your initial sketches. Plan out which construction lines to make and which parts to keep. In addition, plan out your actions ahead of time so that you do the minimum amount of work.

Exercise

Layout of Interior Walls

Continue from the previous exercise. Use figure 12.10 in the following exercise to help you construct some of the interior walls. Remember, the interior walls are 3.5" wide.

You begin by offsetting existing wall lines into the house and then offsetting the new lines 3.5" to make the interior walls.

Command: *Choose* **C**onstruct, *then* **O**ffset	Issues the OFFSET command
Offset distance or Through <Through>: **12'** (Enter)	Sets the distance to 12'
Select object to offset: *Pick* ① (*see fig. 12.10*)	Highlights the line
Side to offset: *Pick above the line*	Duplicates the line
Select object to offset: *Pick* ②	Highlights the line
Side to offset: *Pick to the left*	Duplicates the line
Select object to offset: *Right click once*	Ends the OFFSET command
Command: **F** (Enter)	Issues the FILLET command
Polyline/Radius/<Select first object> *Pick* ③	Selects first entity
Select second object *Pick* ④	Selects second entity and fillets the lines
Command: *Choose* **C**onstruct, *then* **O**ffset	Issues the OFFSET command
Offset distance or Through <Through>: **3.5"** (Enter)	Sets the distance to 3.5"
Select object to offset: *Pick* ③	Highlights the line
Side to offset: *Pick to the left*	Duplicates the line
Select object to offset: *Pick* ⑤	Highlights the line
Side to offset: *Pick to the left*	Duplicates the line
Select object to offset: *Pick* ④	Highlights the line

`Side to offset:` *Pick above the line*	Duplicates the line
`Select object to offset:` *Right click on the mouse twice*	Ends and reissues the OFFSET command
`Offset distance or Through <Through>:` **7'** **[Enter]**	Sets the distance to 7'
`Select object to offset:` *Pick* ④	Highlights the line
`Side to offset:` *Pick below line*	Duplicates the line
`Select object to offset:` *Pick* ⑥	Highlights the line
`Side to offset:` *Pick below line and click on the mouse once*	Duplicates line and ends OFFSET command
`Command:` *Choose the Extend button*	Issues the EXTEND command
`Select boundary edge(s)`	Highlights boundary line
`Select objects:` *Pick* ⑤ *and right click on the mouse once*	
`<Select object to extend> /Undo:` *Pick* ⑦ *and right click on the mouse*	Extends line to boundary and ends the EXTEND command
Save the drawing	

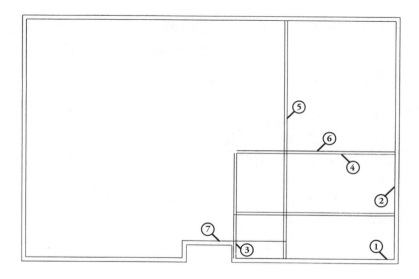

Figure 12.10

The layout of interior walls.

Now that you have seen how to place the lines for the interior walls, continue in the next exercise to trim, fillet, offset, and erase more construction lines. When you are finished, you will look at how to place door and window symbols.

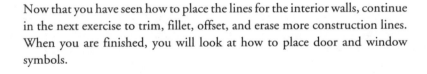

Exercise

Continuing the Layout of Interior Walls

Continue from the previous exercise. Refer to figure 12.11 for the following exercise.

Figure 12.11

The layout of interior walls.

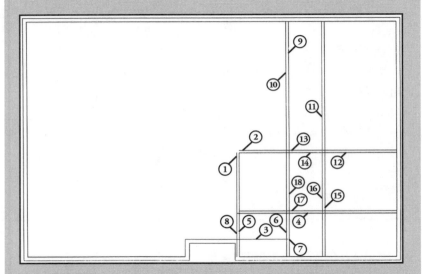

Command: **F** (Enter)	Issues the FILLET command
Polyline/Radius/<Select first object> *Pick* ① *(see fig. 12.11)*	Selects first entity
Select second object *Pick* ②	Fillets the two lines
Command: *Choose the Trim button*	Issues the TRIM command
Select cutting edge(s):	
Select objects: *Pick* ③④ ⑥⑦ *and right click once*	Highlights cutting edge

`<Select object to trim>` `/Undo:` *Pick* ④⑤⑥⑦⑧⑰ *and right click once*	Erases line(s) and ends TRIM command
`Command:` *Choose* **C**onstruct, *then* **O**ffset	Issues offset command
`Offset distance or Through` `<Through>:` **4'** (Enter)	Sets distance to 4'
`Select object to offset:` *Pick* ⑱	Highlights the line
`Side to offset:` *Pick to the right* *and right click on the mouse twice*	Duplicates the line and reissues OFFSET command
`Offset distance or Through` `<Through>:` **5.5"** (Enter)	Sets distance to 5.5"
`Select object to offset:` *Pick* ⑯	Highlights the line
`Side to offset:` *Pick to the right* *and right click once*	Duplicates the line and ends OFFSET command
`Command:` *Choose the Trim button*	Issues the TRIM command
`Select cutting edge(s):`	Highlights cutting edge
`Select objects:` *Pick* ② *and* *right click once*	
`<Select object to trim>`	Erases line(s) and reissues the TRIM command
`/Undo:` *Pick* ⑨⑩ *and right click* *twice*	
`Select cutting edge(s):`	Highlights cutting edge
`Select objects:` *Pick* ⑫ *and* *right click once*	
`Select object to trim>` `/Undo:` *Pick* ⑮⑯ *and right click* *twice*	Erases line(s) and reissues the TRIM command
`Select cutting edge(s):`	Highlights cutting edge
`Select objects:` *Pick* ⑪⑱ *and right click once*	

continues

`Select object to trim>/Undo:` *Pick* ⑬⑭ *and right click once*	Erases line(s) and ends the TRIM command

Save the drawing

In figure 12.12, you can see the construction of some of the interior walls

Use the FILLET or TRIM command as previously discussed to modify the intersections, as shown in figure 12.13

Figure 12.12

The layout of interior walls.

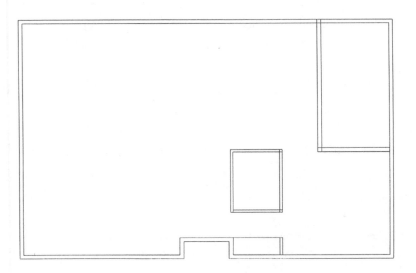

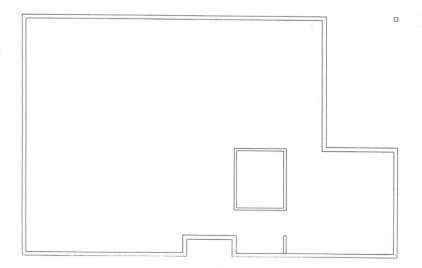

Figure 12.13

The layout of interior walls using FILLET and TRIM commands.

Understanding Door and Windows Symbols

The use of blocks, as you read earlier, is a very efficient tool. Blocks, used along with an extensive and organized "symbols" library, can drastically reduce the amount of drafting time on a particular drawing. Architects use blocks on a regular basis, as do most other drafting professionals. Some of the common blocks used on floor plans are the following:

◆ **Doors.** In any given structure, there may be large quantities of doors, particularly in commercial projects. Even when drafting residential structures, however, it is a great advantage to have access to a library of door symbols. As with so many aspects of architectural drafting, there is no one particular standard for drafting a door. Door blocks can easily be created to your own specifications.

In figure 12.14, you can see three types of specific doors, each drafted two different ways. Your doors can be creative and detailed, or clean and simple. It is important to remember that floor plan scales are typically at 1/4"=1'–0" or 1/8"=1'–0"; if you provide too much detail to your doors, it may not show on the final plot. Again, whichever way you select to draft your door blocks, stay consistent in detail and style.

Figure 12.14

A variety of door blocks.

♦ **Windows.** Many of the concerns discussed regarding door blocks apply here as well. The graphic presentation of a window, however, is not as important as doors, where the type of representation on the plan has a possible impact on the walkway circulation patterns or light switch locations within the room.

Consistency and accuracy are important in all aspects of drafting; however, the actual window type may not be known to you at the time you are drafting the floor plan. In fact, a window block, whether it is *sliding* or *double-hung*, usually will not have any impact on walkway circulation patterns or light switch locations. For this reason, some firms choose to use only one standard graphic representation, in plan, for all windows. This can help during the drafting stages of the project, when perhaps the architect, client, or budget has not allowed for a decision to be made as to the window style or even the manufacturer. The style of the window, however, should be selected prior to completing the exterior elevations, where such information is critical to the overall design and the accuracy of that drawing. Some graphic varieties of the three window types are shown in figure 12.15.

Figure 12.15

A variety of window blocks.

You have worked through the basic exercises for constructing walls. Notice the placement of the door block in figure 12.16. While drafting floor plans, it is often most efficient to complete all of the walls and then move on to additional objects, such as the placement of door and window blocks. In the next exercise, you learn how AutoCAD LT can help make this process even more efficient.

TIP

When making door and window blocks, you may want to make one block and insert it at different X or Y scales to make the door fit different size openings. If you do this, remember that you cannot explode an unequally scaled block. In addition, when designing your block, make sure you include the door or window jams that will cap the wall openings. Not only does this give you a better-looking drawing, it also gives you trim lines for trimming the door or window openings in the walls.

Exercise

Using the MIRROR Command

Continue from the previous exercises. See figure 12.16 for the pick points.

Set Ortho (F8) on

Command: *Choose* Construct, *then* Mirror	Issues the MIRROR command
Select objects: *Use crossing window around the four walls with door, including door*	Highlights line(s)
Select objects: *Right click once*	Issues next prompt
First point of mirror line: *Pick* MID *of* ① *(see fig. 12.16)*	Selects point and issues next prompt
Second point: *Pick above first point*	Duplicates line(s) and issues next prompt
Delete old objects?<N> *Select* No	Displays old and new objects and ends MIRROR command

Figure 12.17 shows the object mirrored. Also notice additional objects that have been added. Please refer to figure 12.18 as a guideline for completing the floor plan. It shows the completed layout of the floor plan.

Figure 12.16

The floor plan with a door.

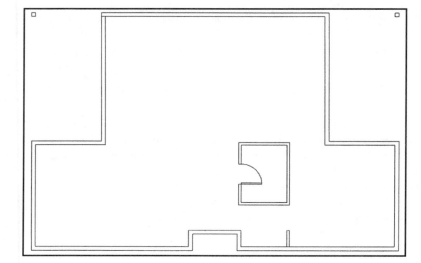

Figure 12.17

Continuation of the floor plan.

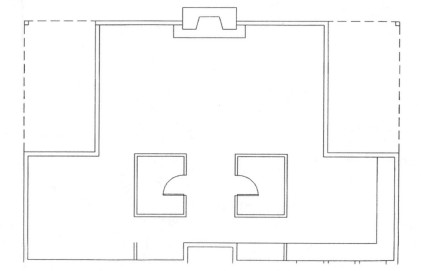

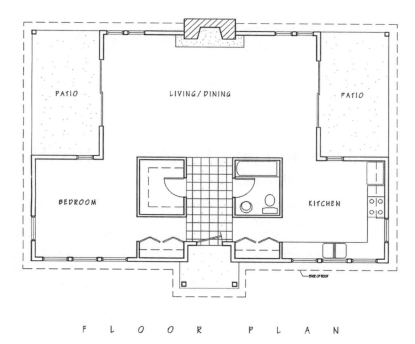

Figure 12.18

*The completed
floor plan.*

You can elect to redesign or enhance any portion of the floor plan, but the process of constructing the floor plan remains the same. Through the use of COPY and MIRROR commands, in addition to blocks, the time you spend creating entities for the floor plan should gradually decrease. It is important to use the software to your advantage.

TIP

Remember to use BHATCH command in addition to symbol blocks when annotating your drawings. Figure 12.18 shows hatching in the entryway, patio, and fireplace.

Developing the Elevation

An *elevation* is a drawing that specifically displays vertical information—it is an orthographic projection that helps communicate additional aspects of the structure. You read earlier in this chapter that the floor plan is often the first area of focus. This is true; however, the elevations of a structure are equally important.

Elevations convey information about a project that other drawings do not, such as door and window shapes, exterior materials, roof lines, and so forth. Because many buildings are rectangular in plan, there are usually four elevations drafted. If a plan of a building or house is irregular in shape, more elevations may be required. The first step, though, is to look at how a basic elevation is constructed.

Using Orthographic Projections to Make Elevation Outlines

The typical process of drafting an elevation is to work with a completed floor plan. The floor plan is used as a basis for translating information from the plan that relates directly to the elevation, such as the location of exterior doors or windows. Because this type of information appears on both sets of drawings, it is most common for the elevations to be drafted at the same scale as the floor plan. Many residential plans and elevations, such as the one in this chapter, are drafted at 1/4"=1'–0".

Typically, the plans would be drafted on a separate sheet from the elevations. In this chapter, the vacation home is quite small. In addition, the following exercise only explores blocking out one of the elevations. This allows both the plan and elevation to be placed on one sheet, which you may have noticed in figure 12.5.

Figure 12.19 shows four basic orthographic projections of the small house used previously in this chapter. This should help give you an understanding of how these drawings are blocked out. The following exercise outlines the steps for drafting the front elevation. You begin by placing the *ground line,* which, as the name suggests, represents the ground on which the house sits.

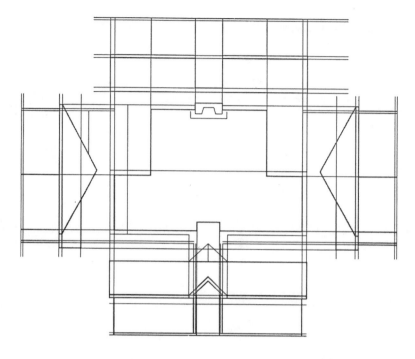

Figure 12.19

Orthographic projections.

Exercise

Blocking Out the Elevation

Continue from the previous exercise. You start by creating the ground line and then using the EXTEND command to create the orthographic projection. See figure 12.20 for pick points and construction lines.

Draft a horizontal line, as seen in figure 12.20. This line should be 44' long, centered directly below floor plan. For now, this line should be about 20 feet away from the floor plan.	Creates the ground line

continues

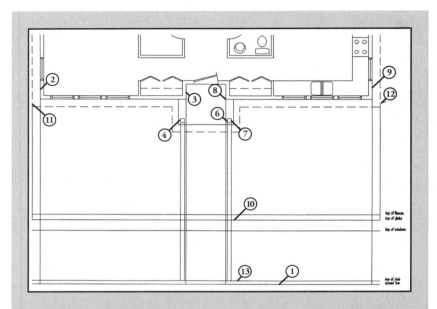

Command: *Choose the Extend button*	Issues the EXTEND command
Select boundary edge(s) Select objects: *Pick ① and right click once on the mouse (see fig. 12.20)*	Highlights the boundary line
<Select object to extend> /Undo:	Extends lines to boundary
Select objects: *Pick ② ③ ④ ⑤ ⑥ ⑦ ⑧ ⑨ and right click on the mouse once*	Ends the EXTEND command
Command: *Choose* **C***onstruct, then* **O***ffset*	Issues the OFFSET command
Offset distance or Through <Through>: **6** **(Enter)**	Sets the distance to 6"
Select object to offset: *Pick ①*	Highlights the line
Side to offset: *Pick above the line*	Duplicates the line

`Select object to offset:` *Right click on the mouse twice*	Ends and reissues the OFFSET command
`Offset distance or Through` `<Through>:` **6'10** (Enter)	Sets distance to 6'–10"
`Select object to offset:` *Pick* ⑬	Highlights the line
`Side to offset:` *Pick above the line*	Duplicates the line
`Select object to offset:` *Right click on the mouse twice*	Ends and reissues the OFFSET command
`Offset distance or Through` `<Through>:` **8'1** (Enter)	Sets distance to 8'–1"
`Select object to offset:` *Pick* ⑬	Highlights the line
`Side to offset:` *Pick above the line*	Duplicates the line
`Select object to offset:` *Right click on the mouse twice*	Ends and reissues the OFFSET command
`Offset distance or Through` `<Through>:` **8** (Enter)	Sets distance to 8"
`Select object to offset:` *Pick* ⑩	Highlights the line
`Side to offset:` *Pick above the line*	Duplicates the line
`Select object to offset:` *Right click on mouse once*	End the OFFSET command
`Command:` *Choose the Line button*	Issues the LINE command
`From point:` *Pick* ENDPOINT *of* ⑪	Sets first point of line
`To point:` *Pick* PER *of* ⑩	Sets second point and draws line
`To point:` *Right click on mouse twice*	Reissues the LINE command
`From point:` *Pick* ENDPOINT *of* ⑫	Sets first point of line
`To point:` *Pick* PER *of* ⑩	Sets second point and draws line
`To point:` *Right click on mouse once*	Ends the LINE command

As you can see in figure 12.20, the initial construction lines for the elevation are now established. The purpose of the horizontal lines is indicated toward the right side of the figure.

Exercise

Continuing the Front Elevation

Continue from the previous exercise. Use figure 12.21 for pick points and constuction lines.

Figure 12.21

Construction lines for front elevation.

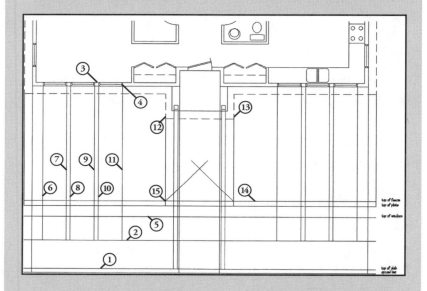

Command: *Choose* **C**onstruct, *then* **O**ffset	Issues OFFSET command
Offset distance or Through <Through>: **3'10** (Enter)	Sets the distance to 3'10
Select object to offset: *Pick* ① *(see fig. 12.21)*	Highlights the line
Side to offset: *Pick above the line*	Duplicates the line
Select object to offset: *Right click on mouse once*	Ends OFFSET command

`Command:` *Choose the Extend button*	Issues the EXTEND command
`Select boundary edge(s)`	Highlights boundary line
`Select objects:` *Pick ② and right click on mouse once*	
`<Select object to extend> /Undo:`	Extends line to boundary
`Select objects:` *Pick ③④ and right click on mouse once*	Ends the EXTEND command
Use the EXTEND *command to construct the left and right side of each window on this elevation.*	
`Command:` *Choose the Trim button*	Issues the TRIM command
`<Select cutting edge(s):`	Highlights cutting edge
`Select objects:` *Pick ⑤ and right click once*	
`<Select object to trim> /Undo:` *Pick ⑥⑦⑧⑨⑩⑪*	Erases line(s) and ends the TRIM command
Use the TRIM command to trim construction lines for each window on the right side of the elevation, or MIRROR the left windows to the right side.	
`Command:` *Choose the Line button*	Issues the LINE command
`From point:` *Pick* ENDPOINT *of ⑫*	Sets first point of line
`To point:` *Pick* PER *of ⑭ and right click on mouse twice*	Draws line and reissues the LINE command.
`From point:` *Pick* ENDPOINT *of ⑬*	Sets first point of line
`To point:` *Pick* PER *of ⑭ and right click on mouse once*	Draws line and ends the LINE command

continues

To construct the gable on the center of the elevation, use the LINE command to start a line from INT of o in a 45° angle. Repeat this process on the opposite side, or use the MIRROR command.

Continue using the TRIM command to eliminate unneeded construction lines. See figure 12.22 for the completed elevation.

Once the elevation has been completed, use the MOVE command to relocate the elevation to the proper place on the sheet. Refer to figures 12.5 and 12.6.

Figure 12.22

The completed front elevation.

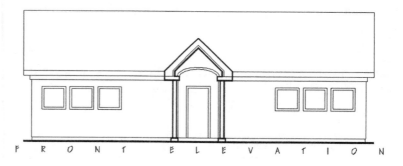

Printing or Plotting the Drawing

Printing or plotting a computer-aided drawing is something that is done as an on-going process throughout any architectural project. Because the final product is on paper, it is important that you periodically look at your work on paper, not just on your computer screen. Looking only at your computer screen to review the drawing does not allow for the same perspective as on paper.

Reviewing the hard copy in stages is a very effective way to work. This enables you to identify errors or omissions in your drawing prior to completing it. You will find that it is easier, and more efficient, to "do it right" the first time, rather than waiting until the end of the project to identify mistakes, which you then must correct.

There are some advantages and disadvantages to both printing and plotting. The following list identifies these issues:

◆ **Printing.** Printing a drawing can mean a few different things within the drafting community. In reference to computer-aided drafting, it most commonly refers to the use of a *laser printer* as an output device for printing your drawing or document on A-size paper. Laser printers can be a very fast and efficient way to see and check your work, primarily due to the fact that most laser printers can produce a physical document for you to look at in a period of seconds, depending on the byte size of the drawing. The small size of the paper also makes it easy to handle and have as a reference.

You may find, however, that the A-size print of your work is too small, particularly if your drawing has a great deal of detail to it, or if it is drafted at a very small scale. One other disadvantage is regarding line weights. Earlier in this chapter, you read that it is common for line weights, or line thicknesses, to be determined by a particular pen assignment on the plotter. Using a laser printer does not always allow for this, especially if the plotter's driver does not allow for variable line widths. The only way a "heavy" line weight can be produced from a laser printer is by creating that line using PLINE, and then varying the width of the line.

Finally, you should be aware that the term *printing* has a different definition when used in reference to drawings that were completed manually on a drafting table. In this case, printing can be used to describe the process of "blue-printing," which was discussed in Chapter 10, "Reproducing Drawings."

◆ **Plotting.** Plotting is a common method used for producing the drawing on larger sheets of vellum or paper such as B, C, D, or E. Plotters can also plot on A-size paper, but if you need your drawing on this size of paper, the laser printer may save you time, depending on the purpose of the print. Some common styles of plotters are pen plotters, inkjet plotters, thermal plotters, and electrostatic plotters; there are even large size laser plotters. Within any given type of plotter, you have options for different line qualities and paper types.

TIP

Before plotting, check with your CAD manager to find out what is the quickest and least expensive way to produce a given plot. Make sure you use the appropriate paper and line qualities. Both of these factors can save you considerable plotting time and money on supplies.

NOTE

For more information on printing and plotting, see Chapter 10.

As you gain experience working with architects, you will discover that there is both a time and a place for using the plotter or the printer. Aside from your preference, or even your supervisor's preference, there is the preference of the consulting engineers, who might request blueprints, laser prints, or plotted prints to scale. The amount of communication that exists during a project between the architect's office and all of the other consultants is vast. On one given project, you may work with many different professionals, such as the landscape architect, structural engineer, mechanical or electrical engineer, interior designer, specifications writer, and so forth.

Commercial projects tend to be complex; however, residential projects can also involve consultants. Whether the project is commercial or residential, the architect's office is often the nucleus or center stage of the project. With this comes the responsibility to communicate with all of the other professionals involved. During the design and drafting of the project, this communication is both verbal and graphic. The verbal communication may take place on the phone or in meetings, but the graphic communication takes place on paper. The importance of constant communication between trades is critical for the success of the project. Therefore, the methods of producing your hardcopy need to vary according to the stage of the project, the people involved at that time, and the purpose of the communication.

The use of any or all of these printing or reproduction methods will be something you encounter on a continuous basis within the architectural profession. Once the project is completely drafted, your hardcopy needs still continue. There are several stages of checking the drawings that must occur prior to obtaining a final construction bid or even a construction permit, as discussed next.

Checking the Drawing

As mentioned at the beginning of this chapter, the process of completing a full set of architectural drawings can be complicated. The production of these working drawings is not really complete, however, until the work has been thoroughly checked. Every architect, designer, or draftsperson should be casually checking the accuracy of his or her own work on a continuous basis. This enables you to discover and correct the errors prior to releasing them for print. Even if you are certain that the working drawings are accurate, it is highly recommended that you allow another person to review the drawings. A new set of eyes on the project is often very helpful, and it is a procedure most firms employ.

One common method, particularly in large architectural firms, is to have a series of people review the drawings. Often a set of prints, sometimes called the *checkset,* is made of the working drawings. The set might be checked by a few people within the firm, or perhaps another draftsperson reviews your work. It may then travel to the project manager or architect for yet another review. In any case, it is important that someone else besides you reviews the work. Whether one or more people check the drawings, the set will inevitably come back to the draftsperson with corrections to be made.

The corrections, which are often marked with red pencil on the prints, need to be addressed prior to releasing them for final printing. This is all done to ensure accuracy and consistency. The more accurate and consistent the working drawings, the more efficient the construction process will be for the contractor.

Prior to construction though, there is yet one more person who will check your work—the plan checker. The *plan checker* is the person at the township, city, county, or state level who checks the entire set of drawings for errors and omissions. The plan checker will sometimes look for simple drafting and spelling errors, and will always be searching for mistakes or omissions that could jeopardize the health, safety, and welfare of the occupants of the structure and the general public. This review of the working drawings is mostly objective and based primarily on codes. Plan checkers virtually always find some form of correction needed. Once the person, or company, responsible for the drawings completes these final corrections, a permit is issued. Upon receipt of a permit, all of your ideas and work on paper can now be transformed into an actual structure.

Summary

The concepts in this chapter only begin to touch on the complexity of the profession of architecture. You should apply these concepts to projects in which you have a direct interest. Once you have practiced drafting a few plans and elevations, you will discover you have a skill that is applicable to practically all plans and elevations. You will discover that a floor plan is always a floor plan, and an elevation is always an elevation, regardless of the design. Some designs are more complex than others, but the drafting principles remain the same. Should your architectural experience never go beyond plans and elevations, you have now acquired a better understanding of the profession of architecture, and a skill that will better diversify your drafting abilities using AutoCAD LT.

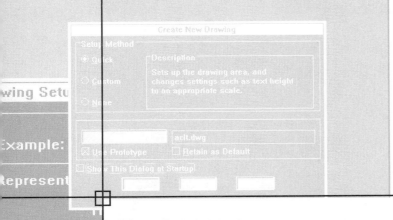

Part Five

Appendix

A. Glossary of CAD and Drafting Terms

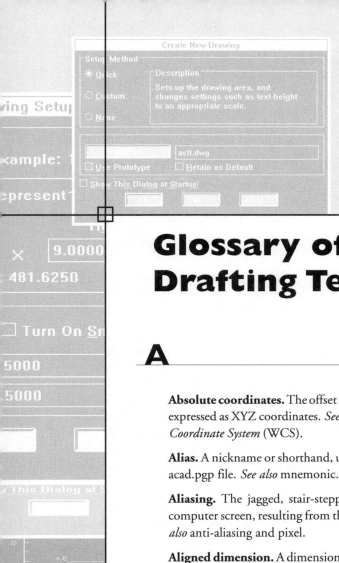

Glossary of CAD and Drafting Terms

A

Absolute coordinates. The offset of a point from the origin of the current UCS, generally expressed as XYZ coordinates. *See also* origin, *User Coordinate System* (UCS), and *World Coordinate System* (WCS).

Alias. A nickname or shorthand, usually an abbreviation, for a command, defined in the acad.pgp file. *See also* mnemonic.

Aliasing. The jagged, stair-stepped appearance of diagonal lines and curves on the computer screen, resulting from their approximation by the bitmapped raster image. *See also* anti-aliasing and pixel.

Aligned dimension. A dimension measuring the distance between any two points at any angle, with the dimension line placed parallel to the angle between the points.

Angular dimension. A dimension measuring the angle between two points and a vertex point, the angle between two lines, or the angle of an arc segment.

Angular unit. The unit of measurement—such as decimal degrees, degrees/minutes/seconds, grads, and radians—for measuring angles.

Annotation. Non-graphic information, such as text, symbols, notes, dimensions, and tolerances, in a drawing.

Anonymous block. An "unnamed" block (which actually has an automatically assigned name like *X17) used by AutoCAD to store data for associative dimensions, hatch patterns, and other objects whose data the user does not directly access.

ANSI. The American National Standards Institute.

ANSI Y14.5m. The ANSI standard for *Geometric Dimensioning and Tolerancing* (GDT). *See also* ANSI and Geometric Dimensioning and Tolerancing (GDT).

Anti-aliasing. Adjustment of the color or intensity of pixels to provide the visual effect of smoothing the jagged appearance of aliasing.

Approximation points. A set of points that approximate a curve. *See also* interpolation and fit points.

Array. A rectangular or polar pattern containing multiple copies of an object.

Arrowhead. Two or more lines or arcs that form a closed or open, filled or unfilled triangular symbol used to terminate a leader or the beginning or end of a dimension line.

ASCII. American Standard Code for Information Interchange.

ASCII character. One of the characters in the 128-character set defined by the American Standard Code for Information Interchange. Each character is assigned a numeric code used in computer programming and data communications. Nonstandard versions of ASCII define 255 characters.

Aspect ratio. The height to width ratio.

Assembly. The set of parts combined to create a completed product. Generally used for mechanical drawings.

Associative dimension. A dimension that changes as the associated feature, object, or geometry is changed.

Associative hatching. Hatching that changes as the associated feature, object, or geometry is changed. *See also* hatching.

Attribute. Predefined, fixed, or user-defined text values associated with blocks. Attributes are often used for creating bills of materials. Other CAD systems sometimes use the term attribute for what are called object or entity properties in AutoCAD.

Attribute definition. An object in AutoCAD that defines a template used to assign predefined, fixed, or user-defined text values to blocks. Attribute definitions are included in block definitions, inserted with the block (at which time the user may enter attribute values), and may be extracted from the drawing database to create external data files. *See also* attribute.

Attribute tag. A text string, defined in the attribute definition, used to identify a particular attribute during data extraction.

Attribute value. The predefined, fixed, or user-defined text value assigned to a particular attribute, and which can be extracted from the drawing database to external data files.

Axis. A direction in space, such as the x (height), y (width), and z (depth) axes.

Axis tripod. A three-vector symbolic representation of the x, y, and z axes, used in viewing three-dimensional AutoCAD drawings.

B

Base point. A reference point for editing operations, relative to which distances and angles for copy, move, and rotate operations are specified. In grip editing, the initial base point is the highlighted grip.

Baseline. The base reference line of a baseline dimension, or the line defined by the bottoms of a string of text characters (ignoring the descenders that drop below the baseline in characters such as *g* and *y*).

Baseline dimension. One of a series of dimensions measured from a common baseline. Also known as a parallel dimension.

Bevel. A cut surface at an angle other than 90 degrees, generally in mechanical parts and machining. *See also* chamfer.

Bézier curve. A non-uniform polynomial curve mathematically defined by a set of control points. A Bézier curve is a special form of a B-spline curve.

Bill of Materials (BOM). A list identifying parts, quantities, lengths, weights, and other information required for an assembly or product. You can extract a BOM from attribute data. Also known as a parts list.

Bind. To permanently import all or part of the data of an xref into the current AutoCAD drawing.

Bitmap. A representation of an image digitally, as bits, displayed as pixels, in which colors are comprised of red, green, and blue components of varying intensities or values. Also known as a raster image. There are many different formats for bitmap images.

Blip marks. Small temporary cross marks displayed on the screen when you specify a point. They are removed from the current viewport when you redraw or regenerate the viewport. Also known as blips.

Block. A term commonly used to mean a block reference.

Block definition. A complex object defined in the AutoCAD drawing database by an assigned name, an insertion base point, and one or more entities or attribute definitions. The block definition is not visible in the drawing, but you can insert visible references to it with the INSERT command.

Block reference. Visible insertions of a reference to a block definition. A block reference groups one or more AutoCAD objects to create a single complex object.

Break line. A line consisting of two straight segments with a zigzag section between them, used to indicate that a portion of an object is not shown in the drawing.

B-spline curve. A non-uniform curve, mathematically defined by a set of control points, which provide more localized control for refining and manipulating the curve than in the case of a Bézier curve. *See also* Bézier curve.

Bubble. *See* reference bubble.

BYBLOCK. An object property that causes an object to inherit the color or linetype of the block containing it, at the time of insertion. If the object is not contained in a block, it is drawn white and continuous.

BYLAYER. An object property that causes an object to inherit the color or linetype assigned to its layer.

C

CAD. Computer-Aided Design (or Drafting).

CADD. Computer-Aided Design and Drafting.

Cartesian coordinate system. A coordinate system defined by the three principal axes (x, y, and z), which are perpendicular to each other. Points in two- or three-dimensional space are specified by their distances along these axes, projecting from the point to the nearest (perpendicular) point on each axis and measuring in a positive or negative direction from the 0,0,0 coordinate where all three axes meet. *See also* polar, spherical, and cylindrical, and World coordinate system.

Center line. A line that marks the center of an arc, circle, or other symmetrical shape. In a dimensioned arc or circle, center lines usually radiate from the center mark and extend to slightly beyond the edge. When drawn outside the edge of a curve, center lines are drawn with alternating long segments and short dashes.

Center mark. A small cross mark at the center of an arc or circle.

Chain dimension. Another term for continued dimension.

Chamfer. An edge or corner between two lines or surfaces at an angle other than 90 degrees. *See also* bevel.

Check print. A preliminary hard copy used for drawing progress review, or marking corrections or changes.

Chord. A straight line segment between two points on the circumference of an arc or circle.

Circular xref. An externally referenced drawing that itself is referenced, either directly or indirectly, via nested xrefs. The xref that causes the circularity is ignored. *See also* xref and external reference.

Color map. The table that defines the intensity of the *red, green, and blue* (RGB) values that determine the appearance of each displayable color.

Command. A predefined, user-initiated instruction to the computer program.

Command lines. A text area that displays user input, prompts, and messages from the program to the user.

Construction plane. A 2D geometric plane in 3D space where 2D objects or the base geometry of 3D objects are constructed by default. This is the XY plane of the current UCS, unless the ELEVATION system variable is set to a non-zero value, in which case that value specifies the elevation of the construction plane above or below the XY plane of the current UCS. *See also* User Coordinate System (UCS).

Continued dimension. A linear dimension that uses the origin of a previous dimension's second extension line, or of a selected extension line of any existing dimension, as its first extension line origin. Continued dimensions are used to dimension a series of features. Also known as a chain dimension.

Contour line. The outline of a 2D shape or edge of a 3D form drawn with a line, generally curved, that represents a specific level relative to a given datum point.

Control point. One of a set of points defining a curve in space. The locations of the control points can be adjusted to alter the shape of the curve. *See also* Bézier curve and B-spline curve.

Coordinate. A set of values that specify the location of a point in space relative to the axes of a coordinate system.

Coordinate filters. Another term for point filters.

Coordinate system. A system for defining the positions of points in space relative to an origin and associated axes. AutoCAD's primary system is the Cartesian coordinate system, with an origin point of 0,0,0 and three orthogonal axes, x, y, and z. *See also* User Coordinate System (UCS), World Coordinate System (WCS), Cartesian coordinate system, and polar coordinates.

Coplanar. All points within the same geometric plane.

CPolygon. A closed, multisided polygon window used to select all objects either fully within or crossing its edges. *See also* WPolygon.

Cross section. The view of a mechanical part resulting from the intersection of the part with an imaginary cutting plane, or a slice through a building or structural member that shows the internal conditions through that area.

Crosshairs. A graphics cursor defined and indicated by the intersection of two lines that extend over the full current viewport. Also known as a graphics cursor.

Crosshatching. An area fill composed of two or more intersecting sets of continuous or broken parallel lines. *See also* hatching.

Crossing window. A rectangular box used to select all objects either fully within or crossing its edges.

Cubic curve. A mathematical curve of the third order, as defined by a parametric polynomial equation. AutoCAD's default spline type is a piecewise cubic curve.

Cursor. A graphic pointer on a video display screen that marks the current location for text, graphics, or command input, and which can be moved around by keys or the pointing device, such as a mouse. The appearance of the pointer varies to indicate the current mode of the program. *See also* crosshairs.

Cursor menu. A menu that pops up in the graphics area at the cursor location when you press the middle button of a three-button mouse, or hold down either the Shift, Ctrl, or Alt key while you press the right button of a two-button mouse.

Curve. The intersection of two surfaces in 3D space. Most curves are unbounded, but some, such as circles, are bounded.

Curve fitting. The definition of a curve by arc or polynomial segments passing through a number of interpolation points.

Cutaway view. Another term for cross section.

Cylindrical coordinates. Coordinates locating a point in space by its z value, distance from the origin or reference point, and angle from the x axis in the XY plane.

D

Datum-line dimension. Another term for ordinate dimension.

Datum point. A reference point, from which elevations or dimensions are typically specified.

Default. A predefined setting that determines graphic or data values in the absence of overriding input from the user. Default values in AutoCAD prompts are indicated by angle brackets <> and accepted by a null response (usually pressing the spacebar or Enter). *See also* null response.

Default drawing. The default prototype drawing, such as acad.dwg and acadiso.dwg, from which a new drawing inherits its initial settings. Any graphic data in the prototype drawing is also inherited.

Definition points. Points used to relate an associative dimension to associated geometry or objects, so that the dimension is modified when the geometry or objects are modified. These points are stored on the special DEFPOINTS layer, which does not plot. Also known as defpoints.

Dependent symbol. A definition, such as a layer, block, text style, or linetype, in the drawing's named object definition tables, resulting from an xref's attachment.

Detail. A large-scale drawing showing a portion of a building, part, or other object in greater detail to clarify structure, assembly, or other relationships between materials or components.

Diameter dimension. A dimension measuring the diameter of an arc or circle.

DIESEL (Direct Interpretively Evaluated String Expression Language). A macro language used to define custom menu items and labels, and for customizing the AutoCAD status line (via the MODEMACRO system variable).

Dimensions. Numerical values used to indicate size and distance.

Dimension line. A line showing the extents of a measurement, between extension lines. The dimension text is typically placed along this line. Architectural and ISO standards place the dimension text above an unbroken dimension line. ANSI standards place the dimension text within a broken dimension line. The dimension line of an angular dimension is an arc.

Dimension style. A set of dimension settings that specify the dimension's behavior and appearance, and enable you to store and recall sets of dimension system variable settings.

Dimension text. The value associated with dimensioned objects, typically a measured distance or angle.

Dimension variables. Numeric values, text strings, or on/off switches that determine the behavior and appearance of AutoCAD dimensioning.

Dimensioning. The creation of dimensions.

Display extents. The limits or extents, in rectangular coordinates, of the current display or graphics area.

Dithering. The varying of the color or intensity of adjacent pixels to give the impression of colors or intensities not directly available to the device.

Documentation drawing. A 2D drawing, such as an orthographic projection, that describes the physical characteristics of an object or part with graphics and annotation.

Draft. A preliminary sketch, subject to revision.

Drag. To move a drawing or interface object on-screen by using the pointer.

Drawing extents. The smallest rectangular area that contains all objects in a drawing.

Drawing file. A file on disk that stores a drawing and its associated data.

Drawing limits. A user-defined rectangular drawing area, defined by the coordinates of its lower left and upper right corners.

Drawing unit. The unit, such as millimeters, feet, inches, or miles, represented by linear distances and measurements in a drawing. *See also* unit.

DXF (Drawing Interchange Format). A standard drawing format for importing and exporting drawings and drawing data between AutoCAD and other programs, including other CAD programs. DXF files can be in either ASCII or binary file formats.

Edit. To change existing graphic or text data, generally by re-specification of parameters or interactive manipulation.

Elevation. The z value of a 3D point from the XY plane, or the front, side, or back view of a part or structure.

Enter button. The mouse button that acts as if the user pressed the Enter key, generally the right button on a mouse. *See also* pick button and pop-up button.

Entity. An object, such as a point, line, arc, circle, polyline, dimension, or text string, which is treated by AutoCAD as a single element. Also known as an object.

Explode. To break a complex object, such as a block, solid, polyline, or dimension, into simpler objects, such as lines, circles, arcs, and text.

Extension lines. The reference lines that extend from or near the measurement points (on an object or objects) to or just beyond the dimension line location. The dimension is calculated between extension lines. Also known as witness lines or projection lines.

External program. An executable program, command, or file, such as an operating system command or application program, which can be executed from within AutoCAD if so defined in the acad.pgp file.

External reference. Another drawing file inserted in the current drawing, much like a block, except the externally referenced drawing's data is not stored with the current drawing. Also known as an xref.

F

Feature. A defining aspect of a part, such as a point, line, arc, axis, surface, slot, hole, fillet, or chamfer.

Fence. A line or connected series of lines used to select all objects that it passes through.

File locking. The automatic flagging of Auto-CAD files to prevent their simultaneous use by more than one user or program. Files can be read and/or write locked. *See also* lock file.

Fill. A solid color covering an enclosed area.

Fillet. A constant radius rounded internal corner joining two curves or surfaces, represented in 2D as an arc.

Filters. *See* point filters.

Fit. The correct clearances and matched designs required for component parts of an assembly to properly mate.

Fit points. *See* approximation and interpolation points.

Flat projection. The line-of-sight projection of 3D objects onto a 2D plane for display or drawing representation.

Floating viewport. A viewport created by the MVIEW command in paper space. Floating (mview) viewports are movable, editable entities within which the model space drawing environment exists. Floating viewports can be created and viewed when TILEMODE is off. Any part of the drawing model may be viewed from within a floating viewport. *See als* viewport and tiled viewports.

Flush. Aligned, flat, even, or level.

Font. A complete set of characters, including numbers, letters, punctuation, and symbols of a given typeface and size.

Freeze. To suppress the regeneration, display, and plotting of objects on specified layers. *See also* thaw.

G

GDT. Geometric Dimensioning and Tolerancing.

Geometric Dimensioning and Tolerancing (GDT). The complete but non-redundant size, placement, and tolerance information required for the unambiguous geometric specification of a part.

Geometric tolerance. Allowable deviation from the design in form, location, orientation, profile, and runout.

Graphics area. The area of the AutoCAD window used for drawing.

Graphics cursor. *See* crosshairs, cursor, and pointer.

Graphics window. The main AutoCAD LT application window, consisting of the title bar, graphics area, menus, docked toolbars, and command line.

Grid. An evenly spaced, adjustable array of dots used as a drawing aid to help with picking points or estimating distances. Grid dots are not plotted.

Grip modes. Interactive editing modes—Stretch, Move, Rotate, Scale, and Mirror—which are available when grips are displayed on selected objects and one or more grips are then selected.

Grips. Small squares that appear at geometric points of selected objects. When you select a grip, you can use grip modes to edit the objects.

Guidelines. Light lines used for preliminary drawing layout. Also known as construction lines.

H

Handle. A unique alphanumeric tag used by AutoCAD and applications programs to identify a particular object in the AutoCAD database.

Hard copy. A print or plot on paper or other media.

Hatching. An area fill composed of one or more sets of continuous or broken parallel lines. A complex pattern of broken lines, usually predefined, may simulate symbols or dots. *See also* crosshatching.

Hidden line. An obscured line not actually visible from the current viewpoint, but generally shown dashed to so indicate. Used to represent lines that exist, but cannot be seen.

Hole. In mechanical parts or solid models, a feature that passes through an object. In hatching, another term for an island.

Hook line. A short horizontal line between the end of a leader line and its text. *See also* leader.

Horizontal dimension. A linear dimension measuring a distance parallel to the x axis.

I

Icon menu. A pictorial menu containing multiple image tiles that display icons. You select an icon menu item by choosing an icon.

Included angle. The inside angular specification or measurement of the extents of a circular path, such as an arc. An included angle in AutoCAD may be positive or negative, in degrees, radians, or surveyor's units, depending on the current angular units.

Initial environment. The settings that a new drawing inherits from its prototype drawing. If no prototype is used, the initial environment is that of the acad.dwg or acadiso.dwg. *See also* prototype drawing.

Instance. One single occurrence or execution, such as a block insertion or the execution of a program.

Interpolation points. A set of defining points through which the resulting curve passes. *See also* approximation and fit points.

Island. An enclosed area within an outer boundary, such as a hole within a hatched area.

Isometric drawing. A drawing in which the x, y, and z axes are equally foreshortened, and therefore spaced 120 degrees apart. The z axis is vertical.

Isometric plane. One of the three orthogonal planes in an isometric projection, XY, XZ, and YZ. In AutoCAD's isometric mode, the grid is aligned with the current isometric plane.

Isometric snap style. In AutoCAD, a snap option that aligns the cursor and grid with the isometric axes and planes.

Isoplane. Another term for isometric plane.

J, K

Key plan. A small-scale plan used to orient other drawings to the overall plan.

L

Lateral tolerance. A dimension component, generally prefixed by the ± symbol, which indicates the maximum allowable plus/minus variance of the dimensioned part. Also known as a plus/minus tolerance. *See also* limits dimension.

Layer. A logical overlay for the grouping of data. You can view layers individually or in any combination, like transparent acetate overlays on a drawing. In AutoCAD, properties can be assigned and visibility can be controlled by layer. *See also* properties.

Leader. A line or series of lines that link annotation to the objects they reference. Also called a leader line. *See also* hook line.

Limits dimension. A dimension with two text strings that indicate the maximum and minimum size (upper and lower limits) of the dimensioned part. The upper limit is placed above the lower limit, unless the dimension is a note, in which case the lower limit precedes the upper limit. Also known as limits. *See also* lateral tolerance.

Linetype. The predefined style or appearance of a line or curve, such as continuous, dashed, dotted, or center. Also known as a line font.

Link. An *Object Linking and Embedding* (OLE) reference to data in another file. Changes in the linked files may be automatically or manually updated.

Lock file. A flag file, automatically created when an AutoCAD drawing or other data file is in use, that prevents the simultaneous accidental use of the drawing or data files by other users.

M

Macro. A group of commands, often including data or pauses for input, which is defined and executed as a single menu item or tool.

Major axis. The longest bisector of an ellipse.

Menu. A list of choices or menu items that, in AutoCAD, generally issue commands, make settings, or open dialog boxes.

Minor axis. The shortest bisector of an ellipse.

Mirror. The creation of a mirror-image copy of an existing object.

Mnemonic. A memory aid, such as an abbreviated alias for a command like BR for BREAK or LT for LINETYPE. *See also* alias.

Mode. The operating state of a program that determines the available commands or options. In dimensioning mode, for example, you can only execute dimensioning commands and certain transparent commands. Also, a setting that determines program or command behavior, such as the SNAPMODE system variable, which controls whether snap is on or off.

Model. The representation of a real-world object as a 2D or 3D drawing.

Model space. One of the two primary spaces in AutoCAD. Typically, model space is where you draw. Model space exists in either tiled viewports or floating (mview) viewports in paper space. *See also* paper space, which is typically where you compose a drawing with specific views and annotations.

N

Named view. A stored viewpoint and zoom magnification, which can be recalled and redisplayed.

Node. A point entity or object.

Normal. At right angles or perpendicular to a line or surface.

Noun/verb. The editing sequence of object selection, then editing operation (as opposed to verb/noun, which is command selection, then object selection).

Null response. Pressing the spacebar or Enter key without making other input, to accept a default value.

Object. One or more simple graphical objects, such as a text string, line, arc, circle, or complex graphical objects such as polylines, blocks, or dimensions, which are treated by AutoCAD as single elements. Also known as an entity.

Object line. A solid line representing a contour of an object.

Object Linking and Embedding (OLE). A Windows protocol that enables you to embed data created by another application as an object in the current application, and to link that object to the application that created it so that it is automatically updated when the source data is changed.

Object snaps. Geometric snap modes that cause the input point to be snapped to a geometric point on an object. The type of geometric point depends on the mode(s) specified. *See also* snap, running object snap, and object snap override.

Object snap override. A single-use object snap specified for a single pick. Any running object snap mode is ignored (overridden) until after the pick. *See also* running object snaps.

Oblique strokes. Another term for tick marks.

Ordinate dimension. A dimension, generally one of a series, measured perpendicular from a common to a dimensioned feature. An ordinate dimension has only a baseline, dimension text, and a leader line that serves as an extension line. Also known as a datum-line dimension.

Origin. The 0-base point at the intersection of the axes in a coordinate system. In the Cartesian coordinate system, for example, the origin is the 0,0,0 coordinate where the x, y, and z axes intersect.

Ortho mode. An AutoCAD setting that constrains the crosshair cursor and point input to points that are orthogonal (relative to the current snap angle in the current UCS) from the current reference or last point. *See also* orthogonal.

Orthogonal. Perpendicular to a line, plane, or tangent at the point of intersection.

Orthographic projection. The orthogonal projection of all points on an object to a 2D viewing plane, resulting in perspective-free 2D representation, typically front, top, and right side views. Also known as parallel projection.

Outline. A line representing the outer boundary or edge of an object.

P

Pan. To shift the drawing image within the current viewport without changing magnification. *See also* zoom and view.

Paper space. An AutoCAD environment intended for arranging floating (mview) viewports. You can work in model space in these viewports, and then use paper space to arrange these viewports and their views to create a finished sheet for printing or plotting. You can annotate the views and place a title block in paper space. *See also* model space and viewport.

Parallel dimension. Another term for baseline dimension.

Parallel projection. Another term for orthographic projection.

Parameter. A value or option that the user enters to control the action or behavior of a command or appearance of an object.

Part. A real-world object that is realistically, not symbolically, represented in the drawing.

Parts list. Another term for bill of materials.

Phantom line. A broken line (long segments and two short dashes) representing an absent object, property line, or alternative position.

Pick button. The mouse button you use to pick points, select interface objects, and specify the cursor position. It is generally the left mouse button. *See also* Enter button and pop-up button.

Pixel. A picture element, the smallest displayable dot in a bitmapped or raster image. Raster devices include computer screens, dot matrix printers, and raster printers and plotters such as laser, ink jet, and PostScript devices. *See also* aliasing and anti-aliasing.

Plan view. An orthogonal view from above an object. *See also* orthogonal and orthographic projection. AutoCAD assumes the plan view to be an orthogonal view of the current UCS.

Plane. A two-dimensional surface at any orientation in space.

Pline. Another term for polyline.

Plus/minus tolerance. Another term for lateral tolerance.

Point. An object having no size or dimension and no geometric properties except location. A geometric location in 3D space.

Point filters. Point input parameters that cause AutoCAD to extract only one or two of the individual x, y, and z coordinate values from the subsequent input point, and then to reprompt for the additional coordinate value or values needed to compose a complete point. Also known as X,Y,Z point filters or coordinate filters.

Polar array. A circular array of objects about a point.

Polar coordinates. Coordinates locating a point in 2D space by its distance from the origin or reference point, and angle from the x axis in the XY plane. *See also* Cartesian coordinate system.

Polygon window. A CPolygon or WPolygon; a multisided selection window for selecting objects. *See also* WPolygon, CPolygon, and crossing window.

polyline. A complex object with one or more connected line and/or circular arc segments, but which is considered a single object by AutoCAD.

Pop-up button. The middle mouse button on a three-button mouse, which pops up a cursor menu in the graphics area at the cursor location. On a two-button mouse, you can hold down either the Shift, Ctrl, or Alt key while you press the right button to simulate the middle button and pop up the menu. *See also* cursor menu.

Pop-up menu. Another term for cursor menu.

Pouche. To darken, or fill with a pattern, an enclosed area of a drawing to make it stand out.

Primitive. The simplest type of AutoCAD entity or object, such as a line, arc, point, or circle.

Profile. A two-dimensional representation of a three-dimensional object showing only those surface edges and silhouette lines visible from a particular viewpoint.

Projection lines. Another term for extension lines.

Prompt. An informative request for input from the computer program to the user, often offering options.

Property. A characteristic attribute of an AutoCAD object, such as its layer, color, linetype, or thickness. Called an attribute in some other CAD systems.

Prototype drawing. A drawing file used to provide the initial settings of a new drawing. If no prototype is used, the initial environment is that of the acad.dwg or acadiso.dwg. Prototype drawings may optionally contain block, layer, and linetype definitions, text and dimension styles, and the definitions of other named items, and any drawing can be used as a prototype. *See also* initial environment.

Q

Quadratic curve. A curve defined by a quadratic equation, a parametric polynomial equation of the form $x = [-b \pm \sqrt{(b^2 - 4ac)}]/2a$.

R

Radius dimension. A dimension measuring the radius of an arc or circle. The text and radius dimension line are drawn either inside or outside the arc or circle.

Real-world scale. The full-size 1:1 scale of objects in the real world. Generally, objects or models representing real-world objects are drawn in real-world scale in CAD, and the drawing is then scaled down to fit a sheet at plot time or in paper space.

Real-world units. Drawing units matching the full-size 1:1 scale units of objects in the real world.

Rectangular coordinates. Another term for coordinates in the Cartesian coordinate system.

Redlining. Marking changes or corrections, generally in red, on a check print, or on-screen. Redlining on-screen generally requires the use of xrefs or third-party redlining software to avoid disturbing the drawing.

Redraw. To update and clean up a drawing image without recalculating the drawing's database. *See also* regenerate.

Reference bubble. A symbol, generally in a circle, which keys an item or portion of a drawing to a detailed drawing or other information.

Regenerate. To recalculate the drawing's database, recompute the virtual screen coordinates, and update and clean up the drawing image. *See also* redraw.

Relative coordinates. The offset of a point from the previous point or coordinates, generally expressed as XYZ or polar coordinates. *See also* polar coordinates.

Resolution. The number of horizontal rows and vertical columns of pixels displayed by a raster device, such as a computer screen, dot matrix printer, and raster printer or plotter. Standard SVGA graphics resolution is 800 columns and 600 rows of pixels. *See also* pixel.

Right-hand rule. A method for determining the positive x, y, and z directions of a Cartesian coordinate system and the positive rotation direction about a given axis.

Rotate. To move around a point in a plane.

Rubber-band line. A dynamic line that stretches from the current reference or last point to the current cursor location as you move the pointer.

Running object snap. One or more object snap modes set to be in effect for point specification except when overridden or until reset. *See also* object snaps and object snap override.

S

Scan lines. The horizontal rows of pixels in a raster video image. The number of scan lines is equal to the vertical resolution.

Schedule. A chart with reference symbols or labels keyed to items in the drawing and specifying related information, such as that used to describe finishes.

Schematic drawing. Initial drawings to plan the general relationships or configuration of rooms or parts, usually drawn freehand and not to any particular scale.

Script file. AutoCAD commands, parameters, and other input in an ASCII text file that can be sequentially executed with the SCRIPT command. Script files have an .scr extension.

Section. Another term for cross section.

Select. To choose one or more drawing or interface objects, or text characters in a list or edit box.

Selection set. A set of one or more AutoCAD objects selected for editing.

Selection window. A rectangular box used for selecting objects. *See also* window and crossing window.

Sight line. The line of sight direction from the eye to the object being viewed.

Slide file. A file containing a vector image matching the image of the viewport current at the time the slide file was made. Slide files have the extension .sld and can be displayed in AutoCAD, Autodesk Animator, Animator Pro, and other programs.

Slide library. A library file created with the slidelib.exe utility and containing a collection of slide files. Slide library files have the extension .slb and make slide management easier.

Snap. To cause an input point to be placed at the nearest invisible grid point defined by the snap grid spacing or nearest geometric point defined by object snap settings.

Snap grid. The invisible grid, defined by the snap grid spacing, to which the crosshairs and input points snap when snap mode is turned on. The snap grid does not necessarily match the visible grid, which is controlled separately by the grid spacing.

Snap grid spacing. The snap resolution or spacing of the invisible points of the snap grid.

Snap mode. A mode that, when on, causes the crosshairs and input points to be placed at the nearest invisible grid point defined by the snap grid spacing.

Snap resolution. The snap grid spacing or spacing of the invisible points of the snap grid.

Solid. A 3D representation of a real-world object. In AutoCAD, a solid entity or object is also a 2D-filled closed primitive defined by three or four points.

Space. In AutoCAD, one of the two drawing environments: model space and paper space. Model space exists in both tiled and floating (mview) viewports. *See also* model space, paper space, and viewports.

Spherical coordinates. Coordinates locating a point in 3D space by its distance from the origin or reference point, and angle from the x axis in the XY plane, and angle in the z direction from the XY plane. *See also* Cartesian coordinate system.

Spline curve. A curve defined by control points near which it passes, and mathematically defined as blended piecewise polynomial curve.

Symbol. A shape or mark that represents an object or action, generally not drawn realistically or to scale. You can create your own symbols by defining and inserting blocks in AutoCAD.

System environment variable. A named variable, generally user-defined, stored in the computer's memory by the operating system. System environment variables are often used to set the initial operating environment of programs such as AutoCAD. AutoCAD can obtain its default path settings from system environment variables.

System variable. A named variable used by AutoCAD to store settings and default parameters that control the behavior of the drawing editor and various commands.

T

Tangent. A line, curve, or surface that touches, but does not intersect, another line, curve, or surface (the AutoCAD INTersection object snap treats a tangent point as an intersection, however).

Target point. A point toward which the view looks, from the eye along the line of sight.

Temporary files. Temporary data files created and used by AutoCAD during a drawing session. Temporary files are normally deleted by AutoCAD at the end of the session, but may remain if AutoCAD is terminated abnormally, such as by a program or system crash or power interruption. Temporary files have names like UNDO.ac$ and Acis.ac$.

Tessellation lines. Evenly spaced, often parallel, lines that help visually indicate 3D surfaces, particularly curved surfaces.

Text style. A named group of settings that modify a font to determine the drawn appearance of text. The effects of these settings may cause the text to be drawn upside-down, backwards, vertically, obliqued, (italicized) stretched or compressed, or with a fixed height.

Thickness. The distance by which a 2D AutoCAD object is extruded above or below its base elevation to simulate 3D. This is sometimes called 2-1/2D.

Tick marks. Slashes at the beginning and end of a dimension line as a terminators. Tick marks are standard in architectural dimensions. Also known as oblique strokes or ticks.

Tiled viewports. One or more non-overlapping viewports that fill the graphics area and within which the model space drawing environment exists. Tiled viewports are the default, when TILEMODE is on. *See also* floating viewports.

TILEMODE. A system variable determines whether floating or tiled viewports are available. *See also* floating viewports and tiled viewports.

Toggle. An on/off setting, such as an on/off system variable.

Tolerance. The total variance allowed for a manufactured part from a specified dimension, indicated as a plus/minus tolerance or a limits tolerance. *See also* lateral, limits, and geometric tolerance, and ANSI Y14.5m.

Toolbar. An interactive interface bar in the graphics window above the drawing area. The toolbar contains icon tools that issue commands and make settings, a coordinate readout, and a layer drop-down list.

Toolbox. A dockable floating window of icon tools that issue commands and make settings.

Tracking. An interactive point specification mode in which you specify a series of temporary relative orthogonal points, then end tracking mode to designate the last point entered as the actual input.

Transparent command. A command that can be used during another command. Transparent commands are entered with a leading apostrophe.

Typeface. See font.

U

UCS. *See* User Coordinate System (UCS).

UCS icon. An icon that indicates the orientation, and optionally the origin, of the current UCS.

Unit. The user-defined distance, such as millimeters, feet, inches, or miles, represented by linear distances and measurements in a drawing. *See also* drawing unit.

User Coordinate System (UCS). A user-defined local Cartesian coordinate system that defines the origin and orientation of the x, y, and z axes relative to the WCS in 3D space. Objects are drawn relative the current UCS. *See also* World Coordinate System (WCS).

V

Variance. Another term for plus/minus tolerance.

Vector. A mathematical quantity completely defined by distance and direction, but without a specific location. *See also* direction vector.

Vertex. The point at which the sides of an angle intersect, such as the vertex where two polyline segments meet. A point on a polyhedron common to three or more sides, such as faces approximating a 3D surface.

Vertical dimension. A linear dimension measuring a distance parallel to the y axis. *See also* horizontal dimension.

View. A graphical representation of a 2D drawing or 3D model from a specific location (viewpoint) in space.

Viewpoint. A point in 3D space from which the view looks, from the eye, along the line of sight to the target point.

Viewport. A subwindow or rectangular viewing area within which model space exists for drawing operations. When TILEMODE is on, one or more fixed viewports fill the drawing window. When TILEMODE is off, you can create one or more floating viewports in paper space. *See also* tiled viewports, floating viewports, and paper space.

Viewport configuration. A named set of tiled viewports that you can save and restore. *See also* tiled viewport.

Virtual screen. The imaginary 32,000 x 32,000 pixel screen that AutoCAD generates in memory when the drawing is regenerated. AutoCAD can quickly transform and scale any portion of the virtual screen to display it in the current viewport, enabling zooming at redraw speed instead of the slower regeneration speed.

W

WCS. *See* World Coordinate System (WCS).

Window. A rectangular box used to select all objects fully within its edges.

Witness line. Another term for extension lines.

Working drawing. A finished drawing from which parts are manufactured or structures constructed.

World Coordinate System (WCS). The default Cartesian coordinate system. The WCS is the base system used to define objects and other coordinate systems.

World coordinates. Coordinates specified relative to the World Coordinate System (WCS). *See also* World Coordinate System (WCS) and User Coordinate System (UCS).

WPolygon. A closed multisided polygon window used to select all objects fully within its edges.

X, Y, Z

X,Y,Z point filters. Another term for point filters.

Xref. See external reference.

Zoom. To change the viewing magnification of the drawing image.

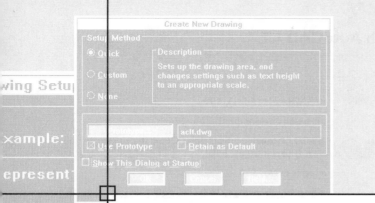

Index

PLUG YOURSELF INTO...

MACMILLAN INFORMATION SUPERLIBRARY™

que · SAMS PUBLISHING · Hayden Books · que COLLEGE · NRP · alpha books · Brady · ADOBE PRESS

THE MACMILLAN INFORMATION SUPERLIBRARY™

Free information and vast computer resources from the world's leading computer book publisher—online!

FIND THE BOOKS THAT ARE RIGHT FOR YOU!

A complete online catalog, plus sample chapters and tables of contents give you an in-depth look at *all* of our books, including hard-to-find titles. It's the best way to find the books you need!

- **STAY INFORMED** with the latest computer industry news through our online newsletter, press releases, and customized Information SuperLibrary Reports.

- **GET FAST ANSWERS** to your questions about MCP books and software.

- **VISIT** our online bookstore for the latest information and editions!

- **COMMUNICATE** with our expert authors through e-mail and conferences.

- **DOWNLOAD SOFTWARE** from the immense MCP library:
    - Source code and files from MCP books
    - The best shareware, freeware, and demos

- **DISCOVER HOT SPOTS** on other parts of the Internet.

- **WIN BOOKS** in ongoing contests and giveaways!

TO PLUG INTO MCP: → **WORLD WIDE WEB: http://www.mcp.com**

GOPHER: gopher.mcp.com

FTP: ftp.mcp.com

Home Page · What's New · Bookstore · Reference Desk · Software Library · Macmillan Overview · Talk to Us

WANT MORE INFORMATION?

CHECK OUT THESE RELATED TOPICS OR SEE YOUR LOCAL BOOKSTORE

CAD and 3D Studio

As the number one CAD publisher in the world, and as a Registered Publisher of Autodesk, New Riders Publishing provides unequaled content on this complex topic. Industry-leading products include AutoCAD and 3D Studio.

Networking

As the leading Novell NetWare publisher, New Riders Publishing delivers cutting-edge products for network professionals. We publish books for all levels of users, from those wanting to gain NetWare Certification, to those administering or installing a network. Leading books in this category include *Inside NetWare 3.12*, *CNE Training Guide: Managing NetWare Systems*, *Inside TCP/IP*, and *NetWare: The Professional Reference*.

Graphics

New Riders provides readers with the most comprehensive product tutorials and references available for the graphics market. Best-sellers include *Inside CorelDRAW! 5*, *Inside Photoshop 3*, and *Adobe Photoshop NOW!*

Internet and Communications

As one of the fastest growing publishers in the communications market, New Riders provides unparalleled information and detail on this ever-changing topic area. We publish international best-sellers such as *New Riders' Official Internet Yellow Pages, 2nd Edition*, a directory of over 10,000 listings of Internet sites and resources from around the world, and *Riding the Internet Highway, Deluxe Edition*.

Operating Systems

Expanding off our expertise in technical markets, and driven by the needs of the computing and business professional, New Riders offers comprehensive references for experienced and advanced users of today's most popular operating systems, including *Understanding Windows 95*, *Inside Unix*, *Inside Windows 3.11 Platinum Edition*, *Inside OS/2 Warp Version 3*, and *Inside MS-DOS 6.22*.

Other Markets

Professionals looking to increase productivity and maximize the potential of their software and hardware should spend time discovering our line of products for Word, Excel, and Lotus 1-2-3. These titles include *Inside Word 6 for Windows*, *Inside Excel 5 for Windows*, *Inside 1-2-3 Release 5*, and *Inside WordPerfect for Windows*.

Orders/Customer Service   **1-800-653-6156**   Source Code   **NRP95**

New Riders Publishing   201 West 103rd Street ◆ Indianapolis, Indiana 46290 USA

Name _____ Title _____

Company _____ Type of
business _____

Address _____

City/State/ZIP _____

Have you used these types of books before? ☐ yes   ☐ no

If yes, which ones? _____

_____

How many computer books do you purchase each year?   ☐ 1–5   ☐ 6 or more

How did you learn about this book? _____

Where did you purchase this book? _____

Which applications do you currently use? _____

_____

Which computer magazines do you subscribe to? _____

_____

What trade shows do you attend? _____

_____

Comments: _____

_____

_____

Would you like to be placed on our preferred mailing list?   ☐ yes   ☐ no

---

☐ **I would like to see my name in print!** You may use my name and quote me in future New Riders products and promotions. My daytime phone number is: _____

New Riders Publishing   201 West 103rd Street ◆ Indianapolis, Indiana 46290 USA

Fax to **317-581-4670**   Orders/Customer Service **1-800-653-6156**   Source Code **NRP95**

Fold Here

- -

BUSINESS REPLY MAIL

FIRST-CLASS MAIL PERMIT NO. 9918 INDIANAPOLIS IN

POSTAGE WILL BE PAID BY THE ADDRESSEE

NEW RIDERS PUBLISHING
201 W 103RD ST
INDIANAPOLIS IN 46290-9058